# A Handbook of Statistical Analyses Using R

SECOND EDITION

# A Handbook of Statistical Analyses Using R

SECOND EDITION

Brian S. Everitt and Torsten Hothorn

CRC Press is an imprint of the
Taylor & Francis Group, an **informa** business
A CHAPMAN & HALL BOOK

Chapman & Hall/CRC
Taylor & Francis Group
6000 Broken Sound Parkway NW, Suite 300
Boca Raton, FL 33487-2742



Printed in the United States of America on acid-free paper
10 9 8 7 6 5 4

International Standard Book Number: 978-1-4200-7933-3 (Paperback)



**Library of Congress Cataloging-in-Publication Data**

Everitt, Brian.
A handbook of statistical analyses using R / Brian S. Everitt and Torsten Hothorn.
-- 2nd ed.
p. cm.
Includes bibliographical references and index.
ISBN 978-1-4200-7933-3 (pbk. : alk. paper)
1. Mathematical statistics--Data processing--Handbooks, manuals, etc. 2. R (Computer program language)--Handbooks, manuals, etc. I. Hothorn, Torsten. II. Title.

QA276.45.R3E94 2010
519.50285'5133--dc22 2009018062

**Visit the Taylor & Francis Web site at**
**http://www.taylorandfrancis.com**

**and the CRC Press Web site at**
**http://www.crcpress.com**

Dedication

To our wives, Mary-Elizabeth and Carolin,
for their constant support and encouragement

# Preface to Second Edition

Like the first edition this book is intended as a guide to data analysis with the R system for statistical computing. New chapters on graphical displays, generalised additive models and simultaneous inference have been added to this second edition and a section on generalised linear mixed models completes the chapter that discusses the analysis of longitudinal data where the response variable does not have a normal distribution. In addition, new examples and additional exercises have been added to several chapters. We have also taken the opportunity to correct a number of errors that were present in the first edition. Most of these errors were kindly pointed out to us by a variety of people to whom we are very grateful, especially Guido Schwarzer, Mike Cheung, Tobias Verbeke, Yihui Xie, Lothar Häberle, and Radoslav Harman.

We learnt that many instructors use our book successfully for introductory courses in applied statistics. We have had the pleasure to give some courses based on the first edition of the book ourselves and we are happy to share slides covering many sections of particular chapters with our readers. LaTeX sources and PDF versions of slides covering several chapters are available from the second author upon request.

A new version of the **HSAUR** package, now called **HSAUR2** for obvious reasons, is available from CRAN. Basically the package vignettes have been updated to cover the new and modified material as well. Otherwise, the technical infrastructure remains as described in the preface to the first edition, with two small exceptions: names of R add-on packages are now printed in bold font and we refrain from showing significance stars in model summaries.

Lastly we would like to thank Thomas Kneib and Achim Zeileis for commenting on the newly added material and again the CRC Press staff, in particular Rob Calver, for their support during the preparation of this second edition.

**Brian S. Everitt and Torsten Hothorn**
London and München, April 2009

# Preface to First Edition

This book is intended as a guide to data analysis with the R system for statistical computing. R is an environment incorporating an implementation of the S programming language, which is powerful and flexible and has excellent graphical facilities (R Development Core Team, 2009b). In the Handbook we aim to give relatively brief and straightforward descriptions of how to conduct a range of statistical analyses using R. Each chapter deals with the analysis appropriate for one or several data sets. A brief account of the relevant statistical background is included in each chapter along with appropriate references, but our prime focus is on how to use R and how to interpret results. We hope the book will provide students and researchers in many disciplines with a self-contained means of using R to analyse their data.

R is an open-source project developed by dozens of volunteers for more than ten years now and is available from the Internet under the General Public Licence. R has become the *lingua franca* of statistical computing. Increasingly, implementations of new statistical methodology first appear as R add-on packages. In some communities, such as in bioinformatics, R already is the primary workhorse for statistical analyses. Because the sources of the R system are open and available to everyone without restrictions and because of its powerful language and graphical capabilities, R has started to become the main computing engine for reproducible statistical research (Leisch, 2002a,b, 2003, Leisch and Rossini, 2003, Gentleman, 2005). For a reproducible piece of research, the original observations, all data preprocessing steps, the statistical analysis as well as the scientific report form a unity and all need to be available for inspection, reproduction and modification by the readers.

Reproducibility is a natural requirement for textbooks such as the *Handbook of Statistical Analyses Using R* and therefore this book is fully reproducible using an R version greater or equal to 2.2.1. All analyses and results, including figures and tables, can be reproduced by the reader without having to retype a single line of R code. The data sets presented in this book are collected in a dedicated add-on package called **HSAUR** accompanying this book. The package can be installed from the Comprehensive R Archive Network (CRAN) via

```
R> install.packages("HSAUR")
```

and its functionality is attached by

```
R> library("HSAUR")
```

The relevant parts of each chapter are available as a *vignette*, basically a document including both the R sources and the rendered output of every analysis contained in the book. For example, the first chapter can be inspected by

```
R> vignette("Ch_introduction_to_R", package = "HSAUR")
```

and the R sources are available for reproducing our analyses by

```
R> edit(vignette("Ch_introduction_to_R", package = "HSAUR"))
```

An overview on all chapter vignettes included in the package can be obtained from

```
R> vignette(package = "HSAUR")
```

We welcome comments on the R package **HSAUR**, and where we think these add to or improve our analysis of a data set we will incorporate them into the package and, hopefully at a later stage, into a revised or second edition of the book.

Plots and tables of results obtained from R are all labelled as 'Figures' in the text. For the graphical material, the corresponding figure also contains the 'essence' of the R code used to produce the figure, although this code may differ a little from that given in the **HSAUR** package, since the latter may include some features, for example thicker line widths, designed to make a basic plot more suitable for publication.

We would like to thank the R Development Core Team for the R system, and authors of contributed add-on packages, particularly Uwe Ligges and Vince Carey for helpful advice on **scatterplot3d** and **gee**. Kurt Hornik, Ludwig A. Hothorn, Fritz Leisch and Rafael Weißbach provided good advice with some statistical and technical problems. We are also very grateful to Achim Zeileis for reading the entire manuscript, pointing out inconsistencies or even bugs and for making many suggestions which have led to improvements. Lastly we would like to thank the CRC Press staff, in particular Rob Calver, for their support during the preparation of the book. Any errors in the book are, of course, the joint responsibility of the two authors.

**Brian S. Everitt and Torsten Hothorn**
London and Erlangen, December 2005

# List of Figures

# List of Tables

# Contents

CHAPTER 1

# An Introduction to R

## 1.1 What is R?

The R system for statistical computing is an environment for data analysis and graphics. The root of R is the S language, developed by John Chambers and colleagues (Becker et al., 1988, Chambers and Hastie, 1992, Chambers, 1998) at Bell Laboratories (formerly AT&T, now owned by Lucent Technologies) starting in the 1960ies. The S language was designed and developed as a programming language for data analysis tasks but in fact it is a full-featured programming language in its current implementations.

The development of the R system for statistical computing is heavily influenced by the open source idea: The base distribution of R and a large number of user contributed extensions are available under the terms of the Free Software Foundation's GNU General Public License in source code form. This licence has two major implications for the data analyst working with R. The complete source code is available and thus the practitioner can investigate the details of the implementation of a special method, can make changes and can distribute modifications to colleagues. As a side-effect, the R system for statistical computing is available to everyone. All scientists, including, in particular, those working in developing countries, now have access to state-of-the-art tools for statistical data analysis without additional costs. With the help of the R system for statistical computing, research really becomes reproducible when both the data and the results of all data analysis steps reported in a paper are available to the readers through an R transcript file. R is most widely used for teaching undergraduate and graduate statistics classes at universities all over the world because students can freely use the statistical computing tools.

The base distribution of R is maintained by a small group of statisticians, the R Development Core Team. A huge amount of additional functionality is implemented in add-on packages authored and maintained by a large group of volunteers. The main source of information about the R system is the world wide web with the official home page of the R project being

`http://www.R-project.org`

All resources are available from this page: the R system itself, a collection of add-on packages, manuals, documentation and more.

The intention of this chapter is to give a rather informal introduction to basic concepts and data manipulation techniques for the R novice. Instead of a rigid treatment of the technical background, the most common tasks are illustrated by practical examples and it is our hope that this will enable readers to get started without too many problems.

## 1.2 Installing R

The R system for statistical computing consists of two major parts: the base system and a collection of user contributed add-on packages. The R language is implemented in the base system. Implementations of statistical and graphical procedures are separated from the base system and are organised in the form of *packages*. A package is a collection of functions, examples and documentation. The functionality of a package is often focused on a special statistical methodology. Both the base system and packages are distributed via the Comprehensive R Archive Network (CRAN) accessible under

`http://CRAN.R-project.org`

### *1.2.1 The Base System and the First Steps*

The base system is available in source form and in precompiled form for various Unix systems, Windows platforms and Mac OS X. For the data analyst, it is sufficient to download the precompiled binary distribution and install it locally. Windows users follow the link

`http://CRAN.R-project.org/bin/windows/base/release.htm`

download the corresponding file (currently named `rw2090.exe`), execute it locally and follow the instructions given by the installer.

Depending on the operating system, R can be started either by typing '`R`' on the shell (Unix systems) or by clicking on the R symbol (as shown left) created by the installer (Windows). R comes without any frills and on start up shows simply a short introductory message including the version number and a prompt '`>`':

```
R : Copyright 2009 The R Foundation for Statistical Computing
Version 2.9.0 (2009-04-17), ISBN 3-900051-07-0

R is free software and comes with ABSOLUTELY NO WARRANTY.
You are welcome to redistribute it under certain conditions.
Type 'license()' or 'licence()' for distribution details.

R is a collaborative project with many contributors.
Type 'contributors()' for more information and
'citation()' on how to cite R or R packages in publications.

Type 'demo()' for some demos, 'help()' for on-line help, or
'help.start()' for an HTML browser interface to help.
Type 'q()' to quit R.
>
```

One can change the appearance of the prompt by

```
> options(prompt = "R> ")
```

and we will use the prompt `R>` for the display of the code examples throughout this book. A `+` sign at the very beginning of a line indicates a continuing command after a newline.

Essentially, the R system evaluates commands typed on the R prompt and returns the results of the computations. The end of a command is indicated by the return key. Virtually all introductory texts on R start with an example using R as a pocket calculator, and so do we:

```
R> x <- sqrt(25) + 2
```

This simple statement asks the R interpreter to calculate $\sqrt{25}$ and then to add 2. The result of the operation is assigned to an R object with variable name `x`. The assignment operator `<-` binds the value of its right hand side to a variable name on the left hand side. The value of the object `x` can be inspected simply by typing

```
R> x
[1] 7
```

which, implicitly, calls the `print` method:

```
R> print(x)
[1] 7
```

### 1.2.2 Packages

The base distribution already comes with some high-priority add-on packages, namely

| | | | |
|---|---|---|---|
| **mgcv** | **KernSmooth** | **MASS** | **base** |
| **boot** | **class** | **cluster** | **codetools** |
| **datasets** | **foreign** | **grDevices** | **graphics** |
| **grid** | **lattice** | **methods** | **nlme** |
| **nnet** | **rcompgen** | **rpart** | **spatial** |
| **splines** | **stats** | **stats4** | **survival** |
| **tcltk** | **tools** | **utils** | |

Some of the packages listed here implement standard statistical functionality, for example linear models, classical tests, a huge collection of high-level plotting functions or tools for survival analysis; many of these will be described and used in later chapters. Others provide basic infrastructure, for example for graphic systems, code analysis tools, graphical user-interfaces or other utilities.

Packages not included in the base distribution can be installed directly from the R prompt. At the time of writing this chapter, 1756 user-contributed packages covering almost all fields of statistical methodology were available. Certain so-called 'task views' for special topics, such as statistics in the social sciences, environmetrics, robust statistics etc., describe important and helpful packages and are available from http://CRAN.R-project.org/web/views/

Given that an Internet connection is available, a package is installed by supplying the name of the package to the function `install.packages`. If, for example, add-on functionality for robust estimation of covariance matrices via sandwich estimators is required (for example in Chapter 13), the **sandwich** package (Zeileis, 2004) can be downloaded and installed via

```
R> install.packages("sandwich")
```

The package functionality is available after *attaching* the package by

```
R> library("sandwich")
```

A comprehensive list of available packages can be obtained from

http://CRAN.R-project.org/web/packages/

Note that on Windows operating systems, precompiled versions of packages are downloaded and installed. In contrast, packages are compiled locally before they are installed on Unix systems.

## 1.3 Help and Documentation

Roughly, three different forms of documentation for the R system for statistical computing may be distinguished: online help that comes with the base distribution or packages, electronic manuals and publications work in the form of books etc.

The help system is a collection of manual pages describing each user-visible function and data set that comes with R. A manual page is shown in a pager or web browser when the name of the function we would like to get help for is supplied to the `help` function

```
R> help("mean")
```

or, for short,

```
R> ?mean
```

Each manual page consists of a general description, the argument list of the documented function with a description of each single argument, information about the return value of the function and, optionally, references, cross-links and, in most cases, executable examples. The function `help.search` is helpful for searching within manual pages. An overview on documented topics in an add-on package is given, for example for the **sandwich** package, by

```
R> help(package = "sandwich")
```

Often a package comes along with an additional document describing the package functionality and giving examples. Such a document is called a *vignette* (Leisch, 2003, Gentleman, 2005). For example, the **sandwich** package vignette is opened using

```
R> vignette("sandwich", package = "sandwich")
```

More extensive documentation is available electronically from the collection of manuals at http://CRAN.R-project.org/manuals.html

For the beginner, at least the first and the second document of the following four manuals (R Development Core Team, 2009a,c,d,e) are mandatory:

**An Introduction to R:** A more formal introduction to data analysis with R than this chapter.

**R Data Import/Export:** A very useful description of how to read and write various external data formats.

**R Installation and Administration:** Hints for installing R on special platforms.

**Writing R Extensions:** The authoritative source on how to write R programs and packages.

Both printed and online publications are available, the most important ones are *Modern Applied Statistics with S* (Venables and Ripley, 2002), *Introductory Statistics with R* (Dalgaard, 2002), *R Graphics* (Murrell, 2005) and the R Newsletter, freely available from

http://CRAN.R-project.org/doc/Rnews/

In case the electronically available documentation and the answers to frequently asked questions (FAQ), available from

http://CRAN.R-project.org/faqs.html

have been consulted but a problem or question remains unsolved, the `r-help` email list is the right place to get answers to well-thought-out questions. It is helpful to read the posting guide

http://www.R-project.org/posting-guide.html

before starting to ask.

## 1.4 Data Objects in R

The data handling and manipulation techniques explained in this chapter will be illustrated by means of a data set of 2000 world leading companies, the Forbes 2000 list for the year 2004 collected by *Forbes Magazine*. This list is originally available from

http://www.forbes.com

and, as an R data object, it is part of the **HSAUR2** package (*Source*: From Forbes.com, New York, New York, 2004. With permission.). In a first step, we make the data available for computations within R. The `data` function searches for data objects of the specified name (`"Forbes2000"`) in the package specified via the `package` argument and, if the search was successful, attaches the data object to the global environment:

```
R> data("Forbes2000", package = "HSAUR2")
R> ls()
[1] "x"          "Forbes2000"
```

The output of the `ls` function lists the names of all objects currently stored in the global environment, and, as the result of the previous command, a variable named `Forbes2000` is available for further manipulation. The variable `x` arises from the pocket calculator example in Subsection 1.2.1.

As one can imagine, printing a list of 2000 companies via

```
R> print(Forbes2000)
```

```
  rank                 name       country       category  sales
1    1            Citigroup United States        Banking  94.71
2    2     General Electric United States Conglomerates 134.19
3    3 American Intl Group United States      Insurance  76.66
  profits  assets marketvalue
1   17.85 1264.03      255.30
2   15.59  626.93      328.54
3    6.46  647.66      194.87
...
```

will not be particularly helpful in gathering some initial information about the data; it is more useful to look at a description of their structure found by using the following command

```
R> str(Forbes2000)
```

```
'data.frame':           2000 obs. of  8 variables:
 $ rank       : int  1 2 3 4 5 ...
 $ name       : chr  "Citigroup" "General Electric" ...
 $ country    : Factor w/ 61 levels "Africa","Australia",...
 $ category   : Factor w/ 27 levels "Aerospace & defense",..
 $ sales      : num  94.7 134.2 ...
 $ profits    : num  17.9 15.6 ...
 $ assets     : num  1264 627 ...
 $ marketvalue: num  255 329 ...
```

The output of the `str` function tells us that `Forbes2000` is an object of class *data.frame*, the most important data structure for handling tabular statistical data in R. As expected, information about 2000 observations, i.e., companies, are stored in this object. For each observation, the following eight variables are available:

`rank`: the ranking of the company,

`name`: the name of the company,

`country`: the country the company is situated in,

`category`: a category describing the products the company produces,

`sales`: the amount of sales of the company in billion US dollars,

`profits`: the profit of the company in billion US dollars,

`assets`: the assets of the company in billion US dollars,

`marketvalue`: the market value of the company in billion US dollars.

A similar but more detailed description is available from the help page for the `Forbes2000` object:

```
R> help("Forbes2000")
```

or

```
R> ?Forbes2000
```

All information provided by `str` can be obtained by specialised functions as well and we will now have a closer look at the most important of these.

The R language is an object-oriented programming language, so every object is an instance of a class. The name of the class of an object can be determined by

```
R> class(Forbes2000)
[1] "data.frame"
```

Objects of class *data.frame* represent data the traditional table-oriented way. Each row is associated with one single observation and each column corresponds to one variable. The dimensions of such a table can be extracted using the `dim` function

```
R> dim(Forbes2000)
[1] 2000    8
```

Alternatively, the numbers of rows and columns can be found using

```
R> nrow(Forbes2000)
[1] 2000
R> ncol(Forbes2000)
[1] 8
```

The results of both statements show that `Forbes2000` has 2000 rows, i.e., observations, the companies in our case, with eight variables describing the observations. The variable names are accessible from

```
R> names(Forbes2000)
[1] "rank"          "name"          "country"       "category"
[5] "sales"         "profits"       "assets"        "marketvalue"
```

The values of single variables can be extracted from the `Forbes2000` object by their names, for example the ranking of the companies

```
R> class(Forbes2000[,"rank"])
[1] "integer"
```

is stored as an integer variable. Brackets `[]` always indicate a subset of a larger object, in our case a single variable extracted from the whole table. Because *data.frames* have two dimensions, observations and variables, the comma is required in order to specify that we want a subset of the second dimension, i.e., the variables. The rankings for all 2000 companies are represented in a *vector* structure the length of which is given by

```
R> length(Forbes2000[,"rank"])
[1] 2000
```

A *vector* is the elementary structure for data handling in R and is a set of simple elements, all being objects of the same class. For example, a simple vector of the numbers one to three can be constructed by one of the following commands

```
R> 1:3
```

```
[1] 1 2 3
```

```
R> c(1,2,3)
```

```
[1] 1 2 3
```

```
R> seq(from = 1, to = 3, by = 1)
```

```
[1] 1 2 3
```

The unique names of all 2000 companies are stored in a character vector

```
R> class(Forbes2000[,"name"])
```

```
[1] "character"
```

```
R> length(Forbes2000[,"name"])
```

```
[1] 2000
```

and the first element of this vector is

```
R> Forbes2000[,"name"][1]
```

```
[1] "Citigroup"
```

Because the companies are ranked, Citigroup is the world's largest company according to the Forbes 2000 list. Further details on vectors and subsetting are given in Section 1.6.

Nominal measurements are represented by *factor* variables in R, such as the category of the company's business segment

```
R> class(Forbes2000[,"category"])
```

```
[1] "factor"
```

Objects of class *factor* and *character* basically differ in the way their values are stored internally. Each element of a vector of class *character* is stored as a *character* variable whereas an integer variable indicating the level of a *factor* is saved for *factor* objects. In our case, there are

```
R> nlevels(Forbes2000[,"category"])
```

```
[1] 27
```

different levels, i.e., business categories, which can be extracted by

```
R> levels(Forbes2000[,"category"])
```

```
[1] "Aerospace & defense"
[2] "Banking"
[3] "Business services & supplies"
...
```

As a simple summary statistic, the frequencies of the levels of such a *factor* variable can be found from

```
R> table(Forbes2000[,"category"])
```

```
         Aerospace & defense                            Banking
                          19                                313
Business services & supplies
                          70
...
```

The sales, assets, profits and market value variables are of type `numeric`, the natural data type for continuous or discrete measurements, for example

```
R> class(Forbes2000[,"sales"])
```

```
[1] "numeric"
```

and simple summary statistics such as the mean, median and range can be found from

```
R> median(Forbes2000[,"sales"])
```

```
[1] 4.365
```

```
R> mean(Forbes2000[,"sales"])
```

```
[1] 9.69701
```

```
R> range(Forbes2000[,"sales"])
```

```
[1]   0.01 256.33
```

The `summary` method can be applied to a numeric vector to give a set of useful summary statistics, namely the minimum, maximum, mean, median and the 25% and 75% quartiles; for example

```
R> summary(Forbes2000[,"sales"])
```

```
   Min. 1st Qu.  Median    Mean 3rd Qu.    Max.
  0.010   2.018   4.365   9.697   9.548 256.300
```

## 1.5 Data Import and Export

In the previous section, the data from the Forbes 2000 list of the world's largest companies were loaded into R from the **HSAUR2** package but we will now explore practically more relevant ways to import data into the R system. The most frequent data formats the data analyst is confronted with are comma separated files, Excel spreadsheets, files in SPSS format and a variety of SQL data base engines. Querying data bases is a nontrivial task and requires additional knowledge about querying languages, and we therefore refer to the *R Data Import/Export* manual – see Section 1.3. We assume that a comma separated file containing the Forbes 2000 list is available as `Forbes2000.csv` (such a file is part of the **HSAUR2** source package in directory `HSAUR2/inst/rawdata`). When the fields are separated by commas and each row begins with a name (a text format typically created by Excel), we can read in the data as follows using the `read.table` function

```
R> csvForbes2000 <- read.table("Forbes2000.csv",
+      header = TRUE, sep = ",", row.names = 1)
```

The argument `header = TRUE` indicates that the entries in the first line of the text file `"Forbes2000.csv"` should be interpreted as variable names. Columns are separated by a comma (`sep = ","`), users of continental versions of Excel should take care of the character symbol coding for decimal points (by default `dec = "."`). Finally, the first column should be interpreted as row names but not as a variable (`row.names = 1`). Alternatively, the function `read.csv` can be used to read comma separated files. The function `read.table` by default guesses the class of each variable from the specified file. In our case, character variables are stored as factors

```
R> class(csvForbes2000[,"name"])

[1] "factor"
```

which is only suboptimal since the names of the companies are unique. However, we can supply the types for each variable to the `colClasses` argument

```
R> csvForbes2000 <- read.table("Forbes2000.csv",
+      header = TRUE, sep = ",", row.names = 1,
+      colClasses = c("character", "integer", "character",
+          "factor", "factor", "numeric", "numeric", "numeric",
+          "numeric"))
R> class(csvForbes2000[,"name"])

[1] "character"
```

and check if this object is identical with our previous Forbes 2000 list object

```
R> all.equal(csvForbes2000, Forbes2000)

[1] TRUE
```

The argument `colClasses` expects a character vector of length equal to the number of columns in the file. Such a vector can be supplied by the `c` function that combines the objects given in the parameter list into a *vector*

```
R> classes <- c("character", "integer", "character", "factor",
+      "factor", "numeric", "numeric", "numeric", "numeric")
R> length(classes)

[1] 9

R> class(classes)

[1] "character"
```

An R interface to the open data base connectivity standard (ODBC) is available in package **RODBC** and its functionality can be used to access Excel and Access files directly:

```
R> library("RODBC")
R> cnct <- odbcConnectExcel("Forbes2000.xls")
R> sqlQuery(cnct, "select * from \"Forbes2000\\$\"")
```

The function odbcConnectExcel opens a connection to the specified Excel or Access file which can be used to send SQL queries to the data base engine and retrieve the results of the query.

Files in SPSS format are read in a way similar to reading comma separated files, using the function read.spss from package **foreign** (which comes with the base distribution).

Exporting data from R is now rather straightforward. A comma separated file readable by Excel can be constructed from a *data.frame* object via

```
R> write.table(Forbes2000, file = "Forbes2000.csv", sep = ",",
+              col.names = NA)
```

The function write.csv is one alternative and the functionality implemented in the **RODBC** package can be used to write data directly into Excel spreadsheets as well.

Alternatively, when data should be saved for later processing in R only, R objects of arbitrary kind can be stored into an external binary file via

```
R> save(Forbes2000, file = "Forbes2000.rda")
```

where the extension .rda is standard. We can get the file names of all files with extension .rda from the working directory

```
R> list.files(pattern = "\\.rda")
```

```
[1] "Forbes2000.rda"
```

and we can load the contents of the file into R by

```
R> load("Forbes2000.rda")
```

## 1.6 Basic Data Manipulation

The examples shown in the previous section have illustrated the importance of *data.frames* for storing and handling tabular data in R. Internally, a *data.frame* is a *list* of vectors of a common length $n$, the number of rows of the table. Each of those vectors represents the measurements of one variable and we have seen that we can access such a variable by its name, for example the names of the companies

```
R> companies <- Forbes2000[,"name"]
```

Of course, the companies vector is of class *character* and of length 2000. A subset of the elements of the vector companies can be extracted using the [] subset operator. For example, the largest of the 2000 companies listed in the Forbes 2000 list is

```
R> companies[1]
```

```
[1] "Citigroup"
```

and the top three companies can be extracted utilising an integer vector of the numbers one to three:

```
R> 1:3
```

```
[1] 1 2 3
```

```
R> companies[1:3]
```

```
[1] "Citigroup"           "General Electric"
[3] "American Intl Group"
```

In contrast to indexing with positive integers, negative indexing returns all elements that are *not* part of the index vector given in brackets. For example, all companies except those with numbers four to two-thousand, i.e., the top three companies, are again

```
R> companies[-(4:2000)]
```

```
[1] "Citigroup"           "General Electric"
[3] "American Intl Group"
```

The complete information about the top three companies can be printed in a similar way. Because *data.frames* have a concept of rows and columns, we need to separate the subsets corresponding to rows and columns by a comma. The statement

```
R> Forbes2000[1:3, c("name", "sales", "profits", "assets")]
```

```
                 name  sales profits  assets
1           Citigroup  94.71   17.85 1264.03
2    General Electric 134.19   15.59  626.93
3 American Intl Group  76.66    6.46  647.66
```

extracts the variables `name`, `sales`, `profits` and `assets` for the three largest companies. Alternatively, a single variable can be extracted from a *data.frame* by

```
R> companies <- Forbes2000$name
```

which is equivalent to the previously shown statement

```
R> companies <- Forbes2000[,"name"]
```

We might be interested in extracting the largest companies with respect to an alternative ordering. The three top selling companies can be computed along the following lines. First, we need to compute the ordering of the companies' sales

```
R> order_sales <- order(Forbes2000$sales)
```

which returns the indices of the ordered elements of the numeric vector `sales`. Consequently the three companies with the lowest sales are

```
R> companies[order_sales[1:3]]
```

```
[1] "Custodia Holding"        "Central European Media"
[3] "Minara Resources"
```

The indices of the three top sellers are the elements 1998, 1999 and 2000 of the integer vector `order_sales`

```
R> Forbes2000[order_sales[c(2000, 1999, 1998)],
+            c("name", "sales", "profits", "assets")]
```

```
                 name  sales profits assets
10 Wal-Mart Stores 256.33    9.05 104.91
5                BP 232.57   10.27 177.57
4        ExxonMobil 222.88   20.96 166.99
```

Another way of selecting vector elements is the use of a logical vector being `TRUE` when the corresponding element is to be selected and `FALSE` otherwise. The companies with assets of more than 1000 billion US dollars are

```
R> Forbes2000[Forbes2000$assets > 1000,
+             c("name", "sales", "profits", "assets")]
                  name sales profits  assets
1            Citigroup 94.71   17.85 1264.03
9           Fannie Mae 53.13    6.48 1019.17
403 Mizuho Financial 24.40  -20.11 1115.90
```

where the expression `Forbes2000$assets > 1000` indicates a logical vector of length 2000 with

```
R> table(Forbes2000$assets > 1000)
FALSE  TRUE 
 1997     3 
```

elements being either `FALSE` or `TRUE`. In fact, for some of the companies the measurement of the `profits` variable are missing. In R, missing values are treated by a special symbol, `NA`, indicating that this measurement is not available. The observations with profit information missing can be obtained via

```
R> na_profits <- is.na(Forbes2000$profits)
R> table(na_profits)
na_profits
FALSE  TRUE 
 1995     5 
R> Forbes2000[na_profits,
+             c("name", "sales", "profits", "assets")]
                        name sales profits assets
772                      AMP  5.40      NA  42.94
1085                     HHG  5.68      NA  51.65
1091                     NTL  3.50      NA  10.59
1425       US Airways Group  5.50      NA   8.58
1909 Laidlaw International  4.48      NA   3.98
```

where the function `is.na` returns a logical vector being `TRUE` when the corresponding element of the supplied vector is `NA`. A more comfortable approach is available when we want to remove all observations with at least one missing value from a *data.frame* object. The function `complete.cases` takes a *data.frame* and returns a logical vector being `TRUE` when the corresponding observation does not contain any missing value:

```
R> table(complete.cases(Forbes2000))
```

```
FALSE  TRUE 
    5  1995 
```

Subsetting *data.frames* driven by logical expressions may induce a lot of typing which can be avoided. The `subset` function takes a *data.frame* as first argument and a logical expression as second argument. For example, we can select a subset of the Forbes 2000 list consisting of all companies situated in the United Kingdom by

```
R> UKcomp <- subset(Forbes2000, country == "United Kingdom")
R> dim(UKcomp)
```

```
[1] 137   8
```

i.e., 137 of the 2000 companies are from the UK. Note that it is not necessary to extract the variable `country` from the *data.frame* `Forbes2000` when formulating the logical expression with `subset`.

## 1.7 Computing with Data

### *1.7.1 Simple Summary Statistics*

Two functions are helpful for getting an overview about R objects: `str` and `summary`, where `str` is more detailed about data types and `summary` gives a collection of sensible summary statistics. For example, applying the `summary` method to the `Forbes2000` data set,

```
R> summary(Forbes2000)
```

results in the following output

```
      rank             name                        country   
 Min.   :   1.0   Length:2000        United States :751  
 1st Qu.: 500.8   Class :character   Japan         :316  
 Median :1000.5   Mode  :character   United Kingdom:137  
 Mean   :1000.5                      Germany       : 65  
 3rd Qu.:1500.2                      France        : 63  
 Max.   :2000.0                      Canada        : 56  
                                     (Other)       :612  
                     category          sales        
 Banking                 : 313   Min.   :  0.010  
 Diversified financials: 158   1st Qu.:  2.018  
 Insurance               : 112   Median :  4.365  
 Utilities               : 110   Mean   :  9.697  
 Materials               :  97   3rd Qu.:  9.547  
 Oil & gas operations    :  90   Max.   :256.330  
 (Other)                 :1120                    
    profits              assets           marketvalue    
 Min.   :-25.8300   Min.   :   0.270   Min.   :  0.02  
 1st Qu.:  0.0800   1st Qu.:   4.025   1st Qu.:  2.72  
 Median :  0.2000   Median :   9.345   Median :  5.15  
 Mean   :  0.3811   Mean   :  34.042   Mean   : 11.88  
 3rd Qu.:  0.4400   3rd Qu.:  22.793   3rd Qu.: 10.60  
```

```
  Max.    : 20.9600    Max.    :1264.030    Max.    :328.54
  NA's    :  5.0000
```

From this output we can immediately see that most of the companies are situated in the US and that most of the companies are working in the banking sector as well as that negative profits, or losses, up to 26 billion US dollars occur.

Internally, `summary` is a so-called *generic function* with methods for a multitude of classes, i.e., `summary` can be applied to objects of different classes and will report sensible results. Here, we supply a *data.frame* object to `summary` where it is natural to apply `summary` to each of the variables in this *data.frame*. Because a *data.frame* is a *list* with each variable being an element of that *list*, the same effect can be achieved by

```
R> lapply(Forbes2000, summary)
```

The members of the `apply` family help to solve recurring tasks for each element of a *data.frame*, *matrix*, *list* or for each level of a *factor*. It might be interesting to compare the profits in each of the 27 categories. To do so, we first compute the median profit for each category from

```
R> mprofits <- tapply(Forbes2000$profits,
+                      Forbes2000$category, median, na.rm = TRUE)
```

a command that should be read as follows. For each level of the factor `category`, determine the corresponding elements of the numeric vector `profits` and supply them to the `median` function with additional argument `na.rm = TRUE`. The latter one is necessary because `profits` contains missing values which would lead to a non-sensible result of the `median` function

```
R> median(Forbes2000$profits)
[1] NA
```

The three categories with highest median profit are computed from the vector of sorted median profits

```
R> rev(sort(mprofits))[1:3]
          Oil & gas operations          Drugs & biotechnology
                          0.35                           0.35
Household & personal products
                          0.31
```

where `rev` rearranges the vector of median profits `sorted` from smallest to largest. Of course, we can replace the `median` function with `mean` or whatever is appropriate in the call to `tapply`. In our situation, `mean` is not a good choice, because the distributions of profits or sales are naturally skewed. Simple graphical tools for the inspection of the empirical distributions are introduced later on and in Chapter 2.

### 1.7.2 Customising Analyses

In the preceding sections we have done quite complex analyses on our data using functions available from R. However, the real power of the system comes to light when writing our own functions for our own analysis tasks. Although R is a full-featured programming language, writing small helper functions for our daily work is not too complicated. We'll study two example cases.

At first, we want to add a robust measure of variability to the location measures computed in the previous subsection. In addition to the median profit, computed via

```
R> median(Forbes2000$profits, na.rm = TRUE)

[1] 0.2
```

we want to compute the inter-quartile range, i.e., the difference between the 3rd and 1st quartile. Although a quick search in the manual pages (via `help("interquartile")`) brings function `IQR` to our attention, we will approach this task without making use of this tool, but using function `quantile` for computing sample quantiles only.

A function in R is nothing but an object, and all objects are created equal. Thus, we 'just' have to assign a *function* object to a variable. A *function* object consists of an argument list, defining arguments and possibly default values, and a body defining the computations. The body starts and ends with braces. Of course, the body is assumed to be valid R code. In most cases we expect a function to return an object, therefore, the body will contain one or more `return` statements the arguments of which define the return values.

Returning to our example, we'll name our function `iqr`. The `iqr` function should operate on numeric vectors, therefore it should have an argument `x`. This numeric vector will be passed on to the `quantile` function for computing the sample quartiles. The required difference between the 3rd and 1st quartile can then be computed using `diff`. The definition of our function reads as follows

```
R> iqr <- function(x) {
+      q <- quantile(x, prob = c(0.25, 0.75), names = FALSE)
+      return(diff(q))
+ }
```

A simple test on simulated data from a standard normal distribution shows that our first function actually works, a comparison with the `IQR` function shows that the result is correct:

```
R> xdata <- rnorm(100)
R> iqr(xdata)

[1] 1.495980

R> IQR(xdata)

[1] 1.495980
```

However, when the numeric vector contains missing values, our function fails as the following example shows:

```
R> xdata[1] <- NA
R> iqr(xdata)
```

```
Error in quantile.default(x, prob = c(0.25, 0.75)):
  missing values and NaN's not allowed if 'na.rm' is FALSE
```

In order to make our little function more flexible it would be helpful to add all arguments of `quantile` to the argument list of `iqr`. The copy-and-paste approach that first comes to mind is likely to lead to inconsistencies and errors, for example when the argument list of `quantile` changes. Instead, the dot argument, a wildcard for any argument, is more appropriate and we redefine our function accordingly:

```
R> iqr <- function(x, ...) {
+      q <- quantile(x, prob = c(0.25, 0.75), names = FALSE,
+                    ...)
+      return(diff(q))
+  }
R> iqr(xdata, na.rm = TRUE)
```

```
[1] 1.503438
```

```
R> IQR(xdata, na.rm = TRUE)
```

```
[1] 1.503438
```

Now, we can assess the variability of the profits using our new `iqr` tool:

```
R> iqr(Forbes2000$profits, na.rm = TRUE)
```

```
[1] 0.36
```

Since there is no difference between functions that have been written by one of the R developers and user-created functions, we can compute the inter-quartile range of profits for each of the business categories by using our `iqr` function inside a `tapply` statement;

```
R> iqr_profits <- tapply(Forbes2000$profits,
+                        Forbes2000$category, iqr, na.rm = TRUE)
```

and extract the categories with the smallest and greatest variability

```
R> levels(Forbes2000$category)[which.min(iqr_profits)]
```

```
[1] "Hotels restaurants & leisure"
```

```
R> levels(Forbes2000$category)[which.max(iqr_profits)]
```

```
[1] "Drugs & biotechnology"
```

We observe less variable profits in tourism enterprises compared with profits in the pharmaceutical industry.

As other members of the `apply` family, `tapply` is very helpful when the same task is to be done more than one time. Moreover, its use is more convenient compared to the usage of `for` loops. For the sake of completeness, we will compute the category-wise inter-quartile range of the profits using a `for` loop.

Like a *function*, a `for` loop consists of a body, i.e., a chain of R commands to be executed. In addition, we need a set of values and a variable that iterates over this set. Here, the set we are interested in is the business categories:

```
R> bcat <- Forbes2000$category
R> iqr_profits2 <- numeric(nlevels(bcat))
R> names(iqr_profits2) <- levels(bcat)
R> for (cat in levels(bcat)) {
+      catprofit <- subset(Forbes2000, category == cat)$profit
+      this_iqr <- iqr(catprofit, na.rm = TRUE)
+      iqr_profits2[levels(bcat) == cat] <- this_iqr
+ }
```

Compared to the usage of `tapply`, the above code is rather complicated. At first, we have to set up a vector for storing the results and assign the appropriate names to it. Next, inside the body of the `for` loop, the `iqr` function has to be called on the appropriate subset of all companies of the current business category `cat`. The corresponding inter-quartile range must then be assigned to the correct vector element in the result vector. Luckily, such complicated constructs will be used in only one of the remaining chapters of the book and are almost always avoidable in practical data analyses.

### *1.7.3 Simple Graphics*

The degree of skewness of a distribution can be investigated by constructing histograms using the `hist` function. (More sophisticated alternatives such as smooth density estimates will be considered in Chapter 8.) For example, the code for producing Figure 1.1 first divides the plot region into two equally spaced rows (the `layout` function) and then plots the histograms of the raw market values in the upper part using the `hist` function. The lower part of the figure depicts the histogram for the log transformed market values which appear to be more symmetric.

Bivariate relationships of two continuous variables are usually depicted as scatterplots. In R, regression relationships are specified by so-called *model formulae* which, in a simple bivariate case, may look like

```
R> fm <- marketvalue ~ sales
R> class(fm)

[1] "formula"
```

with the dependent variable on the left hand side and the independent variable on the right hand side. The tilde separates left and right hand sides. Such a model formula can be passed to a model function (for example to the linear model function as explained in Chapter 6). The `plot` generic function implements a *formula* method as well. Because the distributions of both market value and sales are skewed we choose to depict their logarithms. A raw scatterplot of 2000 data points (Figure 1.2) is rather uninformative due to areas with very high density. This problem can be avoided by choosing a transparent color for the dots as shown in Figure 1.3.

If the independent variable is a factor, a boxplot representation is a natural choice. For four selected countries, the distributions of the logarithms of the

```
R> layout(matrix(1:2, nrow = 2))
R> hist(Forbes2000$marketvalue)
R> hist(log(Forbes2000$marketvalue))
```

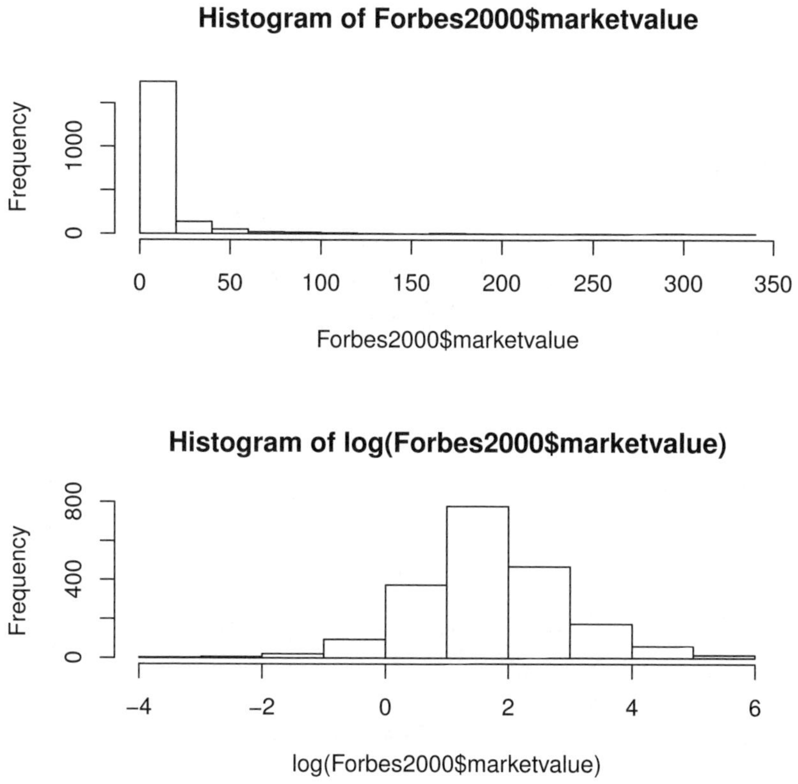

Figure 1.1 Histograms of the market value and the logarithm of the market value for the companies contained in the Forbes 2000 list.

market value may be visually compared in Figure 1.4. Prior to calling the `plot` function on our data, we have to remove empty levels from the `country` variable, because otherwise the $x$-axis would show all and not only the selected countries. This task is most easily performed by subsetting the corresponding factor with additional argument `drop = TRUE`. Here, the width of the boxes are proportional to the square root of the number of companies for each country and extremely large or small market values are depicted by single points. More elaborate graphical methods will be discussed in Chapter 2.

```
R> plot(log(marketvalue) ~ log(sales), data = Forbes2000,
+       pch = ".")
```

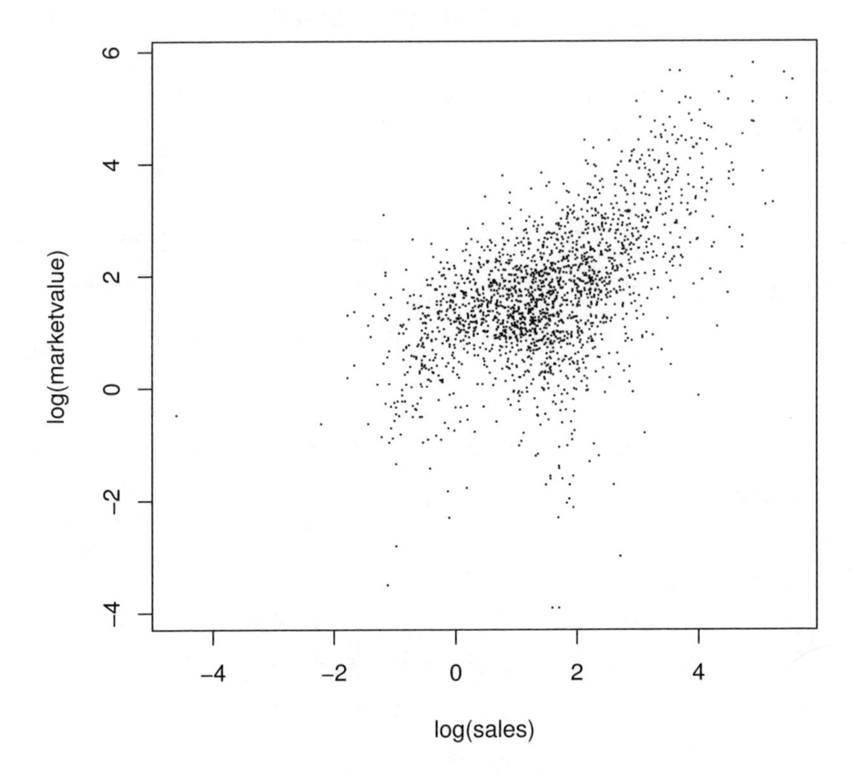


**Figure 1.2** Raw scatterplot of the logarithms of market value and sales.

## 1.8 Organising an Analysis

Although it is possible to perform an analysis typing all commands directly on the R prompt it is much more comfortable to maintain a separate text file collecting all steps necessary to perform a certain data analysis task. Such an R transcript file, for example called `analysis.R` created with your favourite text editor, can be sourced into R using the `source` command

```
R> source("analysis.R", echo = TRUE)
```

When all steps of a data analysis, i.e., data preprocessing, transformations, simple summary statistics and plots, model building and inference as well as reporting, are collected in such an R transcript file, the analysis can be

```
R> plot(log(marketvalue) ~ log(sales), data = Forbes2000,
+       col = rgb(0,0,0,0.1), pch = 16)
```

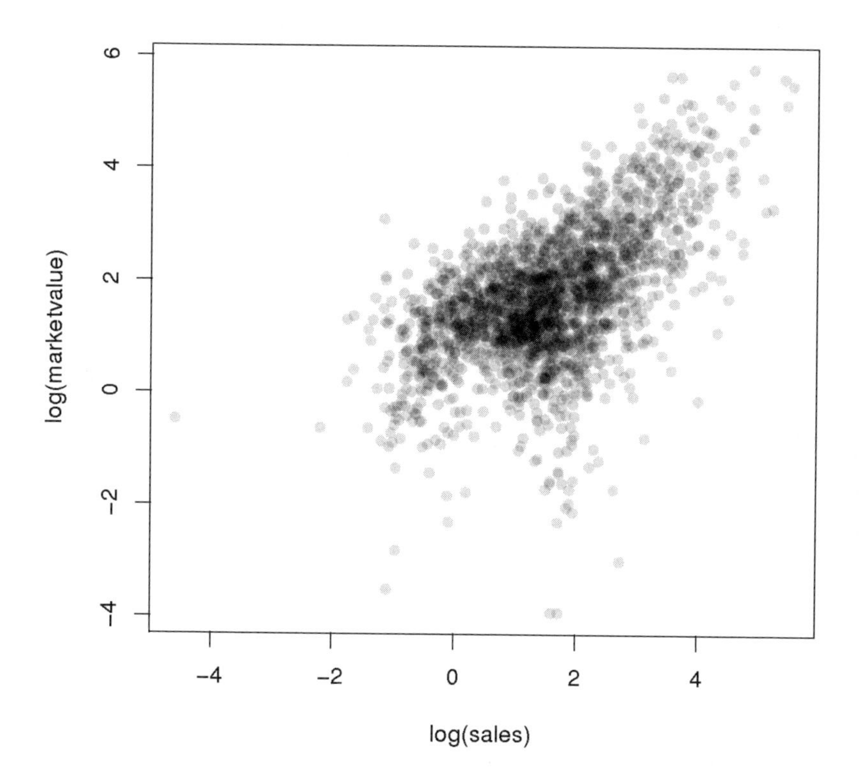


**Figure 1.3** Scatterplot with transparent shading of points of the logarithms of market value and sales.

reproduced at any time, maybe with corrected or updated data as it frequently happens in our consulting practise.

## 1.9 Summary

Reading data into R is possible in many different ways, including direct connections to data base engines. Tabular data are handled by *data.frames* in R, and the usual data manipulation techniques such as sorting, ordering or subsetting can be performed by simple R statements. An overview on data stored in a *data.frame* is given mainly by two functions: `summary` and `str`. Simple graphics such as histograms and scatterplots can be constructed by applying the appropriate R functions (`hist` and `plot`) and we shall give many more

```
R> tmp <- subset(Forbes2000,
+      country %in% c("United Kingdom", "Germany",
+                     "India", "Turkey"))
R> tmp$country <- tmp$country[,drop = TRUE]
R> plot(log(marketvalue) ~ country, data = tmp,
+       ylab = "log(marketvalue)", varwidth = TRUE)
```

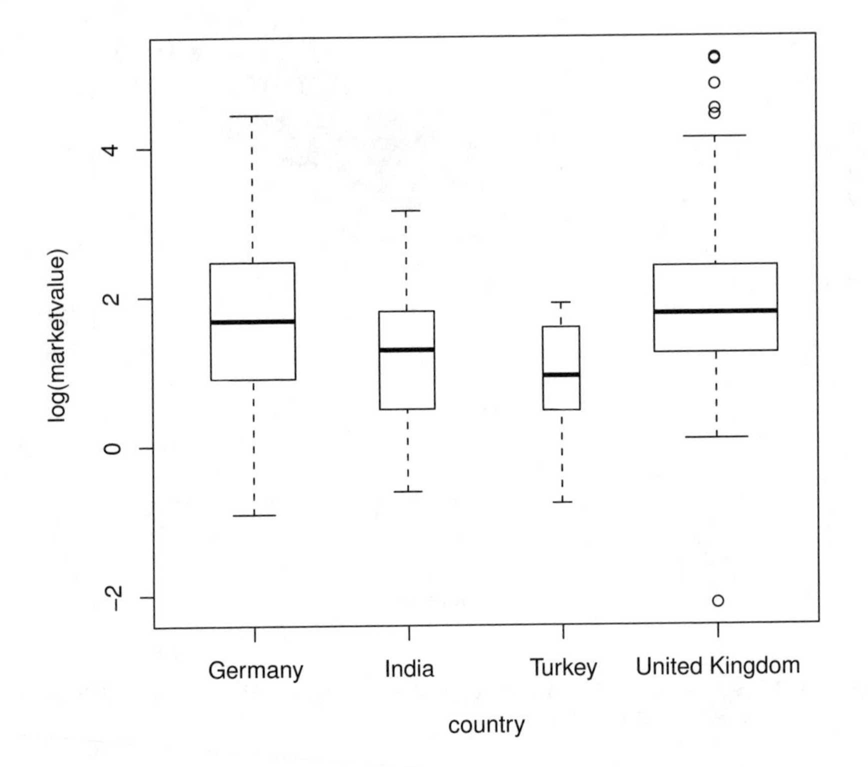


**Figure 1.4** Boxplots of the logarithms of the market value for four selected countries, the width of the boxes is proportional to the square roots of the number of companies.

examples of these functions and those that produce more interesting graphics in later chapters.

**Exercises**

Ex. 1.1 Calculate the median profit for the companies in the US and the median profit for the companies in the UK, France and Germany.

Ex. 1.2 Find all German companies with negative profit.

Ex. 1.3 To which business category do most of the Bermuda island companies belong?

Ex. 1.4 For the 50 companies in the Forbes data set with the highest profits, plot sales against assets (or some suitable transformation of each variable), labelling each point with the appropriate country name which may need to be abbreviated (using `abbreviate`) to avoid making the plot look too 'messy'.

Ex. 1.5 Find the average value of sales for the companies in each country in the Forbes data set, and find the number of companies in each country with profits above 5 billion US dollars.

CHAPTER 2

# Data Analysis Using Graphical Displays: Malignant Melanoma in the USA and Chinese Health and Family Life

## 2.1 Introduction

Fisher and Belle (1993) report mortality rates due to malignant melanoma of the skin for white males during the period 1950–1969, for each state on the US mainland. The data are given in Table 2.1 and include the number of deaths due to malignant melanoma in the corresponding state, the longitude and latitude of the geographic centre of each state, and a binary variable indicating contiguity to an ocean, that is, if the state borders one of the oceans. Questions of interest about these data include: how do the mortality rates compare for ocean and non-ocean states? and how are mortality rates affected by latitude and longitude?

**Table 2.1**: `USmelanoma` data. USA mortality rates for white males due to malignant melanoma.

| | `mortality` | `latitude` | `longitude` | `ocean` |
|---|---|---|---|---|
| Alabama | 219 | 33.0 | 87.0 | yes |
| Arizona | 160 | 34.5 | 112.0 | no |
| Arkansas | 170 | 35.0 | 92.5 | no |
| California | 182 | 37.5 | 119.5 | yes |
| Colorado | 149 | 39.0 | 105.5 | no |
| Connecticut | 159 | 41.8 | 72.8 | yes |
| Delaware | 200 | 39.0 | 75.5 | yes |
| District of Columbia | 177 | 39.0 | 77.0 | no |
| Florida | 197 | 28.0 | 82.0 | yes |
| Georgia | 214 | 33.0 | 83.5 | yes |
| Idaho | 116 | 44.5 | 114.0 | no |
| Illinois | 124 | 40.0 | 89.5 | no |
| Indiana | 128 | 40.2 | 86.2 | no |
| Iowa | 128 | 42.2 | 93.8 | no |
| Kansas | 166 | 38.5 | 98.5 | no |
| Kentucky | 147 | 37.8 | 85.0 | no |
| Louisiana | 190 | 31.2 | 91.8 | yes |

**Table 2.1**: `USmelanoma` data (continued).

| | mortality | latitude | longitude | ocean |
|---|---|---|---|---|
| Maine | 117 | 45.2 | 69.0 | yes |
| Maryland | 162 | 39.0 | 76.5 | yes |
| Massachusetts | 143 | 42.2 | 71.8 | yes |
| Michigan | 117 | 43.5 | 84.5 | no |
| Minnesota | 116 | 46.0 | 94.5 | no |
| Mississippi | 207 | 32.8 | 90.0 | yes |
| Missouri | 131 | 38.5 | 92.0 | no |
| Montana | 109 | 47.0 | 110.5 | no |
| Nebraska | 122 | 41.5 | 99.5 | no |
| Nevada | 191 | 39.0 | 117.0 | no |
| New Hampshire | 129 | 43.8 | 71.5 | yes |
| New Jersey | 159 | 40.2 | 74.5 | yes |
| New Mexico | 141 | 35.0 | 106.0 | no |
| New York | 152 | 43.0 | 75.5 | yes |
| North Carolina | 199 | 35.5 | 79.5 | yes |
| North Dakota | 115 | 47.5 | 100.5 | no |
| Ohio | 131 | 40.2 | 82.8 | no |
| Oklahoma | 182 | 35.5 | 97.2 | no |
| Oregon | 136 | 44.0 | 120.5 | yes |
| Pennsylvania | 132 | 40.8 | 77.8 | no |
| Rhode Island | 137 | 41.8 | 71.5 | yes |
| South Carolina | 178 | 33.8 | 81.0 | yes |
| South Dakota | 86 | 44.8 | 100.0 | no |
| Tennessee | 186 | 36.0 | 86.2 | no |
| Texas | 229 | 31.5 | 98.0 | yes |
| Utah | 142 | 39.5 | 111.5 | no |
| Vermont | 153 | 44.0 | 72.5 | yes |
| Virginia | 166 | 37.5 | 78.5 | yes |
| Washington | 117 | 47.5 | 121.0 | yes |
| West Virginia | 136 | 38.8 | 80.8 | no |
| Wisconsin | 110 | 44.5 | 90.2 | no |
| Wyoming | 134 | 43.0 | 107.5 | no |

*Source*: From Fisher, L. D., and Belle, G. V., *Biostatistics. A Methodology for the Health Sciences*, John Wiley & Sons, Chichester, UK, 1993. With permission.

Contemporary China is on the leading edge of a sexual revolution, with tremendous regional and generational differences that provide unparalleled natural experiments for analysis of the antecedents and outcomes of sexual behaviour. The Chinese Health and Family Life Study, conducted 1999–2000 as a collaborative research project of the Universities of Chicago, Beijing, and

North Carolina, provides a baseline from which to anticipate and track future changes. Specifically, this study produces a baseline set of results on sexual behaviour and disease patterns, using a nationally representative probability sample. The Chinese Health and Family Life Survey sampled 60 villages and urban neighbourhoods chosen in such a way as to represent the full geographical and socioeconomic range of contemporary China excluding Hong Kong and Tibet. Eighty-three individuals were chosen at random for each location from official registers of adults aged between 20 and 64 years to target a sample of 5000 individuals in total. Here, we restrict our attention to women with current male partners for whom no information was missing, leading to a sample of 1534 women with the following variables (see Table 2.2 for example data sets):

`R_edu`: level of education of the responding woman,

`R_income`: monthly income (in yuan) of the responding woman,

`R_health`: health status of the responding woman in the last year,

`R_happy`: how happy was the responding woman in the last year,

`A_edu`: level of education of the woman's partner,

`A_income`: monthly income (in yuan) of the woman's partner.

In the list above the income variables are continuous and the remaining variables are categorical with ordered categories. The income variables are based on (partially) imputed measures. All information, including the partner's income, are derived from a questionnaire answered by the responding woman only. Here, we focus on graphical displays for inspecting the relationship of these health and socioeconomic variables of heterosexual women and their partners.

## 2.2 Initial Data Analysis

According to Chambers et al. (1983), "there is no statistical tool that is as powerful as a well chosen graph". Certainly, the analysis of most (probably all) data sets should begin with an initial attempt to understand the general characteristics of the data by graphing them in some hopefully useful and informative manner. The possible advantages of graphical presentation methods are summarised by Schmid (1954); they include the following

- In comparison with other types of presentation, well-designed charts are more effective in creating interest and in appealing to the attention of the reader.
- Visual relationships as portrayed by charts and graphs are more easily grasped and more easily remembered.
- The use of charts and graphs saves time, since the essential meaning of large measures of statistical data can be visualised at a glance.
- Charts and graphs provide a comprehensive picture of a problem that makes

**Table 2.2**: CHFLS data. Chinese Health and Family Life Survey.

| | R_edu | R_income | R_health | R_happy | A_edu | A_income |
|---|---|---|---|---|---|---|
| 2 | Senior high school | 900 | Good | Somewhat happy | Senior high school | 500 |
| 3 | Senior high school | 500 | Fair | Somewhat happy | Senior high school | 800 |
| 10 | Senior high school | 800 | Good | Somewhat happy | Junior high school | 700 |
| 11 | Junior high school | 300 | Fair | Somewhat happy | Elementary school | 700 |
| 22 | Junior high school | 300 | Fair | Somewhat happy | Junior high school | 400 |
| 23 | Senior high school | 500 | Excellent | Somewhat happy | Junior college | 900 |
| 24 | Junior high school | 0 | Not good | Very happy | Junior high school | 300 |
| 25 | Junior high school | 100 | Good | Not too happy | Senior high school | 800 |
| 26 | Junior high school | 200 | Fair | Not too happy | Junior college | 200 |
| 32 | Senior high school | 400 | Good | Somewhat happy | Senior high school | 600 |
| 33 | Junior high school | 300 | Not good | Not too happy | Junior high school | 200 |
| 35 | Junior high school | 0 | Fair | Somewhat happy | Junior high school | 400 |
| 36 | Junior high school | 200 | Good | Somewhat happy | Junior high school | 500 |
| 37 | Senior high school | 300 | Excellent | Somewhat happy | Senior high school | 200 |
| 38 | Junior college | 3000 | Fair | Somewhat happy | Junior college | 800 |
| 39 | Junior college | 0 | Fair | Somewhat happy | University | 500 |
| 40 | Senior high school | 500 | Excellent | Somewhat happy | Senior high school | 500 |
| 41 | Junior high school | 0 | Not good | Not too happy | Junior high school | 600 |
| 55 | Senior high school | 0 | Excellent | Somewhat happy | Junior high school | 0 |
| 56 | Junior high school | 500 | Not good | Very happy | Junior high school | 200 |
| 57 | ⋮ | ⋮ | ⋮ | ⋮ | ⋮ | ⋮ |

for a more complete and better balanced understanding than could be derived from tabular or textual forms of presentation.

- Charts and graphs can bring out hidden facts and relationships and can stimulate, as well as aid, analytical thinking and investigation.

Graphs are very popular; it has been estimated that between 900 billion ($9 \times 10^{11}$) and 2 trillion ($2 \times 10^{12}$) images of statistical graphics are printed each year. Perhaps one of the main reasons for such popularity is that graphical presentation of data often provides the vehicle for discovering the unexpected; the human visual system is very powerful in detecting patterns, although the following caveat from the late Carl Sagan (in his book *Contact*) should be kept in mind:

> Humans are good at discerning subtle patterns that are really there, but equally so at imagining them when they are altogether absent.

During the last two decades a wide variety of new methods for displaying data graphically have been developed; these will hunt for special effects in data, indicate outliers, identify patterns, diagnose models and generally search for novel and perhaps unexpected phenomena. Large numbers of graphs may be required and computers are generally needed to supply them for the same reasons they are used for numerical analyses, namely that they are fast and they are accurate.

So, because the machine is doing the work the question is no longer "shall we plot?" but rather "what shall we plot?" There are many exciting possibilities including dynamic graphics but graphical exploration of data usually begins, at least, with some simpler, well-known methods, for example, *histograms*, *barcharts*, *boxplots* and *scatterplots*. Each of these will be illustrated in this chapter along with more complex methods such as *spinograms* and *trellis plots*.

## 2.3 Analysis Using R

### *2.3.1 Malignant Melanoma*

We might begin to examine the malignant melanoma data in Table 2.1 by constructing a histogram or boxplot for *all* the mortality rates in Figure 2.1. The `plot`, `hist` and `boxplot` functions have already been introduced in Chapter 1 and we want to produce a plot where both techniques are applied at once. The `layout` function organises two independent plots on one plotting device, for example on top of each other. Using this relatively simple technique (more advanced methods will be introduced later) we have to make sure that the $x$-axis is the same in both graphs. This can be done by computing a plausible range of the data, later to be specified in a plot via the `xlim` argument:

```
R> xr <- range(USmelanoma$mortality) * c(0.9, 1.1)
R> xr
[1]  77.4 251.9
```

Now, plotting both the histogram and the boxplot requires setting up the plotting device with equal space for two independent plots on top of each other.

```
R> layout(matrix(1:2, nrow = 2))
R> par(mar = par("mar") * c(0.8, 1, 1, 1))
R> boxplot(USmelanoma$mortality, ylim = xr, horizontal = TRUE,
+         xlab = "Mortality")
R> hist(USmelanoma$mortality, xlim = xr, xlab = "", main = "",
+      axes = FALSE, ylab = "")
R> axis(1)
```

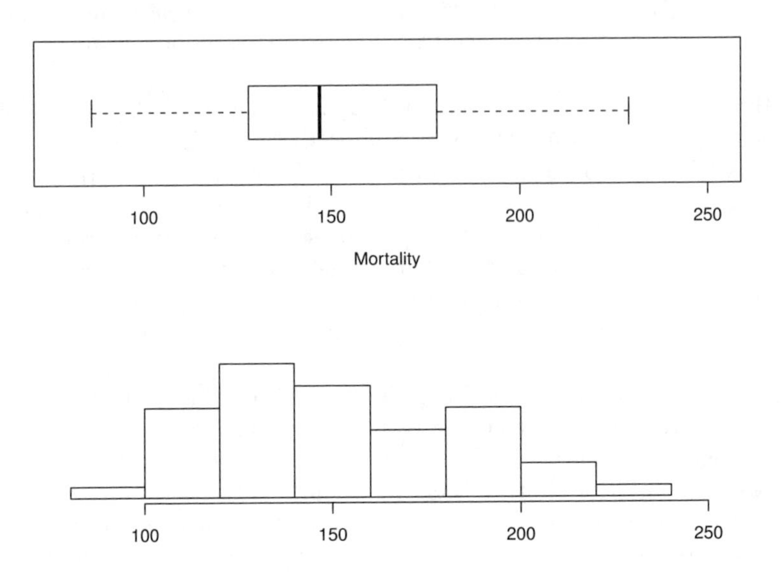


**Figure 2.1** Histogram (top) and boxplot (bottom) of malignant melanoma mortality rates.

Calling the `layout` function on a matrix with two cells in two rows, containing the numbers one and two, leads to such a partitioning. The `boxplot` function is called first on the mortality data and then the `hist` function, where the range of the $x$-axis in both plots is defined by $(77.4, 251.9)$. One tiny problem to solve is the size of the margins; their defaults are too large for such a plot. As with many other graphical parameters, one can adjust their value for a specific plot using function `par`. The R code and the resulting display are given in Figure 2.1.

Both the histogram and the boxplot in Figure 2.1 indicate a certain skewness of the mortality distribution. Looking at the characteristics of all the mortality rates is a useful beginning but for these data we might be more interested in comparing mortality rates for ocean and non-ocean states. So we might construct two histograms or two boxplots. Such a *parallel boxplot*, vi-

```
R> plot(mortality ~ ocean, data = USmelanoma,
+       xlab = "Contiguity to an ocean", ylab = "Mortality")
```

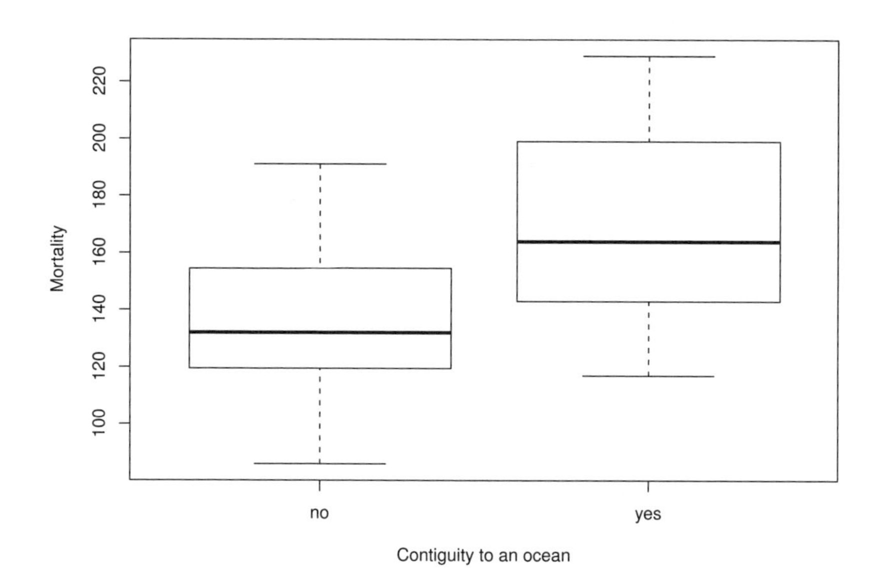


**Figure 2.2** Parallel boxplots of malignant melanoma mortality rates by contiguity to an ocean.

sualising the conditional distribution of a numeric variable in groups as given by a categorical variable, are easily computed using the `boxplot` function. The continuous response variable and the categorical independent variable are specified via a *formula* as described in Chapter 1. Figure 2.2 shows such parallel boxplots, as by default produced the `plot` function for such data, for the mortality in ocean and non-ocean states and leads to the impression that the mortality is increased in east or west coast states compared to the rest of the country.

Histograms are generally used for two purposes: counting and displaying the distribution of a variable; according to Wilkinson (1992), "they are effective for neither". Histograms can often be misleading for displaying distributions because of their dependence on the number of classes chosen. An alternative is to formally estimate the density function of a variable and then plot the resulting estimate; details of density estimation are given in Chapter 8 but for the ocean and non-ocean states the two density estimates can be produced and plotted as shown in Figure 2.3 which supports the impression from Figure 2.2. For more details on such density estimates we refer to Chapter 8.

```
R> dyes <- with(USmelanoma, density(mortality[ocean == "yes"]))
R> dno <- with(USmelanoma, density(mortality[ocean == "no"]))
R> plot(dyes, lty = 1, xlim = xr, main = "", ylim = c(0, 0.018))
R> lines(dno, lty = 2)
R> legend("topleft", lty = 1:2, legend = c("Coastal State",
+         "Land State"), bty = "n")
```

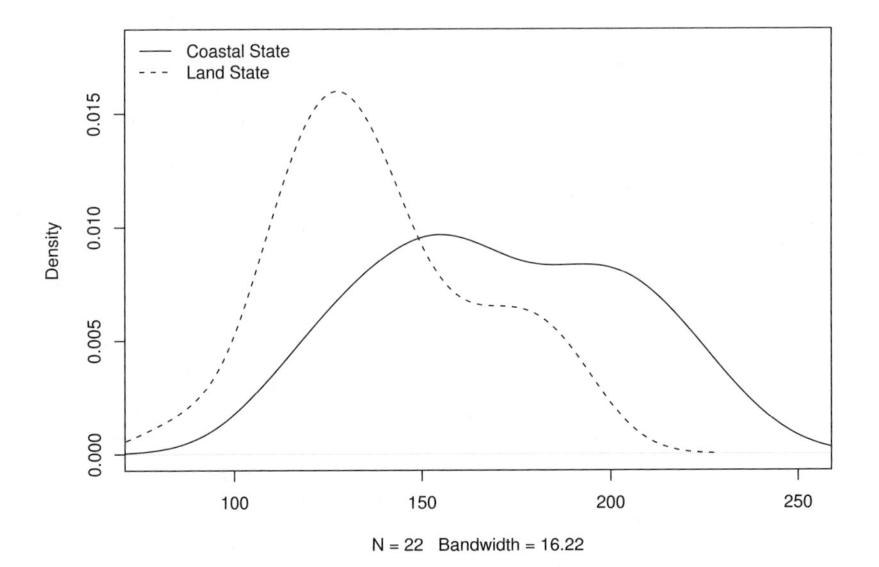


**Figure 2.3** Estimated densities of malignant melanoma mortality rates by contiguity to an ocean.

Now we might move on to look at how mortality rates are related to the geographic location of a state as represented by the latitude and longitude of the centre of the state. Here the main graphic will be the scatterplot. The simple $xy$ scatterplot has been in use since at least the eighteenth century and has many virtues – indeed according to Tufte (1983):

> The relational graphic – in its barest form the scatterplot and its variants – is the greatest of all graphical designs. It links at least two variables, encouraging and even imploring the viewer to assess the possible causal relationship between the plotted variables. It confronts causal theories that $x$ causes $y$ with empirical evidence as to the actual relationship between $x$ and $y$.

Let's begin with simple scatterplots of mortality rate against longitude and mortality rate against latitude which can be produced by the code preceding Figure 2.4. Again, the `layout` function is used for partitioning the plotting device, now resulting in two side by-side-plots. The argument to `layout` is

```
R> layout(matrix(1:2, ncol = 2))
R> plot(mortality ~ longitude, data = USmelanoma)
R> plot(mortality ~ latitude, data = USmelanoma)
```

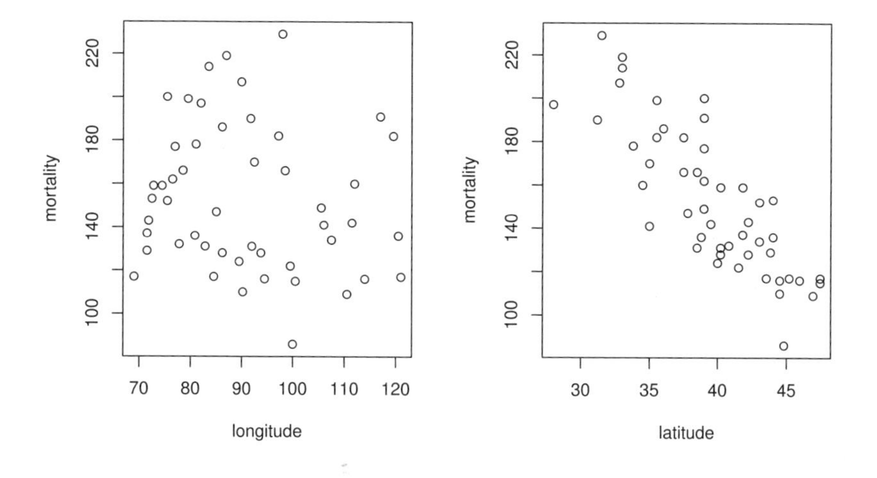


**Figure 2.4** Scatterplot of malignant melanoma mortality rates by geographical location.

now a matrix with only one row but two columns containing the numbers one and two. In each cell, the `plot` function is called for producing a scatterplot of the variables given in the *formula*.

Since mortality rate is clearly related only to latitude we can now produce scatterplots of mortality rate against latitude separately for ocean and non-ocean states. Instead of producing two displays, one can choose different plotting symbols for either states. This can be achieved by specifying a vector of integers or characters to the `pch`, where the $i$th element of this vector defines the plot symbol of the $i$th observation in the data to be plotted. For the sake of simplicity, we convert the `ocean` factor to an *integer* vector containing the numbers one for land states and two for ocean states. As a consequence, land states can be identified by the dot symbol and ocean states by triangles. It is useful to add a legend to such a plot, most conveniently by using the `legend` function. This function takes three arguments: a string indicating the position of the legend in the plot, a character vector of labels to be printed and the corresponding plotting symbols (referred to by integers). In addition, the display of a bounding box is anticipated (`bty = "n"`). The scatterplot in Figure 2.5 highlights that the mortality is lowest in the northern land states. Coastal states show a higher mortality than land states at roughly the same

```
R> plot(mortality ~ latitude, data = USmelanoma,
+       pch = as.integer(USmelanoma$ocean))
R> legend("topright", legend = c("Land state", "Coast state"),
+        pch = 1:2, bty = "n")
```

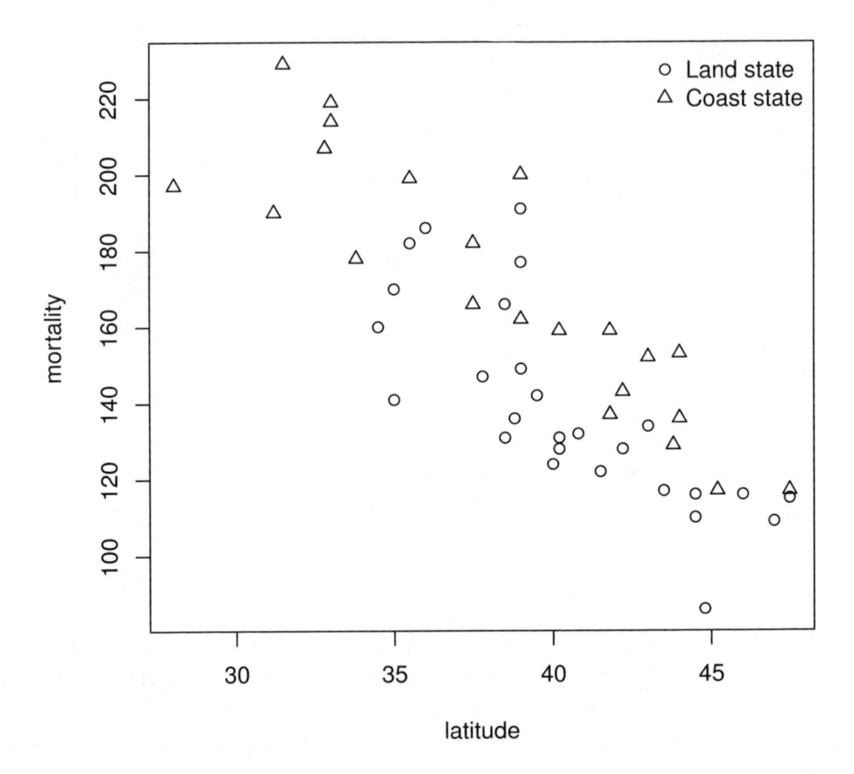


**Figure 2.5** Scatterplot of malignant melanoma mortality rates against latitude.

latitude. The highest mortalities can be observed for the south coastal states with latitude less than 32°, say, that is

```
R> subset(USmelanoma, latitude < 32)
```

```
          mortality latitude longitude ocean
Florida         197     28.0      82.0   yes
Louisiana       190     31.2      91.8   yes
Texas           229     31.5      98.0   yes
```

Up to now we have primarily focused on the visualisation of continuous variables. We now extend our focus to the visualisation of categorical variables.

```
R> barplot(xtabs(~ R_happy, data = CHFLS))
```

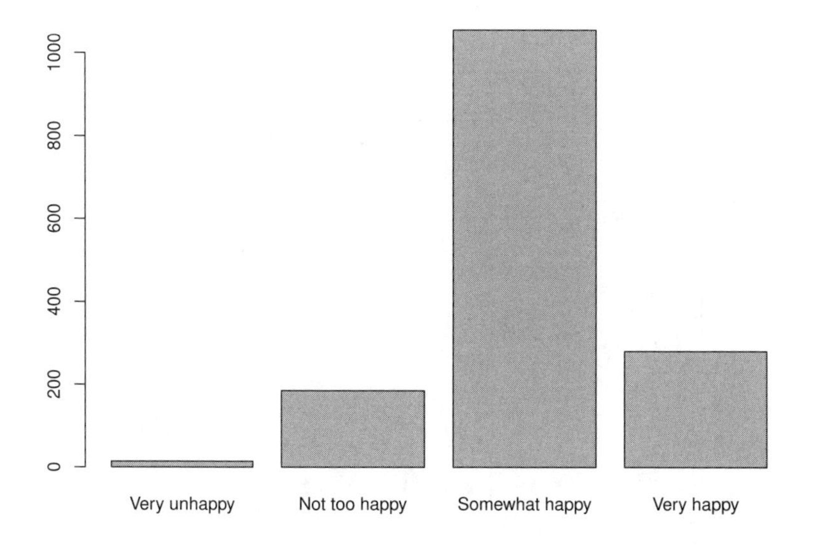


**Figure 2.6** Bar chart of happiness.

### *2.3.2 Chinese Health and Family Life*

One part of the questionnaire the Chinese Health and Family Life Survey focuses on is the self-reported health status. Two questions are interesting for us. The first one is "Generally speaking, do you consider the condition of your health to be excellent, good, fair, not good, or poor?". The second question is "Generally speaking, in the past twelve months, how happy were you?". The distribution of such variables is commonly visualised using barcharts where for each category the total or relative number of observations is displayed. Such a barchart can conveniently be produced by applying the `barplot` function to a tabulation of the data. The empirical density of the variable `R_happy` is computed by the `xtabs` function for producing (contingency) tables; the resulting barchart is given in Figure 2.6.

The visualisation of two categorical variables could be done by conditional barcharts, i.e., barcharts of the first variable within the categories of the second variable. An attractive alternative for displaying such two-way tables are *spineplots* (Friendly, 1994, Hofmann and Theus, 2005, Chen et al., 2008); the meaning of the name will become clear when looking at such a plot in Figure 2.7.

Before constructing such a plot, we produce a two-way table of the health status and self-reported happiness using the `xtabs` function:

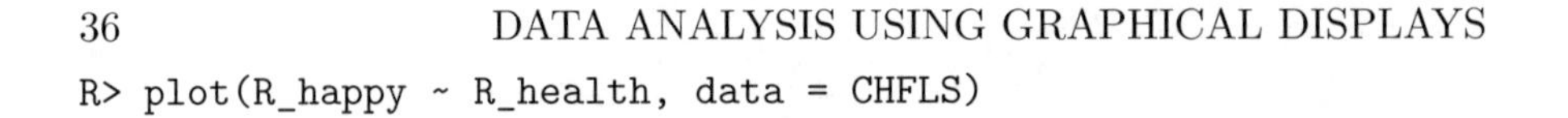

```
R> plot(R_happy ~ R_health, data = CHFLS)
```

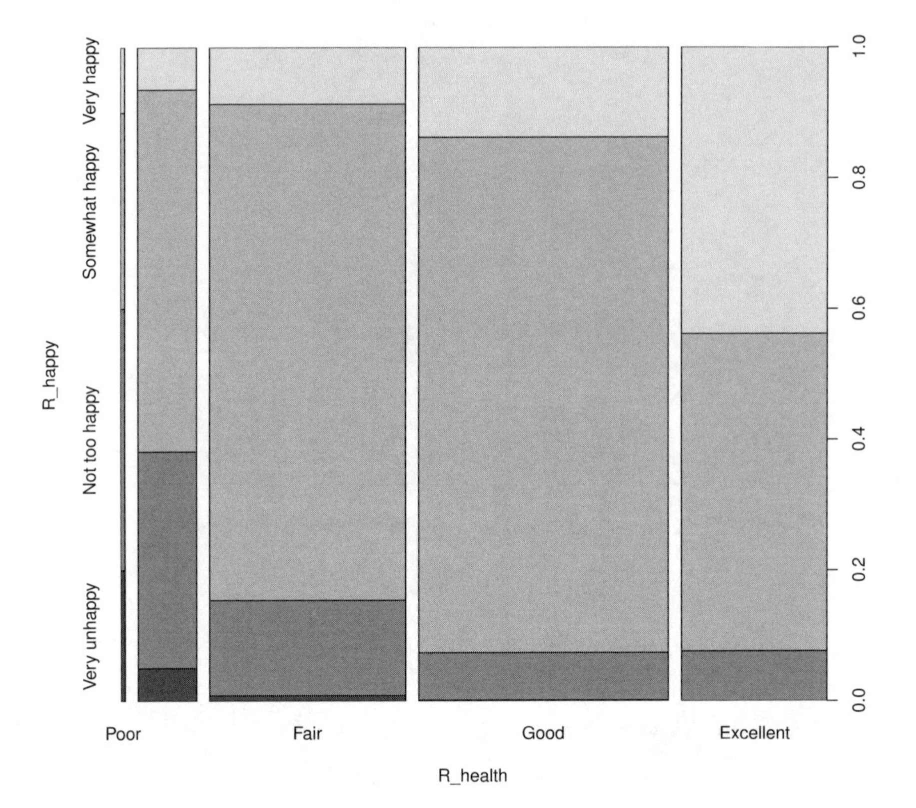


**Figure 2.7** Spineplot of health status and happiness.

```
R> xtabs(~ R_happy + R_health, data = CHFLS)
                R_health
R_happy          Poor Not good Fair Good Excellent
  Very unhappy      2        7    4    1         0
  Not too happy     4       46   67   42        26
  Somewhat happy    3       77  350  459       166
  Very happy        1        9   40   80       150
```

A *spineplot* is a group of rectangles, each representing one cell in the two-way contingency table. The area of the rectangle is proportional with the number of observations in the cell. Here, we produce a mosaic plot of health status and happiness in Figure 2.7.

Consider the right upper cell in Figure 2.7, i.e., the 150 very happy women with excellent health status. The width of the right-most bar corresponds to the frequency of women with excellent health status. The length of the topright rectangle corresponds to the conditional frequency of very happy women given their health status is excellent. Multiplying these two quantities gives the area of this cell which corresponds to the frequency of women who are both very happy and enjoy an excellent health status. The conditional frequency of very happy women increases with increasing health status, whereas the conditional frequency of very unhappy or not too happy women decreases.

When the association of a categorical and a continuous variable is of interest, say the monthly income and self-reported happiness, one might use parallel boxplots to visualise the distribution of the income depending on happiness. If we were studying self-reported happiness as response and income as independent variable, however, this would give a representation of the conditional distribution of income given happiness, but we are interested in the conditional distribution of happiness given income. One possibility to produce a more appropriate plot is called *spinogram*. Here, the continuous $x$-variable is categorised first. Within each of these categories, the conditional frequencies of the response variable are given by stacked barcharts, in a way similar to spineplots. For happiness depending on log-income (since income is naturally skewed we use a log-transformation of the income) it seems that the proportion of unhappy and not too happy women decreases with increasing income whereas the proportion of very happy women stays rather constant. In contrast to spinograms, where bins, as in a histogram, are given on the $x$-axis, a *conditional density plot* uses the original $x$-axis for a display of the conditional density of the categorical response given the independent variable.

For our last example we return to scatterplots for inspecting the association between a woman's monthly income and the income of her partner. Both income variables have been computed and partially imputed from other self-reported variables and are only rough assessments of the real income. Moreover, the data itself is numeric but heavily tied, making it difficult to produce 'correct' scatterplots because points will overlap. A relatively easy trick is to jitter the observation by adding a small random noise to each point in order to avoid overlapping plotting symbols. In addition, we want to study the relationship between both monthly incomes conditional on the woman's education. Such conditioning plots are called *trellis* plots and are implemented in the package **lattice** (Sarkar, 2009, 2008). We utilise the `xyplot` function from package **lattice** to produce a scatterplot. The formula reads as already explained with the exception that a third *conditioning* variable, `R_edu` in our case, is present. For each level of education, a separate scatterplot will be produced. The plots are directly comparable since the axes remain the same for all plots.

The plot reveals several interesting issues. Some observations are positioned on a straight line with slope one, most probably an artifact of missing value imputation by linear models (as described in the data dictionary, see `?CHFLS`). Four constellations can be identified: both partners have zero income, the partner has no income, the woman has no income or both partners have a positive income.

```
R> layout(matrix(1:2, ncol = 2))
R> plot(R_happy ~ log(R_income + 1), data = CHFLS)
R> cdplot(R_happy ~ log(R_income + 1), data = CHFLS)
```

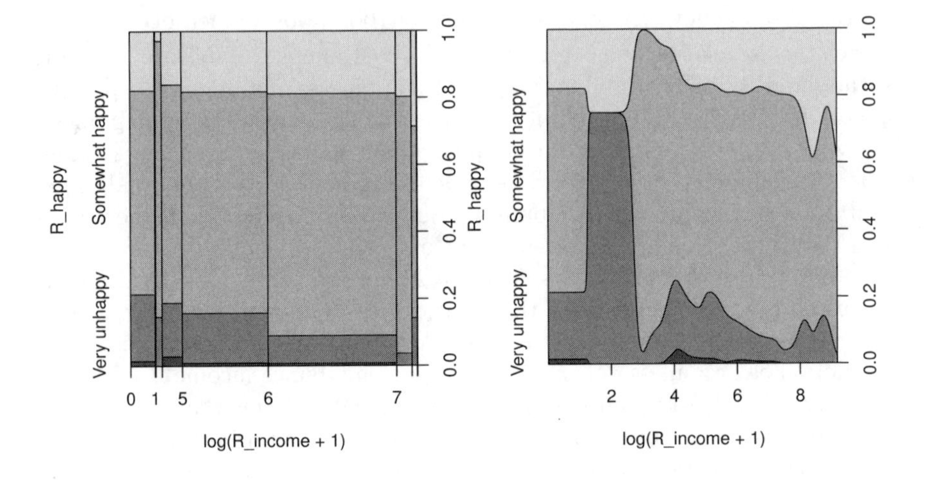


**Figure 2.8** Spinogram (left) and conditional density plot (right) of happiness depending on log-income

For couples where the woman has a university degree, the income of both partners is relatively high (except for two couples where only the woman has income). A small number of former junior college students live in relationships where only the man has income, the income of both partners seems only slightly positively correlated for the remaining couples. For lower levels of education, all four constellations are present. The frequency of couples where only the man has some income seems larger than the other way around. Ignoring the observations on the straight line, there is almost no association between the income of both partners.

## 2.4 Summary

Producing publication-quality graphics is one of the major strengths of the R system and almost anything is possible since graphics are programmable in R. Naturally, this chapter can be only a very brief introduction to some commonly used displays and the reader is referred to specialised books, most important Murrell (2005), Sarkar (2008), and Chen et al. (2008). Interactive 3D-graphics are available from package **rgl** (Adler and Murdoch, 2009).

```
R> xyplot(jitter(log(A_income + 0.5)) ~
+          jitter(log(R_income + 0.5)) | R_edu, data = CHFLS)
```

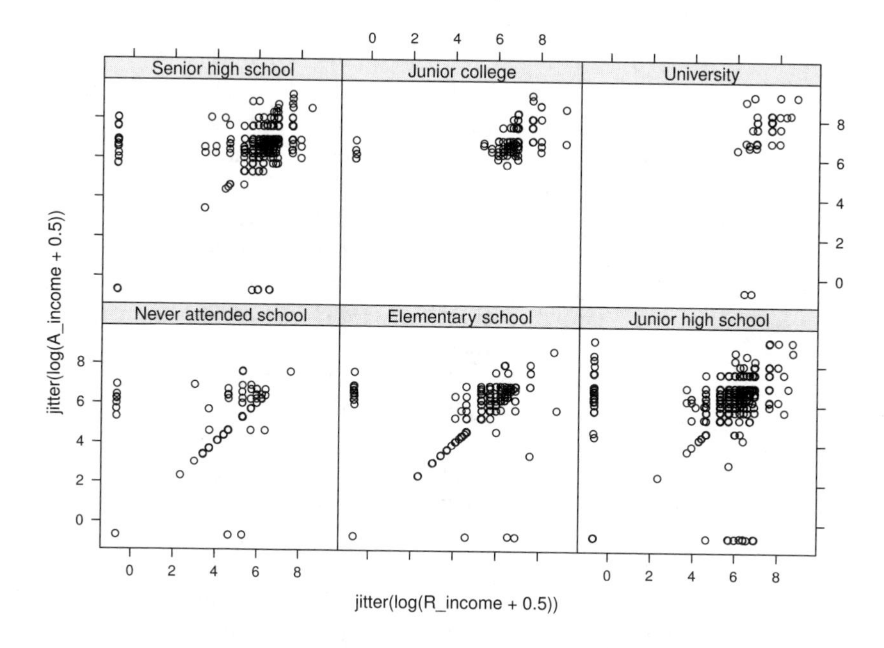


## Exercises

Ex. 2.1 The data in Table 2.3 are part of a data set collected from a survey of household expenditure and give the expenditure of 20 single men and 20 single women on four commodity groups. The units of expenditure are Hong Kong dollars, and the four commodity groups are

`housing`: housing, including fuel and light,

`food`: foodstuffs, including alcohol and tobacco,

`goods`: other goods, including clothing, footwear and durable goods,

`services`: services, including transport and vehicles.

The aim of the survey was to investigate how the division of household expenditure between the four commodity groups depends on total expenditure and to find out whether this relationship differs for men and women. Use appropriate graphical methods to answer these questions and state your conclusions.

**Table 2.3**: `household` data. Household expenditure for single men and women.

| housing | food | goods | service | gender |
|---|---|---|---|---|
| 820 | 114 | 183 | 154 | female |
| 184 | 74 | 6 | 20 | female |
| 921 | 66 | 1686 | 455 | female |
| 488 | 80 | 103 | 115 | female |
| 721 | 83 | 176 | 104 | female |
| 614 | 55 | 441 | 193 | female |
| 801 | 56 | 357 | 214 | female |
| 396 | 59 | 61 | 80 | female |
| 864 | 65 | 1618 | 352 | female |
| 845 | 64 | 1935 | 414 | female |
| 404 | 97 | 33 | 47 | female |
| 781 | 47 | 1906 | 452 | female |
| 457 | 103 | 136 | 108 | female |
| 1029 | 71 | 244 | 189 | female |
| 1047 | 90 | 653 | 298 | female |
| 552 | 91 | 185 | 158 | female |
| 718 | 104 | 583 | 304 | female |
| 495 | 114 | 65 | 74 | female |
| 382 | 77 | 230 | 147 | female |
| 1090 | 59 | 313 | 177 | female |
| 497 | 591 | 153 | 291 | male |
| 839 | 942 | 302 | 365 | male |
| 798 | 1308 | 668 | 584 | male |
| 892 | 842 | 287 | 395 | male |
| 1585 | 781 | 2476 | 1740 | male |
| 755 | 764 | 428 | 438 | male |
| 388 | 655 | 153 | 233 | male |
| 617 | 879 | 757 | 719 | male |
| 248 | 438 | 22 | 65 | male |
| 1641 | 440 | 6471 | 2063 | male |
| 1180 | 1243 | 768 | 813 | male |
| 619 | 684 | 99 | 204 | male |
| 253 | 422 | 15 | 48 | male |
| 661 | 739 | 71 | 188 | male |
| 1981 | 869 | 1489 | 1032 | male |
| 1746 | 746 | 2662 | 1594 | male |
| 1865 | 915 | 5184 | 1767 | male |
| 238 | 522 | 29 | 75 | male |
| 1199 | 1095 | 261 | 344 | male |
| 1524 | 964 | 1739 | 1410 | male |

Ex. 2.2 Mortality rates per 100,000 from male suicides for a number of age groups and a number of countries are given in Table 2.4. Construct side-by-side box plots for the data from different age groups, and comment on what the graphic tells us about the data.

**Table 2.4:** `suicides2` data. Mortality rates per 100,000 from male suicides.

| | `A25.34` | `A35.44` | `A45.54` | `A55.64` | `A65.74` |
|---|---|---|---|---|---|
| Canada | 22 | 27 | 31 | 34 | 24 |
| Israel | 9 | 19 | 10 | 14 | 27 |
| Japan | 22 | 19 | 21 | 31 | 49 |
| Austria | 29 | 40 | 52 | 53 | 69 |
| France | 16 | 25 | 36 | 47 | 56 |
| Germany | 28 | 35 | 41 | 49 | 52 |
| Hungary | 48 | 65 | 84 | 81 | 107 |
| Italy | 7 | 8 | 11 | 18 | 27 |
| Netherlands | 8 | 11 | 18 | 20 | 28 |
| Poland | 26 | 29 | 36 | 32 | 28 |
| Spain | 4 | 7 | 10 | 16 | 22 |
| Sweden | 28 | 41 | 46 | 51 | 35 |
| Switzerland | 22 | 34 | 41 | 50 | 51 |
| UK | 10 | 13 | 15 | 17 | 22 |
| USA | 20 | 22 | 28 | 33 | 37 |

Ex. 2.3 The data set shown in Table 2.5 contains values of seven variables for ten states in the US. The seven variables are

`Population`: population size divided by 1000,

`Income`: average per capita income,

`Illiteracy`: illiteracy rate (% population),

`Life.Expectancy`: life expectancy (years),

`Homicide`: homicide rate (per 1000),

`Graduates`: percentage of high school graduates,

`Freezing`: average number of days per below freezing.

With these data

1. Construct a scatterplot matrix of the data labelling the points by state name (using function `text`).
2. Construct a plot of life expectancy and homicide rate conditional on average per capita income.

**Table 2.5**: `USstates` data. Socio-demographic variables for ten US states.

| Population | Income | Illiteracy | Life.Expectancy | Homicide | Graduates | Freezing |
|---|---|---|---|---|---|---|
| 3615 | 3624 | 2.1 | 69.05 | 15.1 | 41.3 | 20 |
| 21198 | 5114 | 1.1 | 71.71 | 10.3 | 62.6 | 20 |
| 2861 | 4628 | 0.5 | 72.56 | 2.3 | 59.0 | 140 |
| 2341 | 3098 | 2.4 | 68.09 | 12.5 | 41.0 | 50 |
| 812 | 4281 | 0.7 | 71.23 | 3.3 | 57.6 | 174 |
| 10735 | 4561 | 0.8 | 70.82 | 7.4 | 53.2 | 124 |
| 2284 | 4660 | 0.6 | 72.13 | 4.2 | 60.0 | 44 |
| 11860 | 4449 | 1.0 | 70.43 | 6.1 | 50.2 | 126 |
| 681 | 4167 | 0.5 | 72.08 | 1.7 | 52.3 | 172 |
| 472 | 3907 | 0.6 | 71.64 | 5.5 | 57.1 | 168 |

Ex. 2.4 Flury and Riedwyl (1988) report data that give various lengths measurements on 200 Swiss bank notes. The data are available from package **alr3** (Weisberg, 2008); a sample of ten bank notes is given in Table 2.6.

**Table 2.6**: `banknote` data (package **alr3**). Swiss bank note data.

| Length | Left | Right | Bottom | Top | Diagonal |
|---|---|---|---|---|---|
| 214.8 | 131.0 | 131.1 | 9.0 | 9.7 | 141.0 |
| 214.6 | 129.7 | 129.7 | 8.1 | 9.5 | 141.7 |
| 214.8 | 129.7 | 129.7 | 8.7 | 9.6 | 142.2 |
| 214.8 | 129.7 | 129.6 | 7.5 | 10.4 | 142.0 |
| 215.0 | 129.6 | 129.7 | 10.4 | 7.7 | 141.8 |
| 214.4 | 130.1 | 130.3 | 9.7 | 11.7 | 139.8 |
| 214.9 | 130.5 | 130.2 | 11.0 | 11.5 | 139.5 |
| 214.9 | 130.3 | 130.1 | 8.7 | 11.7 | 140.2 |
| 215.0 | 130.4 | 130.6 | 9.9 | 10.9 | 140.3 |
| 214.7 | 130.2 | 130.3 | 11.8 | 10.9 | 139.7 |
| ⋮ | ⋮ | ⋮ | ⋮ | ⋮ | ⋮ |

Use whatever graphical techniques you think are appropriate to investigate whether there is any ‘pattern’ or structure in the data. Do you observe something suspicious?

CHAPTER 3

# Simple Inference: Guessing Lengths, Wave Energy, Water Hardness, Piston Rings, and Rearrests of Juveniles

## 3.1 Introduction

Shortly after metric units of length were officially introduced in Australia in the 1970s, each of a group of 44 students was asked to guess, to the nearest metre, the width of the lecture hall in which they were sitting. Another group of 69 students in the same room was asked to guess the width in feet, to the nearest foot. The data were collected by Professor T. Lewis, and are given here in Table 3.1, which is taken from Hand et al. (1994). The main question is whether estimation in feet and in metres gives different results.

**Table 3.1**: `roomwidth` data. Room width estimates (`width`) in feet and in metres (`unit`).

| unit | width | unit | width | unit | width | unit | width |
|---|---|---|---|---|---|---|---|
| metres | 8 | metres | 16 | feet | 34 | feet | 45 |
| metres | 9 | metres | 16 | feet | 35 | feet | 45 |
| metres | 10 | metres | 17 | feet | 35 | feet | 45 |
| metres | 10 | metres | 17 | feet | 36 | feet | 45 |
| metres | 10 | metres | 17 | feet | 36 | feet | 45 |
| metres | 10 | metres | 17 | feet | 36 | feet | 46 |
| metres | 10 | metres | 18 | feet | 37 | feet | 46 |
| metres | 10 | metres | 18 | feet | 37 | feet | 47 |
| metres | 11 | metres | 20 | feet | 40 | feet | 48 |
| metres | 11 | metres | 22 | feet | 40 | feet | 48 |
| metres | 11 | metres | 25 | feet | 40 | feet | 50 |
| metres | 11 | metres | 27 | feet | 40 | feet | 50 |
| metres | 12 | metres | 35 | feet | 40 | feet | 50 |
| metres | 12 | metres | 38 | feet | 40 | feet | 51 |
| metres | 13 | metres | 40 | feet | 40 | feet | 54 |
| metres | 13 | feet | 24 | feet | 40 | feet | 54 |
| metres | 13 | feet | 25 | feet | 40 | feet | 54 |
| metres | 14 | feet | 27 | feet | 41 | feet | 55 |
| metres | 14 | feet | 30 | feet | 41 | feet | 55 |
| metres | 14 | feet | 30 | feet | 42 | feet | 60 |

**Table 3.1**: `roomwidth` data (continued).

| unit | width | unit | width | unit | width | unit | width |
|---|---|---|---|---|---|---|---|
| metres | 15 | feet | 30 | feet | 42 | feet | 60 |
| metres | 15 | feet | 30 | feet | 42 | feet | 63 |
| metres | 15 | feet | 30 | feet | 42 | feet | 70 |
| metres | 15 | feet | 30 | feet | 43 | feet | 75 |
| metres | 15 | feet | 32 | feet | 43 | feet | 80 |
| metres | 15 | feet | 32 | feet | 44 | feet | 94 |
| metres | 15 | feet | 33 | feet | 44 | | |
| metres | 15 | feet | 34 | feet | 44 | | |
| metres | 16 | feet | 34 | feet | 45 | | |

In a design study for a device to generate electricity from wave power at sea, experiments were carried out on scale models in a wave tank to establish how the choice of mooring method for the system affected the bending stress produced in part of the device. The wave tank could simulate a wide range of sea states and the model system was subjected to the same sample of sea states with each of two mooring methods, one of which was considerably cheaper than the other. The resulting data (from Hand et al., 1994, giving root mean square bending moment in Newton metres) are shown in Table 3.2. The question of interest is whether bending stress differs for the two mooring methods.

**Table 3.2**: `waves` data. Bending stress (root mean squared bending moment in Newton metres) for two mooring methods in a wave energy experiment.

| method1 | method2 | method1 | method2 | method1 | method2 |
|---|---|---|---|---|---|
| 2.23 | 1.82 | 8.98 | 8.88 | 5.91 | 6.44 |
| 2.55 | 2.42 | 0.82 | 0.87 | 5.79 | 5.87 |
| 7.99 | 8.26 | 10.83 | 11.20 | 5.50 | 5.30 |
| 4.09 | 3.46 | 1.54 | 1.33 | 9.96 | 9.82 |
| 9.62 | 9.77 | 10.75 | 10.32 | 1.92 | 1.69 |
| 1.59 | 1.40 | 5.79 | 5.87 | 7.38 | 7.41 |

The data shown in Table 3.3 were collected in an investigation of environmental causes of disease and are taken from Hand et al. (1994). They show the annual mortality per 100,000 for males, averaged over the years 1958–1964, and the calcium concentration (in parts per million) in the drinking water for 61 large towns in England and Wales. The higher the calcium concentration, the harder the water. Towns at least as far north as Derby are identified in the table. Here there are several questions that might be of interest including: are mortality and water hardness related, and do either or both variables differ between northern and southern towns?

**Table 3.3**: `water` data. Mortality (per 100,000 males per year, `mortality`) and water hardness for 61 cities in England and Wales.

| location | town | mortality | hardness |
|---|---|---|---|
| South | Bath | 1247 | 105 |
| North | Birkenhead | 1668 | 17 |
| South | Birmingham | 1466 | 5 |
| North | Blackburn | 1800 | 14 |
| North | Blackpool | 1609 | 18 |
| North | Bolton | 1558 | 10 |
| North | Bootle | 1807 | 15 |
| South | Bournemouth | 1299 | 78 |
| North | Bradford | 1637 | 10 |
| South | Brighton | 1359 | 84 |
| South | Bristol | 1392 | 73 |
| North | Burnley | 1755 | 12 |
| South | Cardiff | 1519 | 21 |
| South | Coventry | 1307 | 78 |
| South | Croydon | 1254 | 96 |
| North | Darlington | 1491 | 20 |
| North | Derby | 1555 | 39 |
| North | Doncaster | 1428 | 39 |
| South | East Ham | 1318 | 122 |
| South | Exeter | 1260 | 21 |
| North | Gateshead | 1723 | 44 |
| North | Grimsby | 1379 | 94 |
| North | Halifax | 1742 | 8 |
| North | Huddersfield | 1574 | 9 |
| North | Hull | 1569 | 91 |
| South | Ipswich | 1096 | 138 |
| North | Leeds | 1591 | 16 |
| South | Leicester | 1402 | 37 |
| North | Liverpool | 1772 | 15 |
| North | Manchester | 1828 | 8 |
| North | Middlesbrough | 1704 | 26 |
| North | Newcastle | 1702 | 44 |
| South | Newport | 1581 | 14 |
| South | Northampton | 1309 | 59 |
| South | Norwich | 1259 | 133 |
| North | Nottingham | 1427 | 27 |
| North | Oldham | 1724 | 6 |

**Table 3.3**: `water` data (continued).

| `location` | `town` | `mortality` | `hardness` |
|---|---|---|---|
| South | Oxford | 1175 | 107 |
| South | Plymouth | 1486 | 5 |
| South | Portsmouth | 1456 | 90 |
| North | Preston | 1696 | 6 |
| South | Reading | 1236 | 101 |
| North | Rochdale | 1711 | 13 |
| North | Rotherham | 1444 | 14 |
| North | St Helens | 1591 | 49 |
| North | Salford | 1987 | 8 |
| North | Sheffield | 1495 | 14 |
| South | Southampton | 1369 | 68 |
| South | Southend | 1257 | 50 |
| North | Southport | 1587 | 75 |
| North | South Shields | 1713 | 71 |
| North | Stockport | 1557 | 13 |
| North | Stoke | 1640 | 57 |
| North | Sunderland | 1709 | 71 |
| South | Swansea | 1625 | 13 |
| North | Wallasey | 1625 | 20 |
| South | Walsall | 1527 | 60 |
| South | West Bromwich | 1627 | 53 |
| South | West Ham | 1486 | 122 |
| South | Wolverhampton | 1485 | 81 |
| North | York | 1378 | 71 |

The two-way contingency table in Table 3.4 shows the number of piston-ring failures in each of three legs of four steam-driven compressors located in the same building (Haberman, 1973). The compressors have identical design and are oriented in the same way. The question of interest is whether the two categorical variables (compressor and leg) are independent.

The data in Table 3.5 (taken from Agresti, 1996) arise from a sample of juveniles convicted of felony in Florida in 1987. Matched pairs were formed using criteria such as age and the number of previous offences. For each pair, one subject was handled in the juvenile court and the other was transferred to the adult court. Whether or not the juvenile was rearrested by the end of 1988 was then noted. Here the question of interest is whether the true proportions rearrested were identical for the adult and juvenile court assignments?

**Table 3.4:** `pistonrings` data. Number of piston ring failures for three legs of four compressors.

| compressor | leg | | |
|---|---|---|---|
| | North | Centre | South |
| C1 | 17 | 17 | 12 |
| C2 | 11 | 9 | 13 |
| C3 | 11 | 8 | 19 |
| C4 | 14 | 7 | 28 |

*Source*: From Haberman, S. J., *Biometrics*, 29, 205–220, 1973. With permission.

**Table 3.5:** `rearrests` data. Rearrests of juvenile felons by type of court in which they were tried.

| Adult court | Juvenile court | |
|---|---|---|
| | Rearrest | No rearrest |
| Rearrest | 158 | 515 |
| No rearrest | 290 | 1134 |

*Source*: From Agresti, A., *An Introduction to Categorical Data Analysis*, John Wiley & Sons, New York, 1996. With permission.

## 3.2 Statistical Tests

Inference, the process of drawing conclusions about a population on the basis of measurements or observations made on a sample of individuals from the population, is central to statistics. In this chapter we shall use the data sets described in the introduction to illustrate both the application of the most common statistical tests, and some simple graphics that may often be used to aid in understanding the results of the tests. Brief descriptions of each of the tests to be used follow.

### *3.2.1 Comparing Normal Populations: Student's t-Tests*

The $t$-test is used to assess hypotheses about two population means where the measurements are assumed to be sampled from a normal distribution. We shall describe two types of $t$-tests, the independent samples test and the paired test.

The independent samples $t$-test is used to test the null hypothesis that the means of two populations are the same, $H_0 : \mu_1 = \mu_2$, when a sample of observations from each population is available. The subjects of one population must not be individually matched with subjects from the other population and the subjects within each group should not be related to each other. The variable to be compared is assumed to have a normal distribution with the same standard deviation in both populations. The test statistic is essentially a standardised difference of the two sample means,

$$t = \frac{\bar{y}_1 - \bar{y}_2}{s\sqrt{1/n_1 + 1/n_2}} \tag{3.1}$$

where $\bar{y}_i$ and $n_i$ are the means and sample sizes in groups $i = 1$ and 2, respectively. The pooled standard deviation $s$ is given by

$$s = \sqrt{\frac{(n_1 - 1)s_1^2 + (n_2 - 1)s_2^2}{n_1 + n_2 - 2}}$$

where $s_1$ and $s_2$ are the standard deviations in the two groups.

Under the null hypothesis, the $t$-statistic has a Student's $t$-distribution with $n_1 + n_2 - 2$ degrees of freedom. A $100(1 - \alpha)\%$ confidence interval for the difference between two means is useful in giving a plausible range of values for the differences in the two means and is constructed as

$$\bar{y}_1 - \bar{y}_2 \pm t_{\alpha, n_1+n_2-2}\, s\sqrt{n_1^{-1} + n_2^{-1}}$$

where $t_{\alpha, n_1+n_2-2}$ is the percentage point of the $t$-distribution such that the cumulative distribution function, $\mathsf{P}(t \leq t_{\alpha, n_1+n_2-2})$, equals $1 - \alpha/2$.

If the two populations are suspected of having different variances, a modified form of the $t$ statistic, known as the Welch test, may be used, namely

$$t = \frac{\bar{y}_1 - \bar{y}_2}{\sqrt{s_1^2/n_1 + s_2^2/n_2}}.$$

In this case, $t$ has a Student's $t$-distribution with $\nu$ degrees of freedom, where

$$\nu = \left(\frac{c}{n_1 - 1} + \frac{(1 - c)^2}{n_2 - 1}\right)^{-1}$$

with

$$c = \frac{s_1^2/n_1}{s_1^2/n_1 + s_2^2/n_2}.$$

A paired $t$-test is used to compare the means of two populations when samples from the populations are available, in which each individual in one sample is paired with an individual in the other sample or each individual in the sample is observed twice. Examples of the former are anorexic girls and their healthy sisters and of the latter the same patients observed before and after treatment.

If the values of the variable of interest, $y$, for the members of the $i$th pair in groups 1 and 2 are denoted as $y_{1i}$ and $y_{2i}$, then the differences $d_i = y_{1i} - y_{2i}$ are assumed to have a normal distribution with mean $\mu$ and the null hypothesis here is that the mean difference is zero, i.e., $H_0 : \mu = 0$. The paired $t$-statistic is

$$t = \frac{\bar{d}}{s/\sqrt{n}}$$

where $\bar{d}$ is the mean difference between the paired measurements and $s$ is its standard deviation. Under the null hypothesis, $t$ follows a $t$-distribution with $n - 1$ degrees of freedom. A $100(1 - \alpha)\%$ confidence interval for $\mu$ can be constructed by

$$\bar{d} \pm t_{\alpha,n-1} s/\sqrt{n}$$

where $\mathsf{P}(t \leq t_{\alpha,n-1}) = 1 - \alpha/2$.

### *3.2.2 Non-parametric Analogues of Independent Samples and Paired t-Tests*

One of the assumptions of both forms of $t$-test described above is that the data have a normal distribution, i.e., are unimodal and symmetric. When departures from those assumptions are extreme enough to give cause for concern, then it might be advisable to use the non-parametric analogues of the $t$-tests, namely the *Wilcoxon Mann-Whitney rank sum test* and the *Wilcoxon signed rank test*. In essence, both procedures throw away the original measurements and only retain the rankings of the observations.

For two independent groups, the Wilcoxon Mann-Whitney rank sum test applies the $t$-statistic to the joint ranks of all measurements in both groups instead of the original measurements. The null hypothesis to be tested is that the two populations being compared have identical distributions. For two normally distributed populations with common variance, this would be equivalent to the hypothesis that the means of the two populations are the same. The alternative hypothesis is that the population distributions differ in location, i.e., the median.

The test is based on the joint ranking of the observations from the two samples (as if they were from a single sample). The test statistic is the sum of the ranks of one sample (the lower of the two rank sums is generally used). A version of this test applicable in the presence of ties is discussed in Chapter 4.

For small samples, $p$-values for the test statistic can be assigned relatively simply. A large sample approximation is available that is suitable when the two sample sizes are greater and there are no ties. In R, the large sample approximation is used by default when the sample size in one group exceeds 50 observations.

In the paired situation, we first calculate the differences $d_i = y_{1i} - y_{2i}$ between each pair of observations. To compute the Wilcoxon signed-rank statistic, we rank the absolute differences $|d_i|$. The statistic is defined as the sum of the ranks associated with positive difference $d_i > 0$. Zero differences are discarded, and the sample size $n$ is altered accordingly. Again, $p$-values for small sample sizes can be computed relatively simply and a large sample approximation is available. It should be noted that this test is valid only when the differences $d_i$ are symmetrically distributed.

### *3.2.3 Testing Independence in Contingency Tables*

When a sample of $n$ observations in two nominal (categorical) variables are available, they can be arranged into a cross-classification (see Table 3.6) in which the number of observations falling in each cell of the table is recorded. Table 3.6 is an example of such a contingency table, in which the observations for a sample of individuals or objects are cross-classified with respect to two categorical variables. Testing for the independence of the two variables $x$ and $y$ is of most interest in general and details of the appropriate test follow.

**Table 3.6**: The general $r \times c$ table.

| | | $y$ | | | |
|---|---|---|---|---|---|
| | | 1 | ... | $c$ | |
| | 1 | $n_{11}$ | ... | $n_{1c}$ | $n_{1\cdot}$ |
| | 2 | $n_{21}$ | ... | $n_{2c}$ | $n_{2\cdot}$ |
| $x$ | $\vdots$ | $\vdots$ | ... | $\vdots$ | $\vdots$ |
| | $r$ | $n_{r1}$ | ... | $n_{rc}$ | $n_{r\cdot}$ |
| | | $n_{\cdot 1}$ | ... | $n_{\cdot c}$ | $n$ |

Under the null hypothesis of independence of the row variable $x$ and the column variable $y$, estimated expected values $E_{jk}$ for cell $(j, k)$ can be computed from the corresponding margin totals $E_{jk} = n_{j\cdot} n_{\cdot k}/n$. The test statistic for assessing independence is

$$X^2 = \sum_{j=1}^{r} \sum_{k=1}^{c} \frac{(n_{jk} - E_{jk})^2}{E_{jk}}.$$

Under the null hypothesis of independence, the test statistic $X^2$ is asymptotically distributed according to a $\chi^2$-distribution with $(r-1)(c-1)$ degrees of freedom, the corresponding test is usually known as *chi-squared test.*

### *3.2.4 McNemar's Test*

The chi-squared test on categorical data described previously assumes that the observations are independent. Often, however, categorical data arise from *paired* observations, for example, cases matched with controls on variables such as gender, age and so on, or observations made on the same subjects on two occasions (cf. paired $t$-test). For this type of paired data, the required procedure is McNemar's test. The general form of such data is shown in Table 3.7.

**Table 3.7**: Frequencies in matched samples data.

| | | Sample 1 | |
|---|---|---|---|
| | | present | absent |
| Sample 2 | present | $a$ | $b$ |
| | absent | $c$ | $d$ |

Under the hypothesis that the two populations do not differ in their probability of having the characteristic present, the test statistic

$$X^2 = \frac{(c-b)^2}{c+b}$$

has a $\chi^2$-distribution with a single degree of freedom.

## 3.3 Analysis Using R

### 3.3.1 Estimating the Width of a Room

The data shown in Table 3.1 are available as `roomwidth` *data.frame* from the **HSAUR2** package and can be attached by using

```
R> data("roomwidth", package = "HSAUR2")
```

If we convert the estimates of the room width in metres into feet by multiplying each by 3.28 then we would like to test the hypothesis that the mean of the population of 'metre' estimates is equal to the mean of the population of 'feet' estimates. We shall do this first by using an independent samples $t$-test, but first it is good practise to check, informally at least, the normality and equal variance assumptions. Here we can use a combination of numerical and graphical approaches. The first step should be to convert the metre estimates into feet by a factor

```
R> convert <- ifelse(roomwidth$unit == "feet", 1, 3.28)
```

which equals one for all feet measurements and 3.28 for the measurements in metres. Now, we get the usual summary statistics and standard deviations of each set of estimates using

```
R> tapply(roomwidth$width * convert, roomwidth$unit, summary)
$feet
   Min. 1st Qu.  Median    Mean 3rd Qu.    Max. 
   24.0    36.0    42.0    43.7    48.0    94.0 

$metres
   Min. 1st Qu.  Median    Mean 3rd Qu.    Max. 
  26.24   36.08   49.20   52.55   55.76  131.20 
```

```
R> tapply(roomwidth$width * convert, roomwidth$unit, sd)

    feet   metres
12.49742 23.43444
```

where tapply applies summary, or sd, to the converted widths for both groups of measurements given by roomwidth$unit. A boxplot of each set of estimates might be useful and is depicted in Figure 3.1. The layout function (line 1 in Figure 3.1) divides the plotting area in three parts. The boxplot function produces a boxplot in the upper part and the two qqnorm statements in lines 8 and 11 set up the normal probability plots that can be used to assess the normality assumption of the $t$-test.

The boxplots indicate that both sets of estimates contain a number of outliers and also that the estimates made in metres are skewed and more variable than those made in feet, a point underlined by the numerical summary statistics above. Both normal probability plots depart from linearity, suggesting that the distributions of both sets of estimates are not normal. The presence of outliers, the apparently different variances and the evidence of non-normality all suggest caution in applying the $t$-test, but for the moment we shall apply the usual version of the test using the t.test function in R.

The two-sample test problem is specified by a *formula*, here by

```
I(width * convert) ~ unit
```

where the response, width, on the left hand side needs to be converted first and, because the star has a special meaning in formulae as will be explained in Chapter 5, the conversion needs to be embedded by I. The factor unit on the right hand side specifies the two groups to be compared.

From the output shown in Figure 3.2 we see that there is considerable evidence that the estimates made in feet are lower than those made in metres by between about 2 and 15 feet. The test statistic $t$ from 3.1 is $-2.615$ and, with 111 degrees of freedom, the two-sided $p$-value is 0.01. In addition, a 95% confidence interval for the difference of the estimated widths between feet and metres is reported.

But this form of $t$-test assumes both normality and equality of population variances, both of which are suspect for these data. Departure from the equality of variance assumption can be accommodated by the modified $t$-test described above and this can be applied in R by choosing var.equal = FALSE (note that var.equal = FALSE is the default in R). The result shown in Figure 3.3 as well indicates that there is strong evidence for a difference in the means of the two types of estimate.

But there remains the problem of the outliers and the possible non-normality; consequently we shall apply the Wilcoxon Mann-Whitney test which, since it is based on the ranks of the observations, is unlikely to be affected by the outliers, and which does not assume that the data have a normal distribution. The test can be applied in R using the wilcox.test function.

Figure 3.4 shows a two-sided $p$-value of 0.028 confirming the difference in location of the two types of estimates of room width. Note that, due to ranking

```
R> layout(matrix(c(1,2,1,3), nrow = 2, ncol = 2, byrow = FALSE))
R> boxplot(I(width * convert) ~ unit, data = roomwidth,
+          ylab = "Estimated width (feet)",
+          varwidth = TRUE, names = c("Estimates in feet",
+          "Estimates in metres (converted to feet)"))
R> feet <- roomwidth$unit == "feet"
R> qqnorm(roomwidth$width[feet],
+         ylab = "Estimated width (feet)")
R> qqline(roomwidth$width[feet])
R> qqnorm(roomwidth$width[!feet],
+         ylab = "Estimated width (metres)")
R> qqline(roomwidth$width[!feet])
```

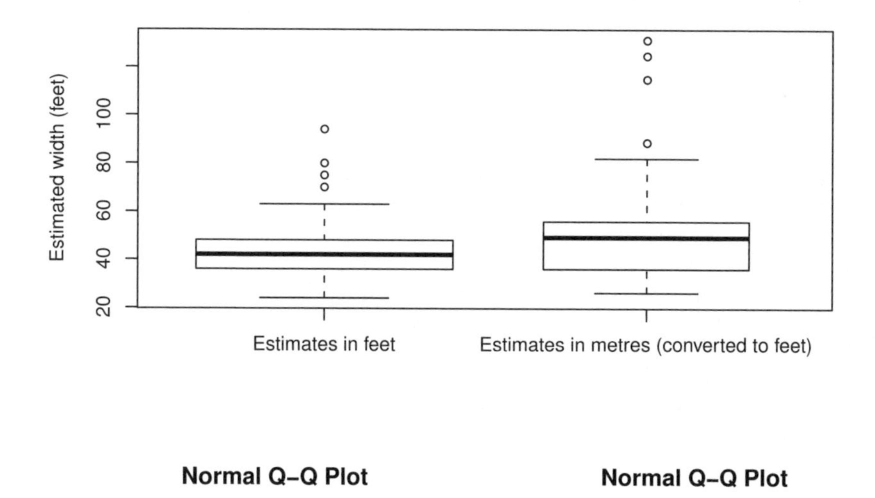


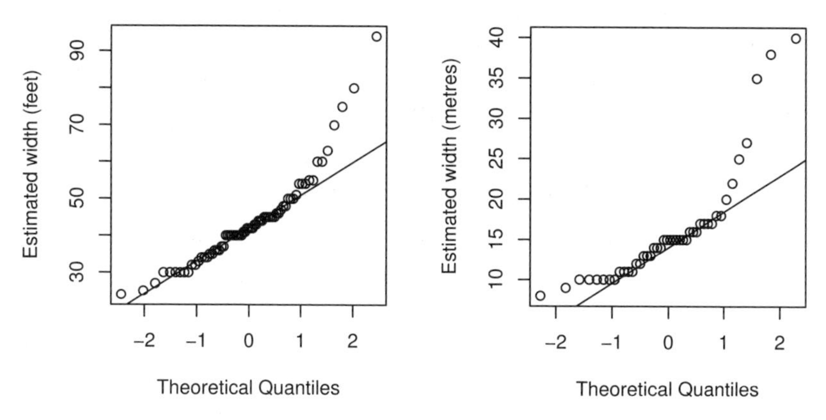


**Figure 3.1** Boxplots of estimates of room width in feet and metres (after conversion to feet) and normal probability plots of estimates of room width made in feet and in metres.

```
R> t.test(I(width * convert) ~ unit, data = roomwidth,
+         var.equal = TRUE)

        Two Sample t-test

data:  I(width * convert) by unit
t = -2.6147, df = 111, p-value = 0.01017
95 percent confidence interval:
 -15.572734  -2.145052
sample estimates:
  mean in group feet mean in group metres
            43.69565             52.55455
```

**Figure 3.2** R output of the independent samples $t$-test for the `roomwidth` data.

```
R> t.test(I(width * convert) ~ unit, data = roomwidth,
+         var.equal = FALSE)

        Welch Two Sample t-test

data:  I(width * convert) by unit
t = -2.3071, df = 58.788, p-value = 0.02459
95 percent confidence interval:
 -16.54308  -1.17471
sample estimates:
  mean in group feet mean in group metres
            43.69565             52.55455
```

**Figure 3.3** R output of the independent samples Welch test for the `roomwidth` data.

the observations, the confidence interval for the median difference reported here is much smaller than the confidence interval for the difference in means as shown in Figures 3.2 and 3.3. Further possible analyses of the data are considered in Exercise 3.1 and in Chapter 4.

### *3.3.2 Wave Energy Device Mooring*

The data from Table 3.2 are available as *data.frame* `waves`

```
R> data("waves", package = "HSAUR2")
```

and requires the use of a matched pairs $t$-test to answer the question of interest. This test assumes that the differences between the matched observations have a normal distribution so we can begin by checking this assumption by constructing a boxplot and a normal probability plot – see Figure 3.5.

The boxplot indicates a possible outlier, and the normal probability plot gives little cause for concern about departures from normality, although with

```
R> wilcox.test(I(width * convert) ~ unit, data = roomwidth,
+              conf.int = TRUE)

        Wilcoxon rank sum test with continuity correction

data:  I(width * convert) by unit
W = 1145, p-value = 0.02815
95 percent confidence interval:
 -9.3599953 -0.8000423
sample estimates:
difference in location
             -5.279955
```

**Figure 3.4** R output of the Wilcoxon rank sum test for the `roomwidth` data.

only 18 observations it is perhaps difficult to draw any convincing conclusion. We can now apply the paired $t$-test to the data again using the `t.test` function. Figure 3.6 shows that there is no evidence for a difference in the mean bending stress of the two types of mooring device. Although there is no real reason for applying the non-parametric analogue of the paired $t$-test to these data, we give the R code for interest in Figure 3.7. The associated $p$-value is 0.316 confirming the result from the $t$-test.

### 3.3.3 Mortality and Water Hardness

There is a wide range of analyses we could apply to the data in Table 3.3 available from

```
R> data("water", package = "HSAUR2")
```

But to begin we will construct a scatterplot of the data enhanced somewhat by the addition of information about the marginal distributions of water hardness (calcium concentration) and mortality, and by adding the estimated linear regression fit (see Chapter 6) for mortality on hardness. The plot and the required R code is given along with Figure 3.8. In line 1 of Figure 3.8, we divide the plotting region into four areas of different size. The scatterplot (line 3) uses a plotting symbol depending on the location of the city (by the `pch` argument); a legend for the location is added in line 6. We add a least squares fit (see Chapter 6) to the scatterplot and, finally, depict the marginal distributions by means of a boxplot and a histogram. The scatterplot shows that as hardness increases mortality decreases, and the histogram for the water hardness shows it has a rather skewed distribution.

We can both calculate the Pearson's correlation coefficient between the two variables and test whether it differs significantly for zero by using the `cor.test` function in R. The test statistic for assessing the hypothesis that the population correlation coefficient is zero is

$$r/\sqrt{(1-r^2)/(n-2)}$$

```
R> mooringdiff <- waves$method1 - waves$method2
R> layout(matrix(1:2, ncol = 2))
R> boxplot(mooringdiff, ylab = "Differences (Newton metres)",
+         main = "Boxplot")
R> abline(h = 0, lty = 2)
R> qqnorm(mooringdiff, ylab = "Differences (Newton metres)")
R> qqline(mooringdiff)
```

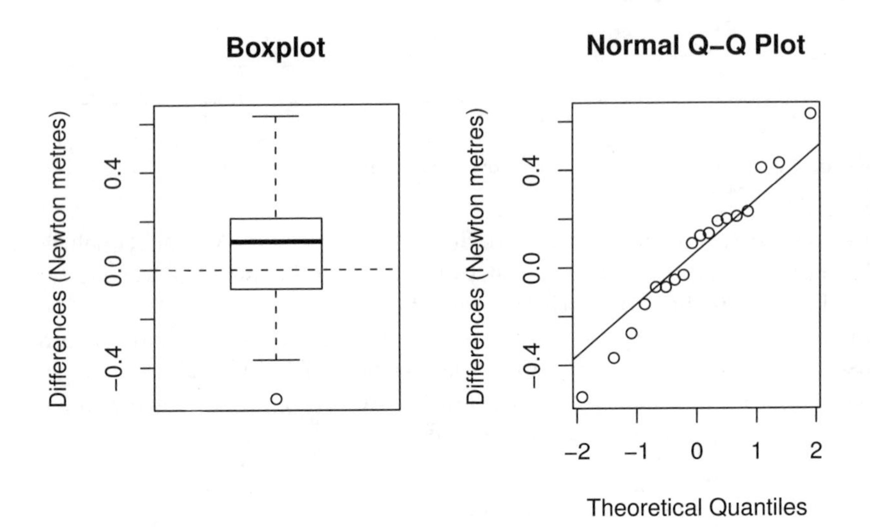


**Figure 3.5** Boxplot and normal probability plot for differences between the two mooring methods.

where $r$ is the sample correlation coefficient and $n$ is the sample size. If the population correlation is zero and assuming the data have a bivariate normal distribution, then the test statistic has a Student's $t$ distribution with $n - 2$ degrees of freedom.

The estimated correlation shown in Figure 3.9 is -0.655 and is highly significant. We might also be interested in the correlation between water hardness and mortality in each of the regions North and South but we leave this as an exercise for the reader (see Exercise 3.2).

### *3.3.4 Piston-ring Failures*

The first step in the analysis of the `pistonrings` data is to apply the chi-squared test for independence. This we can do in R using the `chisq.test` function. The output of the chi-squared test, see Figure 3.10, shows a value of the $X^2$ test statistic of 11.722 with 6 degrees of freedom and an associated

```
R> t.test(mooringdiff)

        One Sample t-test

data:  mooringdiff
t = 0.9019, df = 17, p-value = 0.3797
95 percent confidence interval:
 -0.08258476  0.20591810
sample estimates:
 mean of x
0.06166667
```

**Figure 3.6** R output of the paired $t$-test for the `waves` data.

```
R> wilcox.test(mooringdiff)

        Wilcoxon signed rank test with continuity correction

data:  mooringdiff
V = 109, p-value = 0.3165
```

**Figure 3.7** R output of the Wilcoxon signed rank test for the `waves` data.

$p$-value of 0.068. The evidence for departure from independence of compressor and leg is not strong, but it may be worthwhile taking the analysis a little further by examining the estimated expected values and the differences of these from the corresponding observed value.

Rather than looking at the simple differences of observed and expected values for each cell which would be unsatisfactory since a difference of fixed size is clearly more important for smaller samples, it is preferable to consider a *standardised residual* given by dividing the observed minus the expected difference by the square root of the appropriate expected value. The $X^2$ statistic for assessing independence is simply the sum, over all the cells in the table, of the squares of these terms. We can find these values extracting the `residuals` element of the object returned by the `chisq.test` function

```
R> chisq.test(pistonrings)$residuals

          leg
compressor      North     Centre      South
        C1  0.6036154  1.6728267 -1.7802243
        C2  0.1429031  0.2975200 -0.3471197
        C3 -0.3251427 -0.4522620  0.6202463
        C4 -0.4157886 -1.4666936  1.4635235
```

A graphical representation of these residuals is called an *association plot* and is available via the `assoc` function from package **vcd** (Meyer et al., 2009) applied to the contingency table of the two categorical variables. Figure 3.11

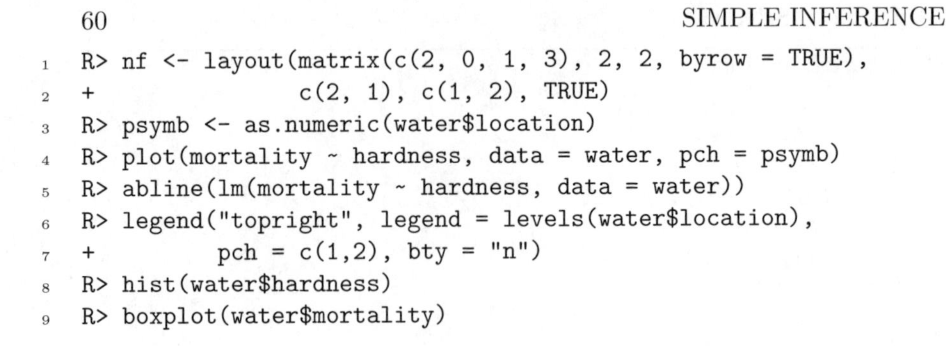

```
R> nf <- layout(matrix(c(2, 0, 1, 3), 2, 2, byrow = TRUE),
+              c(2, 1), c(1, 2), TRUE)
R> psymb <- as.numeric(water$location)
R> plot(mortality ~ hardness, data = water, pch = psymb)
R> abline(lm(mortality ~ hardness, data = water))
R> legend("topright", legend = levels(water$location),
+        pch = c(1,2), bty = "n")
R> hist(water$hardness)
R> boxplot(water$mortality)
```

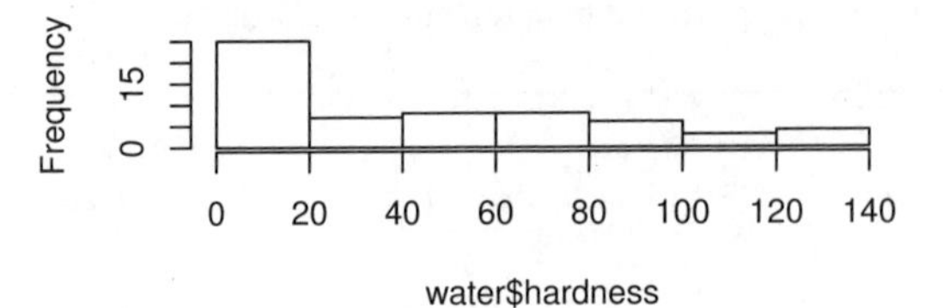


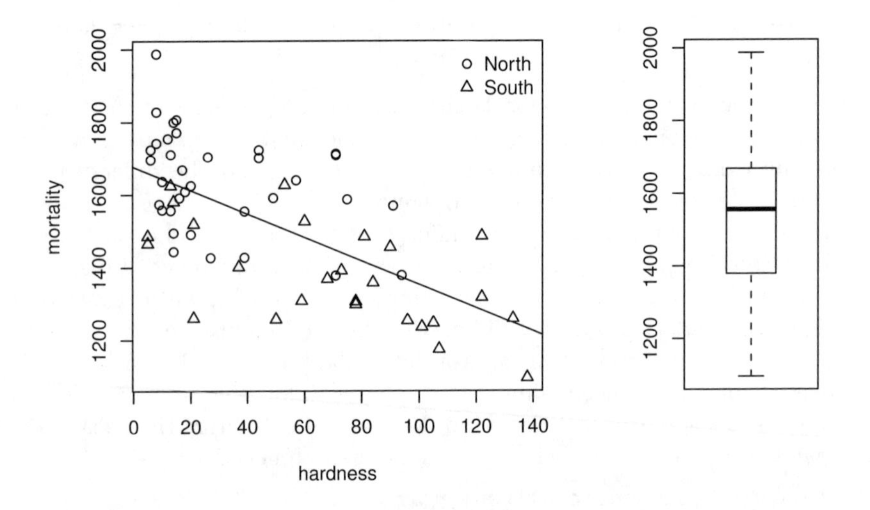


**Figure 3.8** Enhanced scatterplot of water hardness and mortality, showing both the joint and the marginal distributions and, in addition, the location of the city by different plotting symbols.

```
R> cor.test(~ mortality + hardness, data = water)

        Pearson's product-moment correlation

data:  mortality and hardness
t = -6.6555, df = 59, p-value = 1.033e-08
95 percent confidence interval:
 -0.7783208 -0.4826129
sample estimates:
       cor
-0.6548486
```

**Figure 3.9** R output of Pearsons' correlation coefficient for the water data.

```
R> data("pistonrings", package = "HSAUR2")
R> chisq.test(pistonrings)

        Pearson's Chi-squared test

data:  pistonrings
X-squared = 11.7223, df = 6, p-value = 0.06846
```

**Figure 3.10** R output of the chi-squared test for the pistonrings data.

depicts the residuals for the piston ring data. The deviations from independence are largest for C1 and C4 compressors in the centre and south leg.

It is tempting to think that the size of these residuals may be judged by comparison with standard normal percentage points (for example greater than 1.96 or less than 1.96 for significance level $\alpha = 0.05$). Unfortunately it can be shown that the variance of a standardised residual is always less than or equal to one, and in some cases considerably less than one, however, the residuals are asymptotically normal. A more satisfactory 'residual' for contingency table data is considered in Exercise 3.3.

### 3.3.5 Rearrests of Juveniles

The data in Table 3.5 are available as *table* object via

```
R> data("rearrests", package = "HSAUR2")
R> rearrests

             Juvenile court
Adult court   Rearrest No rearrest
  Rearrest         158         515
  No rearrest      290        1134
```

and in rearrests the counts in the four cells refer to the matched pairs of subjects; for example, in 158 pairs both members of the pair were rearrested.

```
R> library("vcd")
R> assoc(pistonrings)
```

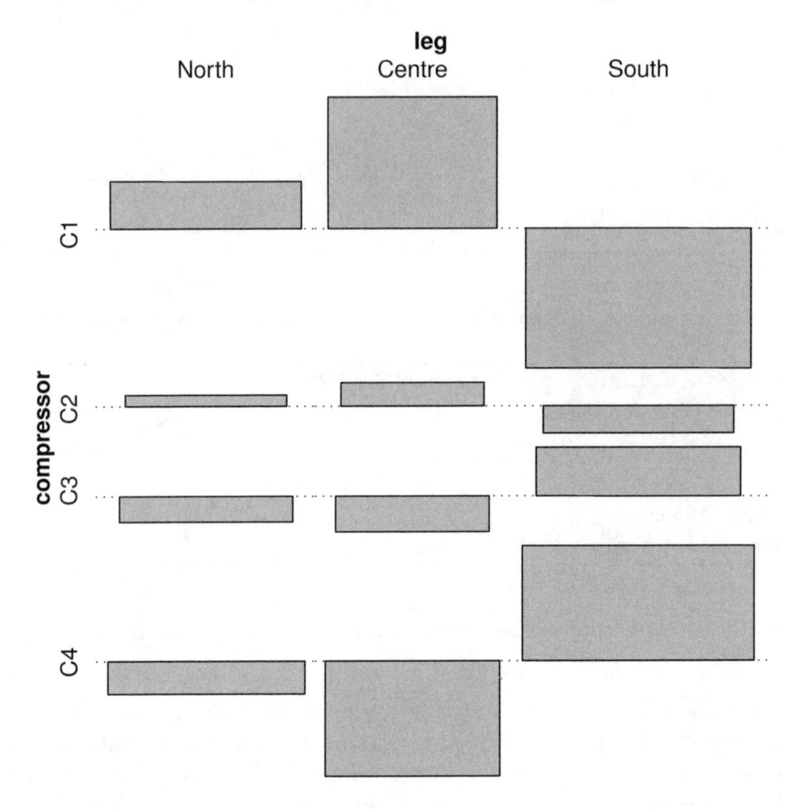


**Figure 3.11** Association plot of the residuals for the `pistonrings` data.

Here we need to use McNemar's test to assess whether rearrest is associated with the type of court where the juvenile was tried. We can use the R function `mcnemar.test`. The test statistic shown in Figure 3.12 is 62.89 with a single degree of freedom – the associated $p$-value is extremely small and there is strong evidence that type of court and the probability of rearrest are related. It appears that trial at a juvenile court is less likely to result in rearrest (see Exercise 3.4). An exact version of McNemar's test can be obtained by testing whether $b$ and $c$ are equal using a binomial test (see Figure 3.13).

```
R> mcnemar.test(rearrests, correct = FALSE)

        McNemar's Chi-squared test

data:  rearrests
McNemar's chi-squared = 62.8882, df = 1, p-value =
2.188e-15
```

**Figure 3.12** R output of McNemar's test for the `rearrests` data.

```
R> binom.test(rearrests[2], n = sum(rearrests[c(2,3)]))

        Exact binomial test

data:  rearrests[2] and sum(rearrests[c(2, 3)])
number of successes = 290, number of trials = 805,
p-value = 1.918e-15
95 percent confidence interval:
 0.3270278 0.3944969
sample estimates:
probability of success
             0.3602484
```

**Figure 3.13** R output of an exact version of McNemar's test for the `rearrests` data computed via a binomial test.

## 3.4 Summary

Significance tests are widely used and they can easily be applied using the corresponding functions in R. But they often need to be accompanied by some graphical material to aid in interpretation and to assess whether assumptions are met. In addition, $p$-values are never as useful as confidence intervals.

### Exercises

Ex. 3.1 After the students had made the estimates of the width of the lecture hall the room width was accurately measured and found to be 13.1 metres (43.0 feet). Use this additional information to determine which of the two types of estimates was more precise.

Ex. 3.2 For the mortality and water hardness data calculate the correlation between the two variables in each region, north and south.

Ex. 3.3 The standardised residuals calculated for the piston ring data are not entirely satisfactory for the reasons given in the text. An alternative residual suggested by Haberman (1973) is defined as the ratio of the standardised residuals and an adjustment:

$$\frac{\sqrt{(n_{jk} - E_{jk})^2 / E_{jk}}}{\sqrt{(1 - n_{j\cdot}/n)(1 - n_{\cdot k}/n)}}.$$

When the variables forming the contingency table are independent, the adjusted residuals are approximately normally distributed with mean zero and standard deviation one. Write a general R function to calculate both standardised and adjusted residuals for any $r \times c$ contingency table and apply it to the piston ring data.

Ex. 3.4 For the data in table `rearrests` estimate the difference between the probability of being rearrested after being tried in an adult court and in a juvenile court, and find a 95% confidence interval for the population difference.

CHAPTER 4

# Conditional Inference: Guessing Lengths, Suicides, Gastrointestinal Damage, and Newborn Infants

## 4.1 Introduction

There are many experimental designs or studies where the subjects are not a random sample from some well-defined population. For example, subjects recruited for a clinical trial are hardly ever a random sample from the set of all people suffering from a certain disease but are a selection of patients showing up for examination in a hospital participating in the trial. Usually, the subjects are randomly assigned to certain groups, for example a control and a treatment group, and the analysis needs to take this randomisation into account. In this chapter, we discuss such test procedures usually known as *(re)-randomisation* or *permutation tests.*

In the room width estimation experiment reported in Chapter 3, 40 of the estimated widths (in feet) of 69 students and 26 of the estimated widths (in metres) of 44 students are tied. In fact, this violates one assumption of the *unconditional* test procedures applied in Chapter 3, namely that the measurements are drawn from a continuous distribution. In this chapter, the data will be reanalysed using conditional test procedures, i.e., statistical tests where the distribution of the test statistics under the null hypothesis is determined *conditionally* on the data at hand. A number of other data sets will also be considered in this chapter and these will now be described.

Mann (1981) reports a study carried out to investigate the causes of jeering or baiting behaviour by a crowd when a person is threatening to commit suicide by jumping from a high building. A hypothesis is that baiting is more likely to occur in warm weather. Mann (1981) classified 21 accounts of threatened suicide by two factors, the time of year and whether or not baiting occurred. The data are given in Table 4.1 and the question is whether they give any evidence to support the hypothesis? The data come from the northern hemisphere, so June–September are the warm months.

**Table 4.1**: `suicides` data. Crowd behaviour at threatened suicides.

| NA | | NA | |
|---|---|---|---|
| | | Baiting | Nonbaiting |
| | June–September | 8 | 4 |
| | October–May | 2 | 7 |

*Source*: From Mann, L., *J. Pers. Soc. Psy.*, 41, 703–709, 1981. With permission.

The administration of non-steroidal anti-inflammatory drugs for patients suffering from arthritis induces gastrointestinal damage. Lanza (1987) and Lanza et al. (1988a,b, 1989) report the results of placebo-controlled randomised clinical trials investigating the prevention of gastrointestinal damage by the application of Misoprostol. The degree of the damage is determined by endoscopic examinations and the response variable is defined as the classification described in Table 4.2. Further details of the studies as well as the data can be found in Whitehead and Jones (1994). The data of the four studies are given in Tables 4.3, 4.4, 4.5 and 4.6.

**Table 4.2**: Classification system for the response variable.

| Classification | Endoscopy Examination |
|---|---|
| 1 | No visible lesions |
| 2 | One haemorrhage or erosion |
| 3 | 2-10 haemorrhages or erosions |
| 4 | 11-25 haemorrhages or erosions |
| 5 | More than 25 haemorrhages or erosions or an invasive ulcer of any size |

*Source*: From Whitehead, A. and Jones, N. M. B., *Stat. Med.*, 13, 2503–2515, 1994. With permission.

**Table 4.3**: `Lanza` data. Misoprostol randomised clinical trial from Lanza (1987).

| | | classification | | | | |
|---|---|---|---|---|---|---|
| treatment | | 1 | 2 | 3 | 4 | 5 |
| | Misoprostol | 21 | 2 | 4 | 2 | 0 |
| | Placebo | 2 | 2 | 4 | 9 | 13 |

**Table 4.4**: `Lanza` data. Misoprostol randomised clinical trial from Lanza et al. (1988a).

| treatment | classification 1 | 2 | 3 | 4 | 5 |
|---|---|---|---|---|---|
| Misoprostol | 20 | 4 | 6 | 0 | 0 |
| Placebo | 8 | 4 | 9 | 4 | 5 |

**Table 4.5**: `Lanza` data. Misoprostol randomised clinical trial from Lanza et al. (1988b).

| treatment | classification 1 | 2 | 3 | 4 | 5 |
|---|---|---|---|---|---|
| Misoprostol | 20 | 4 | 3 | 1 | 2 |
| Placebo | 0 | 2 | 5 | 5 | 17 |

**Table 4.6**: `Lanza` data. Misoprostol randomised clinical trial from Lanza et al. (1989).

| treatment | classification 1 | 2 | 3 | 4 | 5 |
|---|---|---|---|---|---|
| Misoprostol | 1 | 4 | 5 | 0 | 0 |
| Placebo | 0 | 0 | 0 | 4 | 6 |

Newborn infants exposed to antiepileptic drugs in utero have a higher risk of major and minor abnormalities of the face and digits. The inter-rater agreement in the assessment of babies with respect to the number of minor physical features was investigated by Carlin et al. (2000). In their paper, the agreement on total number of face anomalies for 395 newborn infants examined by a paediatrician and a research assistant is reported (see Table 4.7). One is interested in investigating whether the paediatrician and the research assistant agree above a chance level.

**Table 4.7:** `anomalies` data. Abnormalities of the face and digits of newborn infants exposed to antiepileptic drugs as assessed by a paediatrician (`MD`) and a research assistant (`RA`).

| MD | RA | | | |
|---|---|---|---|---|
| | 0 | 1 | 2 | 3 |
| 0 | 235 | 41 | 20 | 2 |
| 1 | 23 | 35 | 11 | 1 |
| 2 | 3 | 8 | 11 | 3 |
| 3 | 0 | 0 | 1 | 1 |

*Source*: From Carlin, J. B., et al., *Teratology*, 62, 406-412, 2000. With permission.

## 4.2 Conditional Test Procedures

The statistical test procedures applied in Chapter 3 all are defined for samples randomly drawn from a well-defined population. In many experiments however, this model is far from being realistic. For example in clinical trials, it is often impossible to draw a random sample from all patients suffering a certain disease. Commonly, volunteers and patients are recruited from hospital staff, relatives or people showing up for some examination. The test procedures applied in this chapter make no assumptions about random sampling or a specific model. Instead, the null distribution of the test statistics is computed conditionally on all random permutations of the data. Therefore, the procedures shown in the sequel are known as *permutation tests* or *(re)-randomisation tests.* For a general introduction we refer to the text books of Edgington (1987) and Pesarin (2001).

### *4.2.1 Testing Independence of Two Variables*

Based on $n$ pairs of measurements $(x_i, y_i)$ recorded for $n$ observational units we want to test the null hypothesis of the independence of $x$ and $y$. We may distinguish three situations: both variables $x$ and $y$ are continuous, one is continuous and the other one is a factor or both $x$ and $y$ are factors. The special case of paired observations is treated in Section 4.2.2.

One class of test procedures for the above three situations are randomisation and permutation tests whose basic principles have been described by Fisher (1935) and Pitman (1937) and are best illustrated for the case of continuous measurements $y$ in two groups, i.e., the $x$ variable is a factor that can take values $x = 1$ or $x = 2$. The difference of the means of the $y$ values in both groups is an appropriate statistic for the assessment of the association of $y$

and $x$

$$T = \frac{\sum_{i=1}^{n} I(x_i = 1) y_i}{\sum_{i=1}^{n} I(x_i = 1)} - \frac{\sum_{i=1}^{n} I(x_i = 2) y_i}{\sum_{i=1}^{n} I(x_i = 2)}.$$

Here $I(x_i = 1)$ is the indication function which is equal to one if the condition $x_i = 1$ is true and zero otherwise. Clearly, under the null hypothesis of independence of $x$ and $y$ we expect the distribution of $T$ to be centred about zero.

Suppose that the group labels $x = 1$ or $x = 2$ have been assigned to the observational units by randomisation. When the result of the randomisation procedure is independent of the $y$ measurements, we are allowed to fix the $x$ values and shuffle the $y$ values randomly over and over again. Thus, we can compute, or at least approximate, the distribution of the test statistic $T$ under the conditions of the null hypothesis directly from the data $(x_i, y_i), i = 1, \dots, n$ by the so called *randomisation principle*. The test statistic $T$ is computed for a reasonable number of shuffled $y$ values and we can determine how many of the shuffled differences are at least as large as the test statistic $T$ obtained from the original data. If this proportion is small, smaller than $\alpha = 0.05$ say, we have good evidence that the assumption of independence of $x$ and $y$ is not realistic and we therefore can reject the null hypothesis. The proportion of larger differences is usually referred to as $p$-value.

A special approach is based on ranks assigned to the continuous $y$ values. When we replace the raw measurements $y_i$ by their corresponding ranks in the computation of $T$ and compare this test statistic with its null distribution we end up with the Wilcoxon Mann-Whitney rank sum test. The conditional distribution and the unconditional distribution of the Wilcoxon Mann-Whitney rank sum test as introduced in Chapter 3 coincide when the $y$ values are not tied. Without ties in the $y$ values, the ranks are simply the integers $1, 2, \dots, n$ and the unconditional (Chapter 3) and the conditional view on the Wilcoxon Mann-Whitney test coincide.

In the case that both variables are nominal, the test statistic can be computed from the corresponding contingency table in which the observations $(x_i, y_i)$ are cross-classified. A general $r \times c$ contingency table may be written in the form of Table 3.6 where each cell $(j, k)$ is the number $n_{ij} = \sum_{i=1}^{n} I(x_i = j) I(y_i = k)$, see Chapter 3 for more details.

Under the null hypothesis of independence of $x$ and $y$, estimated expected values $E_{jk}$ for cell $(j, k)$ can be computed from the corresponding margin totals $E_{jk} = n_{j \cdot} n_{\cdot k} / n$ which are fixed for each randomisation of the data. The test statistic for assessing independence is

$$X^2 = \sum_{j=1}^{r} \sum_{k=1}^{c} \frac{(n_{jk} - E_{jk})^2}{E_{jk}}.$$

The exact distribution based on all permutations of the $y$ values for a similar test statistic can be computed by means of Fisher's exact test (Freeman and Halton, 1951). This test procedure is based on the hyper-geometric probability of the observed contingency table. All possible tables can be ordered with respect to this metric and $p$-values are computed from the fraction of tables more extreme than the observed one.

When both the $x$ and the $y$ measurements are numeric, the test statistic can be formulated as the product, i.e., by the sum of all $x_i y_i, i = 1, \dots, n$. Again, we can fix the $x$ values and shuffle the $y$ values in order to approximate the distribution of the test statistic under the laws of the null hypothesis of independence of $x$ and $y$.

### *4.2.2 Testing Marginal Homogeneity*

In contrast to the independence problem treated above the data analyst is often confronted with situations where two (or more) measurements of one variable taken from the same observational unit are to be compared. In this case one assumes that the measurements are independent between observations and the test statistics are aggregated over all observations. Where two nominal variables are taken for each observation (for example see the case of McNemar's test for binary variables as discussed in Chapter 3), the measurement of each observation can be summarised by a $k \times k$ matrix with cell $(i, j)$ being equal to one if the first measurement is the $i$th level and the second measurement is the $j$th level. All other entries are zero. Under the null hypothesis of independence of the first and second measurement, all $k \times k$ matrices with exactly one non-zero element are equally likely. The test statistic is now based on the elementwise sum of all $n$ matrices.

## 4.3 Analysis Using R

### *4.3.1 Estimating the Width of a Room Revised*

The unconditional analysis of the room width estimated by two groups of students in Chapter 3 led to the conclusion that the estimates in metres are slightly larger than the estimates in feet. Here, we reanalyse these data in a conditional framework. First, we convert metres into feet and store the vector of observations in a variable y:

```
R> data("roomwidth", package = "HSAUR2")
R> convert <- ifelse(roomwidth$unit == "feet", 1, 3.28)
R> feet <- roomwidth$unit == "feet"
R> metre <- !feet
R> y <- roomwidth$width * convert
```

The test statistic is simply the difference in means

```
R> T <- mean(y[feet]) - mean(y[metre])
R> T
[1] -8.858893
```

```
R> hist(meandiffs)
R> abline(v = T, lty = 2)
R> abline(v = -T, lty = 2)
```

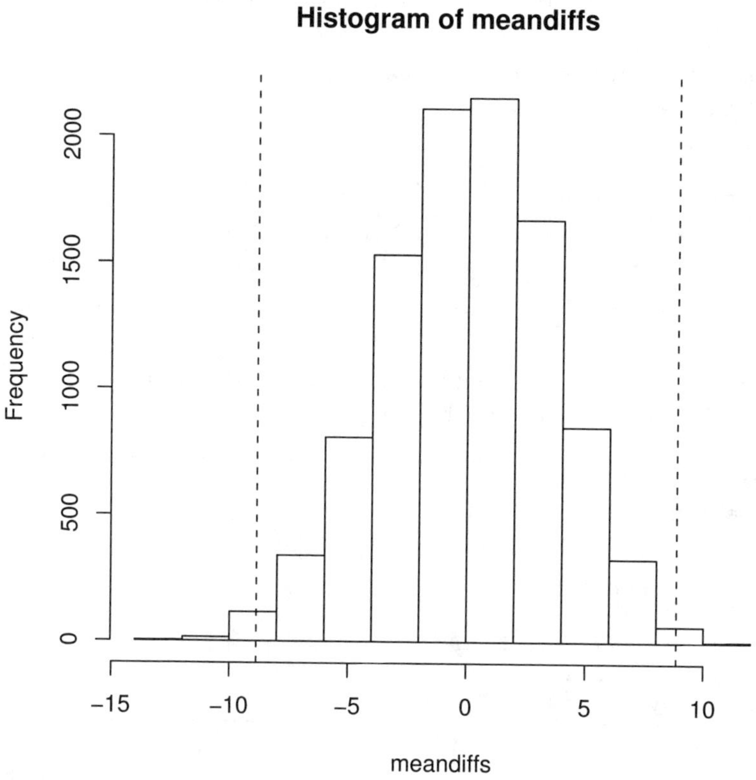

**Figure 4.1** An approximation for the conditional distribution of the difference of mean `roomwidth` estimates in the feet and metres group under the null hypothesis. The vertical lines show the negative and positive absolute value of the test statistic $T$ obtained from the original data.

In order to approximate the conditional distribution of the test statistic $T$ we compute 9999 test statistics for shuffled $y$ values. A permutation of the $y$ vector can be obtained from the `sample` function.

```
R> meandiffs <- double(9999)
R> for (i in 1:length(meandiffs)) {
+      sy <- sample(y)
+      meandiffs[i] <- mean(sy[feet]) - mean(sy[metre])
+ }
```

The distribution of the test statistic $T$ under the null hypothesis of independence of room width estimates and groups is depicted in Figure 4.1. Now, the value of the test statistic $T$ for the original unshuffled data can be compared with the distribution of $T$ under the null hypothesis (the vertical lines in Figure 4.1). The $p$-value, i.e., the proportion of test statistics $T$ larger than 8.859 or smaller than -8.859, is

```
R> greater <- abs(meandiffs) > abs(T)
R> mean(greater)

[1] 0.0080008
```

with a confidence interval of

```
R> binom.test(sum(greater), length(greater))$conf.int

[1] 0.006349087 0.009947933
attr(,"conf.level")
[1] 0.95
```

Note that the approximated conditional $p$-value is roughly the same as the $p$-value reported by the $t$-test in Chapter 3.

```
R> library("coin")
R> independence_test(y ~ unit, data = roomwidth,
+                    distribution = exact())

        Exact General Independence Test

data:  y by unit (feet, metres)
Z = -2.5491, p-value = 0.008492
alternative hypothesis: two.sided
```

**Figure 4.2** R output of the exact permutation test applied to the `roomwidth` data.

For some situations, including the analysis shown here, it is possible to compute the *exact* $p$-value, i.e., the $p$-value based on the distribution evaluated on all possible randomisations of the $y$ values. The function `independence_test` (package **coin**, Hothorn et al., 2006a, 2008b) can be used to compute the exact $p$-value as shown in Figure 4.2. Similarly, the exact conditional distribution of the Wilcoxon Mann-Whitney rank sum test can be computed by a function implemented in package **coin** as shown in Figure 4.3.

One should note that the $p$-values of the permutation test and the $t$-test coincide rather well and that the $p$-values of the Wilcoxon Mann-Whitney rank sum tests in their conditional and unconditional version are roughly three times as large due to the loss of information induced by taking only the ranking of the measurements into account. However, based on the results of the permutation test applied to the `roomwidth` data we can conclude that the estimates in metres are, on average, larger than the estimates in feet.

```
R> wilcox_test(y ~ unit, data = roomwidth,
+              distribution = exact())

        Exact Wilcoxon Mann-Whitney Rank Sum Test

data:  y by unit (feet, metres)
Z = -2.1981, p-value = 0.02763
alternative hypothesis: true mu is not equal to 0
```

**Figure 4.3** R output of the exact conditional Wilcoxon rank sum test applied to the `roomwidth` data.

### *4.3.2 Crowds and Threatened Suicide*

The data in this case are in the form of a $2 \times 2$ contingency table and it might be thought that the chi-squared test could again be applied to test for the independence of crowd behaviour and time of year. However, the $\chi^2$-distribution as an approximation to the independence test statistic is bad when the expected frequencies are rather small. The problem is discussed in detail in Everitt (1992) and Agresti (1996). One solution is to use a conditional test procedure such as Fisher's exact test as described above. We can apply this test procedure using the R function `fisher.test` to the *table* `suicides` (see Figure 4.4).

```
R> data("suicides", package = "HSAUR2")
R> fisher.test(suicides)

        Fisher's Exact Test for Count Data

data:  suicides
p-value = 0.0805
alternative hypothesis: true odds ratio is not equal to 1
95 percent confidence interval:
  0.7306872 91.0288231
sample estimates:
odds ratio
  6.302622
```

**Figure 4.4** R output of Fisher's exact test for the `suicides` data.

The resulting $p$-value obtained from the hypergeometric distribution is 0.08 (the asymptotic $p$-value associated with the $X^2$ statistic for this table is 0.115). There is no strong evidence of crowd behaviour being associated with time of year of threatened suicide, but the sample size is low and the test lacks power. Fisher's exact test can also be applied to larger than $2 \times 2$ tables, especially when there is concern that the cell frequencies are low (see Exercise 4.1).

### *4.3.3 Gastrointestinal Damage*

Here we are interested in the comparison of two groups of patients, where one group received a placebo and the other one Misoprostol. In the trials shown here, the response variable is measured on an ordered scale – see Table 4.2. Data from four clinical studies are available and thus the observations are naturally grouped together. From the *data.frame* `Lanza` we can construct a three-way table as follows:

```
R> data("Lanza", package = "HSAUR2")
R> xtabs(~ treatment + classification + study, data = Lanza)
```

```
, , study = I

             classification
treatment      1  2  3  4  5
  Misoprostol 21  2  4  2  0
  Placebo      2  2  4  9 13

, , study = II

             classification
treatment      1  2  3  4  5
  Misoprostol 20  4  6  0  0
  Placebo      8  4  9  4  5

, , study = III

             classification
treatment      1  2  3  4  5
  Misoprostol 20  4  3  1  2
  Placebo      0  2  5  5 17

, , study = IV

             classification
treatment      1  2  3  4  5
  Misoprostol  1  4  5  0  0
  Placebo      0  0  0  4  6
```

We will first analyse each study separately and then show how one can investigate the effect of Misoprostol for all four studies simultaneously. Because the response is ordered, we take this information into account by assigning a score to each level of the response. Since the classifications are defined by the number of haemorrhages or erosions, the midpoint of the interval for each level is a reasonable choice, i.e., 0, 1, 6, 17 and 30 – compare those scores to the definitions given in Table 4.2. The corresponding linear-by-linear association tests extending the general Cochran-Mantel-Haenszel statistics (see Agresti, 2002, for further details) are implemented in package **coin**.

For the first study, the null hypothesis of independence of treatment and gastrointestinal damage, i.e., of no treatment effect of Misoprostol, is tested by

```
R> library("coin")
R> cmh_test(classification ~ treatment, data = Lanza,
+           scores = list(classification = c(0, 1, 6, 17, 30)),
+           subset = Lanza$study == "I")

        Asymptotic Linear-by-Linear Association Test

data:  classification (ordered) by
         treatment (Misoprostol, Placebo)
chi-squared = 28.8478, df = 1, p-value = 7.83e-08
```

and, by default, the conditional distribution is approximated by the corresponding limiting distribution. The $p$-value indicates a strong treatment effect. For the second study, the asymptotic $p$-value is a little bit larger:

```
R> cmh_test(classification ~ treatment, data = Lanza,
+           scores = list(classification = c(0, 1, 6, 17, 30)),
+           subset = Lanza$study == "II")

        Asymptotic Linear-by-Linear Association Test

data:  classification (ordered) by
         treatment (Misoprostol, Placebo)
chi-squared = 12.0641, df = 1, p-value = 0.000514
```

and we make sure that the implied decision is correct by calculating a confidence interval for the exact $p$-value:

```
R> p <- cmh_test(classification ~ treatment, data = Lanza,
+           scores = list(classification = c(0, 1, 6, 17, 30)),
+           subset = Lanza$study == "II", distribution =
+           approximate(B = 19999))
R> pvalue(p)

[1] 5.00025e-05
99 percent confidence interval:
 2.506396e-07 3.714653e-04
```

The third and fourth study indicate a strong treatment effect as well:

```
R> cmh_test(classification ~ treatment, data = Lanza,
+           scores = list(classification = c(0, 1, 6, 17, 30)),
+           subset = Lanza$study == "III")

        Asymptotic Linear-by-Linear Association Test

data:  classification (ordered) by
         treatment (Misoprostol, Placebo)
chi-squared = 28.1587, df = 1, p-value = 1.118e-07
```

```
R> cmh_test(classification ~ treatment, data = Lanza,
+          scores = list(classification = c(0, 1, 6, 17, 30)),
+          subset = Lanza$study == "IV")

        Asymptotic Linear-by-Linear Association Test

data:  classification (ordered) by
         treatment (Misoprostol, Placebo)
chi-squared = 15.7414, df = 1, p-value = 7.262e-05
```

At the end, a separate analysis for each study is unsatisfactory. Because the design of the four studies is the same, we can use **study** as a block variable and perform a global linear-association test investigating the treatment effect of Misoprostol in all four studies. The block variable can be incorporated into the *formula* by the | symbol.

```
R> cmh_test(classification ~ treatment | study, data = Lanza,
+          scores = list(classification = c(0, 1, 6, 17, 30)))

        Asymptotic Linear-by-Linear Association Test

data:  classification (ordered) by
         treatment (Misoprostol, Placebo)
         stratified by study
chi-squared = 83.6188, df = 1, p-value < 2.2e-16
```

Based on this result, a strong treatment effect can be established.

### 4.3.4 Teratogenesis

In this example, the medical doctor (MD) and the research assistant (RA) assessed the number of anomalies $(0, 1, 2$ or $3)$ for each of 395 babies:

```
R> anomalies <- c(235, 23, 3, 0, 41, 35, 8, 0,
+                  20, 11, 11, 1, 2, 1, 3, 1)
R> anomalies <- as.table(matrix(anomalies,
+      ncol = 4, dimnames = list(MD = 0:3, RA = 0:3)))
R> anomalies
   RA
MD    0   1   2   3
  0 235  41  20   2
  1  23  35  11   1
  2   3   8  11   3
  3   0   0   1   1
```

We are interested in testing whether the number of anomalies assessed by the medical doctor differs structurally from the number reported by the research assistant. Because we compare *paired* observations, i.e., one pair of measurements for each newborn, a test of marginal homogeneity (a generalisation of McNemar's test, Chapter 3) needs to be applied:

```
R> mh_test(anomalies)

        Asymptotic Marginal-Homogeneity Test

data:  response by
         groups (MD, RA)
         stratified by block
chi-squared = 21.2266, df = 3, p-value = 9.446e-05
```

The $p$-value indicates a deviation from the null hypothesis. However, the levels of the response are not treated as ordered. Similar to the analysis of the gastrointestinal damage data above, we can take this information into account by the definition of an appropriate score. Here, the number of anomalies is a natural choice:

```
R> mh_test(anomalies, scores = list(c(0, 1, 2, 3)))

        Asymptotic Marginal-Homogeneity Test for Ordered Data

data:  response (ordered) by
         groups (MD, RA)
         stratified by block
chi-squared = 21.0199, df = 1, p-value = 4.545e-06
```

In our case, both versions coincide and one can conclude that the assessment of the number of anomalies differs between the medical doctor and the research assistant.

## 4.4 Summary

The analysis of randomised experiments, for example the analysis of randomised clinical trials such as the Misoprostol trial presented in this chapter, requires the application of conditional inferences procedures. In such experiments, the observations might not have been sampled from well-defined populations but are assigned to treatment groups, say, by a random procedure which is reiterated when randomisation tests are applied.

## Exercises

Ex. 4.1 Although in the past Fisher's test has been largely applied to sparse $2 \times 2$ tables, it can also be applied to larger tables, especially when there is concern about small values in some cells. Using the data displayed in Table 4.8 (taken from Mehta and Patel, 2003) which gives the distribution of the oral lesion site found in house-to-house surveys in three geographic regions of rural India, find the $p$-value from Fisher's test and the corresponding $p$-value from applying the usual chi-square test to the data. What are your conclusions?

**Table 4.8**: `orallesions` data. Oral lesions found in house-to-house surveys in three geographic regions of rural India.

| `site of lesion` | `region` Kerala | Gujarat | Andhra |
|---|---|---|---|
| Buccal mucosa | 8 | 1 | 8 |
| Commissure | 0 | 1 | 0 |
| Gingiva | 0 | 1 | 0 |
| Hard palate | 0 | 1 | 0 |
| Soft palate | 0 | 1 | 0 |
| Tongue | 0 | 1 | 0 |
| Floor of mouth | 1 | 0 | 1 |
| Alveolar ridge | 1 | 0 | 1 |

*Source*: From Mehta, C. and Patel, N., *StatXact-6: Statistical Software for Exact Nonparametric Inference*, Cytel Software Corporation, Cambridge, MA, 2003. With permission.

Ex. 4.2 Use the `mosaic` and `assoc` functions from the **vcd** package (Meyer et al., 2009) to create a graphical representation of the deviations from independence in the $2 \times 2$ contingency table shown in Table 4.1.

Ex. 4.3 Generate two groups with measurements following a normal distribution having different means. For multiple replications of this experiment (1000, say), compare the $p$-values of the Wilcoxon Mann-Whitney rank sum test and a permutation test (using `independence_test`). Where do the differences come from?

CHAPTER 5

# Analysis of Variance: Weight Gain, Foster Feeding in Rats, Water Hardness and Male Egyptian Skulls

## 5.1 Introduction

The data in Table 5.1 (from Hand et al., 1994) arise from an experiment to study the gain in weight of rats fed on four different diets, distinguished by amount of protein (low and high) and by source of protein (beef and cereal). Ten rats are randomised to each of the four treatments and the weight gain in grams recorded. The question of interest is how diet affects weight gain.

**Table 5.1**: `weightgain` data. Rat weight gain for diets differing by the amount of protein (`type`) and source of protein (`source`).

| source | type | weightgain | source | type | weightgain |
|---|---|---|---|---|---|
| Beef | Low | 90 | Cereal | Low | 107 |
| Beef | Low | 76 | Cereal | Low | 95 |
| Beef | Low | 90 | Cereal | Low | 97 |
| Beef | Low | 64 | Cereal | Low | 80 |
| Beef | Low | 86 | Cereal | Low | 98 |
| Beef | Low | 51 | Cereal | Low | 74 |
| Beef | Low | 72 | Cereal | Low | 74 |
| Beef | Low | 90 | Cereal | Low | 67 |
| Beef | Low | 95 | Cereal | Low | 89 |
| Beef | Low | 78 | Cereal | Low | 58 |
| Beef | High | 73 | Cereal | High | 98 |
| Beef | High | 102 | Cereal | High | 74 |
| Beef | High | 118 | Cereal | High | 56 |
| Beef | High | 104 | Cereal | High | 111 |
| Beef | High | 81 | Cereal | High | 95 |
| Beef | High | 107 | Cereal | High | 88 |
| Beef | High | 100 | Cereal | High | 82 |
| Beef | High | 87 | Cereal | High | 77 |
| Beef | High | 117 | Cereal | High | 86 |
| Beef | High | 111 | Cereal | High | 92 |

The data in Table 5.2 are from a foster feeding experiment with rat mothers and litters of four different genotypes: A, B, I and J (Hand et al., 1994). The measurement is the litter weight (in grams) after a trial feeding period. Here the investigator's interest lies in uncovering the effect of genotype of mother and litter on litter weight.

**Table 5.2**: `foster` data. Foster feeding experiment for rats with different genotypes of the litter (`litgen`) and mother (`motgen`).

| litgen | motgen | weight | litgen | motgen | weight |
|---|---|---|---|---|---|
| A | A | 61.5 | B | J | 40.5 |
| A | A | 68.2 | I | A | 37.0 |
| A | A | 64.0 | I | A | 36.3 |
| A | A | 65.0 | I | A | 68.0 |
| A | A | 59.7 | I | B | 56.3 |
| A | B | 55.0 | I | B | 69.8 |
| A | B | 42.0 | I | B | 67.0 |
| A | B | 60.2 | I | I | 39.7 |
| A | I | 52.5 | I | I | 46.0 |
| A | I | 61.8 | I | I | 61.3 |
| A | I | 49.5 | I | I | 55.3 |
| A | I | 52.7 | I | I | 55.7 |
| A | J | 42.0 | I | J | 50.0 |
| A | J | 54.0 | I | J | 43.8 |
| A | J | 61.0 | I | J | 54.5 |
| A | J | 48.2 | J | A | 59.0 |
| A | J | 39.6 | J | A | 57.4 |
| B | A | 60.3 | J | A | 54.0 |
| B | A | 51.7 | J | A | 47.0 |
| B | A | 49.3 | J | B | 59.5 |
| B | A | 48.0 | J | B | 52.8 |
| B | B | 50.8 | J | B | 56.0 |
| B | B | 64.7 | J | I | 45.2 |
| B | B | 61.7 | J | I | 57.0 |
| B | B | 64.0 | J | I | 61.4 |
| B | B | 62.0 | J | J | 44.8 |
| B | I | 56.5 | J | J | 51.5 |
| B | I | 59.0 | J | J | 53.0 |
| B | I | 47.2 | J | J | 42.0 |
| B | I | 53.0 | J | J | 54.0 |
| B | J | 51.3 | | | |

The data in Table 5.3 (from Hand et al., 1994) give four measurements made on Egyptian skulls from five epochs. The data has been collected with a view to deciding if there are any differences between the skulls from the five epochs. The measurements are:

`mb`: maximum breadths of the skull,

`bh`: basibregmatic heights of the skull,

`bl`: basialiveolar length of the skull, and

`nh`: nasal heights of the skull.

Non-constant measurements of the skulls over time would indicate interbreeding with immigrant populations.

**Table 5.3**: `skulls` data. Measurements of four variables taken from Egyptian skulls of five periods.

| epoch | mb | bh | bl | nh |
|---|---|---|---|---|
| c4000BC | 131 | 138 | 89 | 49 |
| c4000BC | 125 | 131 | 92 | 48 |
| c4000BC | 131 | 132 | 99 | 50 |
| c4000BC | 119 | 132 | 96 | 44 |
| c4000BC | 136 | 143 | 100 | 54 |
| c4000BC | 138 | 137 | 89 | 56 |
| c4000BC | 139 | 130 | 108 | 48 |
| c4000BC | 125 | 136 | 93 | 48 |
| c4000BC | 131 | 134 | 102 | 51 |
| c4000BC | 134 | 134 | 99 | 51 |
| c4000BC | 129 | 138 | 95 | 50 |
| c4000BC | 134 | 121 | 95 | 53 |
| c4000BC | 126 | 129 | 109 | 51 |
| c4000BC | 132 | 136 | 100 | 50 |
| c4000BC | 141 | 140 | 100 | 51 |
| c4000BC | 131 | 134 | 97 | 54 |
| c4000BC | 135 | 137 | 103 | 50 |
| c4000BC | 132 | 133 | 93 | 53 |
| c4000BC | 139 | 136 | 96 | 50 |
| c4000BC | 132 | 131 | 101 | 49 |
| c4000BC | 126 | 133 | 102 | 51 |
| c4000BC | 135 | 135 | 103 | 47 |
| c4000BC | 134 | 124 | 93 | 53 |
| ⋮ | ⋮ | ⋮ | ⋮ | ⋮ |

## 5.2 Analysis of Variance

For each of the data sets described in the previous section, the question of interest involves assessing whether certain populations differ in mean value for, in Tables 5.1 and 5.2, a single variable, and in Table 5.3, for a set of four variables. In the first two cases we shall use *analysis of variance* (ANOVA) and in the last *multivariate analysis of variance* (MANOVA) method for the analysis of this data. Both Tables 5.1 and 5.2 are examples of *factorial designs*, with the factors in the first data set being amount of protein with two levels, and source of protein also with two levels. In the second, the factors are the genotype of the mother and the genotype of the litter, both with four levels. The analysis of each data set can be based on the same model (see below) but the two data sets differ in that the first is *balanced*, i.e., there are the same number of observations in each cell, whereas the second is *unbalanced* having different numbers of observations in the 16 cells of the design. This distinction leads to complications in the analysis of the unbalanced design that we will come to in the next section. But the model used in the analysis of each is

$$y_{ijk} = \mu + \gamma_i + \beta_j + (\gamma\beta)_{ij} + \varepsilon_{ijk}$$

where $y_{ijk}$ represents the $k$th measurement made in cell $(i, j)$ of the factorial design, $\mu$ is the overall mean, $\gamma_i$ is the main effect of the first factor, $\beta_j$ is the main effect of the second factor, $(\gamma\beta)_{ij}$ is the interaction effect of the two factors and $\varepsilon_{ijk}$ is the residual or error term assumed to have a normal distribution with mean zero and variance $\sigma^2$. In R, the model is specified by a model *formula*. The *two-way layout with interactions* specified above reads

```
y ~ a + b + a:b
```

where the variable a is the first and the variable b is the second *factor*. The interaction term $(\gamma\beta)_{ij}$ is denoted by a:b. An equivalent model *formula* is

```
y ~ a * b
```

Note that the mean $\mu$ is implicitly defined in the *formula* shown above. In case $\mu = 0$, one needs to remove the intercept term from the *formula* explicitly, i.e.,

```
y ~ a + b + a:b - 1
```

For a more detailed description of model formulae we refer to R Development Core Team (2009a) and help("lm").

The model as specified above is overparameterised, i.e., there are infinitely many solutions to the corresponding estimation equations, and so the parameters have to be constrained in some way, commonly by requiring them to sum to zero – see Everitt (2001) for a full discussion. The analysis of the rat weight gain data below explains some of these points in more detail (see also Chapter 6).

The model given above leads to a partition of the variation in the observations into parts due to main effects and interaction plus an error term that enables a series of $F$-tests to be calculated that can be used to test hypotheses about the main effects and the interaction. These calculations are generally set out in the familiar *analysis of variance table*. The assumptions made in deriving the $F$-tests are:

- The observations are independent of each other,
- The observations in each cell arise from a population having a normal distribution, and
- The observations in each cell are from populations having the same variance.

The multivariate analysis of variance, or MANOVA, is an extension of the univariate analysis of variance to the situation where a set of variables are measured on each individual or object observed. For the data in Table 5.3 there is a single factor, `epoch`, and four measurements taken on each skull; so we have a *one-way* MANOVA design. The linear model used in this case is

$$y_{ijh} = \mu_h + \gamma_{jh} + \varepsilon_{ijh}$$

where $\mu_h$ is the overall mean for variable $h$, $\gamma_{jh}$ is the effect of the $j$th level of the single factor on the $h$th variable, and $\varepsilon_{ijh}$ is a random error term. The vector $\varepsilon_{ij}^\top = (\varepsilon_{ij1}, \varepsilon_{ij2}, \dots, \varepsilon_{ijq})$ where $q$ is the number of response variables (four in the skull example) is assumed to have a multivariate normal distribution with null mean vector and covariance matrix, $\Sigma$, assumed to be the same in each level of the grouping factor. The hypothesis of interest is that the population mean vectors for the different levels of the grouping factor are the same.

In the multivariate situation, when there are more than two levels of the grouping factor, no single test statistic can be derived which is always the most powerful, for *all* types of departures from the null hypothesis of the equality of mean vector. A number of different test statistics are available which may give different results when applied to the same data set, although the final conclusion is often the same. The principal test statistics for the multivariate analysis of variance are *Hotelling-Lawley trace*, *Wilks' ratio of determinants*, *Roy's greatest root*, and the *Pillai trace*. Details are given in Morrison (2005).

## 5.3 Analysis Using R

### 5.3.1 Weight Gain in Rats

Before applying analysis of variance to the data in Table 5.1 we should try to summarise the main features of the data by calculating means and standard deviations and by producing some hopefully informative graphs. The data is available in the *data.frame* `weightgain`. The following R code produces the required summary statistics

```
R> data("weightgain", package = "HSAUR2")
R> tapply(weightgain$weightgain,
+         list(weightgain$source, weightgain$type), mean)
```

```
R> plot.design(weightgain)
```

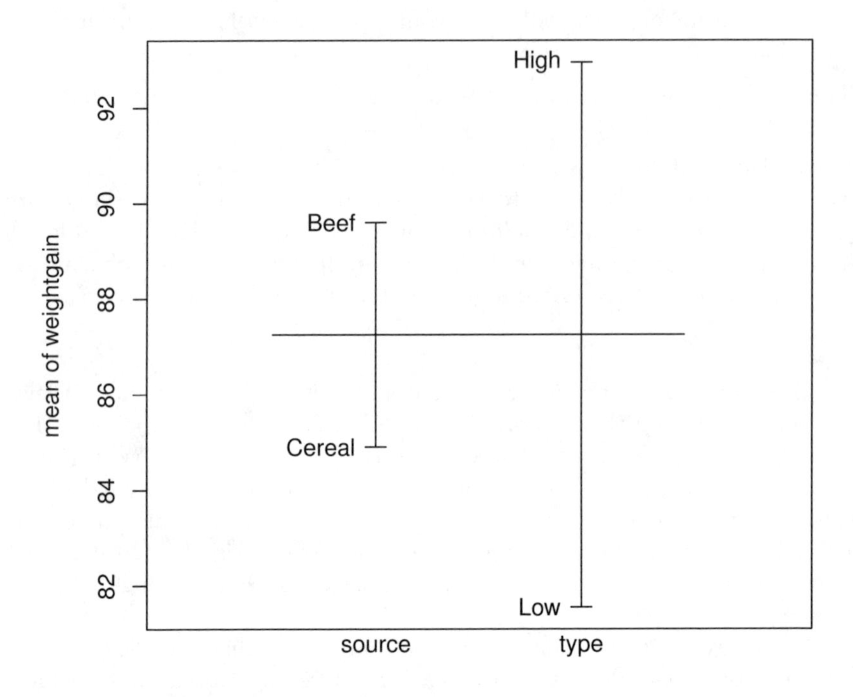


**Figure 5.1** Plot of mean weight gain for each level of the two factors.

```
        High  Low
Beef   100.0 79.2
Cereal  85.9 83.9
```

```
R> tapply(weightgain$weightgain,
+        list(weightgain$source, weightgain$type), sd)
```

```
           High      Low
Beef   15.13642 13.88684
Cereal 15.02184 15.70881
```

The cell variances are relatively similar and there is no apparent relationship between cell mean and cell variance so the homogeneity assumption of the analysis of variance looks like it is reasonable for these data. The plot of cell means in Figure 5.1 suggests that there is a considerable difference in weight gain for the amount of protein factor with the gain for the high-protein diet being far more than for the low-protein diet. A smaller difference is seen for the source factor with beef leading to a higher gain than cereal.

To apply analysis of variance to the data we can use the aov function in R and then the summary method to give us the usual analysis of variance table. The model *formula* specifies a two-way layout with interaction terms, where the first factor is source, and the second factor is type.

```
R> wg_aov <- aov(weightgain ~ source * type, data = weightgain)
```

```
R> summary(wg_aov)

            Df Sum Sq Mean Sq F value  Pr(>F)
source       1  220.9   220.9  0.9879 0.32688
type         1 1299.6  1299.6  5.8123 0.02114
source:type  1  883.6   883.6  3.9518 0.05447
Residuals   36 8049.4   223.6
```

**Figure 5.2** R output of the ANOVA fit for the weightgain data.

The resulting analysis of variance table in Figure 5.2 shows that the main effect of type is highly significant confirming what was seen in Figure 5.1. The main effect of source is not significant. But interpretation of both these main effects is complicated by the type $\times$ source interaction which approaches significance at the 5% level. To try to understand this interaction effect it will be useful to plot the mean weight gain for low- and high-protein diets for each level of source of protein, beef and cereal. The required R code is given with Figure 5.3. From the resulting plot we see that for low-protein diets, the use of cereal as the source of the protein leads to a greater weight gain than using beef. For high-protein diets the reverse is the case with the beef/high diet leading to the highest weight gain.

The estimates of the intercept and the main and interaction effects can be extracted from the model fit by

```
R> coef(wg_aov)

        (Intercept)          sourceCereal               typeLow
              100.0                 -14.1                 -20.8
sourceCereal:typeLow
               18.8
```

Note that the model was fitted with the restrictions $\gamma_1 = 0$ (corresponding to Beef) and $\beta_1 = 0$ (corresponding to High) because treatment contrasts were used as default as can be seen from

```
R> options("contrasts")

$contrasts
        unordered           ordered
"contr.treatment"      "contr.poly"
```

Thus, the coefficient for source of $-14.1$ can be interpreted as an estimate of the difference $\gamma_2 - \gamma_1$. Alternatively, we can use the restriction $\sum_i \gamma_i = 0$ by

```
R> interaction.plot(weightgain$type, weightgain$source,
+                   weightgain$weightgain)
```

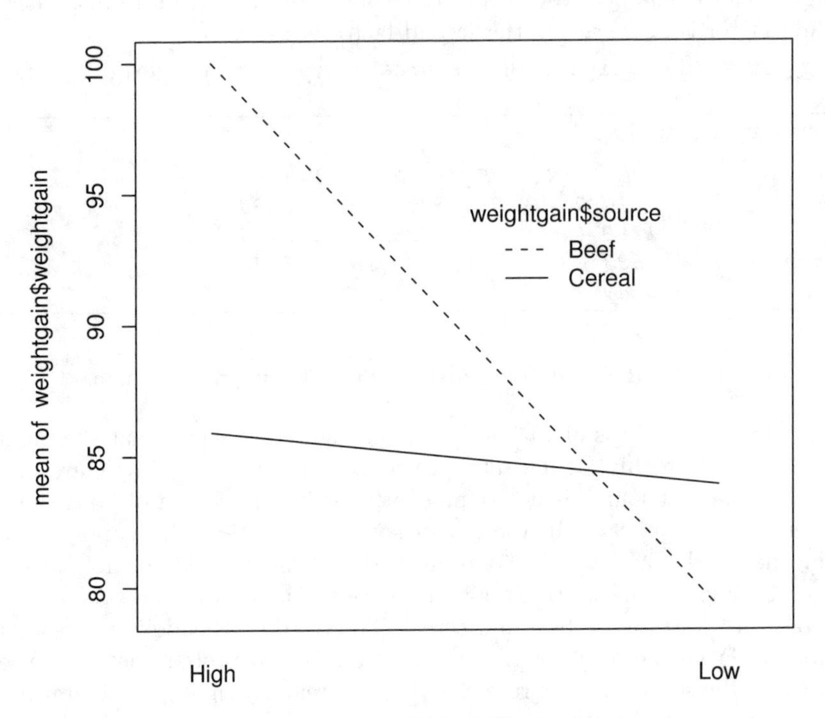


**Figure 5.3** Interaction plot of type and source.

```
R> coef(aov(weightgain ~ source + type + source:type,
+      data = weightgain, contrasts = list(source = contr.sum)))

     (Intercept)           source1           typeLow
           92.95              7.05            -11.40
source1:typeLow
           -9.40
```

*5.3.2 Foster Feeding of Rats of Different Genotype*

As in the previous subsection we will begin the analysis of the foster feeding data in Table 5.2 with a plot of the mean litter weight for the different geno-

```
R> plot.design(foster)
```

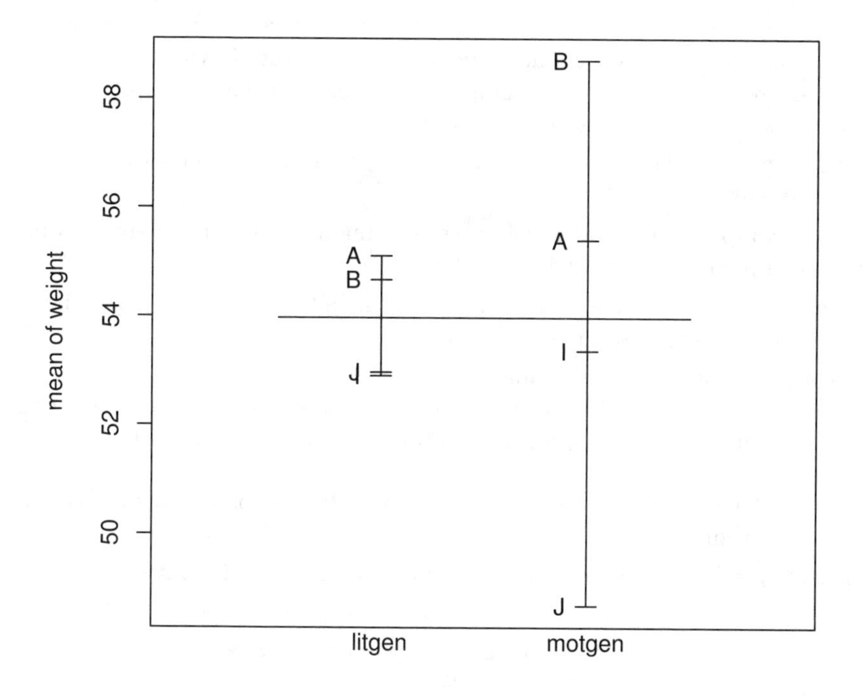


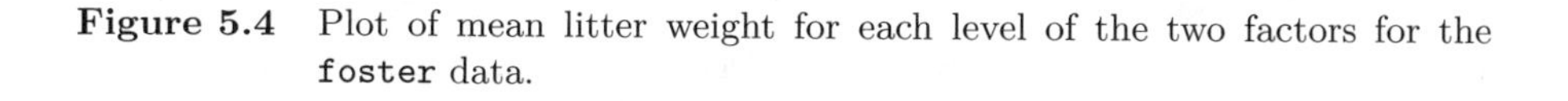

**Figure 5.4** Plot of mean litter weight for each level of the two factors for the `foster` data.

types of mother and litter (see Figure 5.4). The data are in the *data.frame* `foster`

```
R> data("foster", package = "HSAUR2")
```

Figure 5.4 indicates that differences in litter weight for the four levels of mother's genotype are substantial; the corresponding differences for the genotype of the litter are much smaller.

As in the previous example we can now apply analysis of variance using the `aov` function, but there is a complication caused by the unbalanced nature of the data. Here where there are unequal numbers of observations in the 16 cells of the two-way layout, it is no longer possible to partition the variation in the data into *non-overlapping* or *orthogonal* sums of squares representing main effects and interactions. In an unbalanced two-way layout with factors

$A$ and $B$ there is a proportion of the variance of the response variable that can be attributed to either $A$ or $B$. The consequence is that $A$ and $B$ together explain less of the variation of the dependent variable than the sum of which each explains alone. The result is that the sum of squares corresponding to a factor depends on which other terms are currently in the model for the observations, so the sums of squares depend on the order in which the factors are considered and represent a comparison of models. For example, for the order $a, b, a \times b$, the sums of squares are such that

- $SSa$: compares the model containing only the $a$ main effect with one containing only the overall mean.
- $SSb|a$: compares the model including both main effects, but no interaction, with one including only the main effect of $a$.
- $SSab|a, b$: compares the model including an interaction and main effects with one including only main effects.

The use of these sums of squares (sometimes known as *Type I sums of squares*) in a series of tables in which the effects are considered in different orders provides the most appropriate approach to the analysis of unbalanced designs.

We can derive the two analyses of variance tables for the foster feeding example by applying the R code

```
R> summary(aov(weight ~ litgen * motgen, data = foster))
```

to give

```
              Df  Sum Sq Mean Sq F value   Pr(>F)
litgen         3   60.16   20.05  0.3697 0.775221
motgen         3  775.08  258.36  4.7632 0.005736
litgen:motgen  9  824.07   91.56  1.6881 0.120053
Residuals     45 2440.82   54.24
```

and then the code

```
R> summary(aov(weight ~ motgen * litgen, data = foster))
```

to give

```
              Df  Sum Sq Mean Sq F value   Pr(>F)
motgen         3  771.61  257.20  4.7419 0.005869
litgen         3   63.63   21.21  0.3911 0.760004
motgen:litgen  9  824.07   91.56  1.6881 0.120053
Residuals     45 2440.82   54.24
```

There are (small) differences in the sum of squares for the two main effects and, consequently, in the associated $F$-tests and $p$-values. This would not be true if in the previous example in Subsection 5.3.1 we had used the code

```
R> summary(aov(weightgain ~ type * source, data = weightgain))
```

instead of the code which produced Figure 5.2 (readers should confirm that this is the case).

Although for the foster feeding data the differences in the two analyses of variance with different orders of main effects are very small, this may not always be the case and care is needed in dealing with unbalanced designs. For a more complete discussion see Nelder (1977) and Aitkin (1978).

Both ANOVA tables indicate that the main effect of mother's genotype is highly significant and that genotype B leads to the greatest litter weight and genotype J to the smallest litter weight.

We can investigate the effect of genotype B on litter weight in more detail by the use of *multiple comparison procedures* (see Everitt, 1996, and Chapter 14). Such procedures allow a comparison of all pairs of levels of a factor whilst maintaining the nominal significance level at its specified value and producing adjusted confidence intervals for mean differences. One such procedure is called *Tukey honest significant differences* suggested by Tukey (1953); see Hochberg and Tamhane (1987) also. Here, we are interested in simultaneous confidence intervals for the weight differences between all four genotypes of the mother. First, an ANOVA model is fitted

```
R> foster_aov <- aov(weight ~ litgen * motgen, data = foster)
```

which serves as the basis of the multiple comparisons, here with all pair-wise differences by

```
R> foster_hsd <- TukeyHSD(foster_aov, "motgen")
R> foster_hsd
```

```
  Tukey multiple comparisons of means
    95% family-wise confidence level

Fit: aov(formula = weight ~ litgen * motgen, data = foster)

$motgen
         diff        lwr        upr     p adj
B-A  3.330369  -3.859729 10.5204672 0.6078581
I-A -1.895574  -8.841869  5.0507207 0.8853702
J-A -6.566168 -13.627285  0.4949498 0.0767540
I-B -5.225943 -12.416041  1.9641552 0.2266493
J-B -9.896537 -17.197624 -2.5954489 0.0040509
J-I -4.670593 -11.731711  2.3905240 0.3035490
```

A convenient `plot` method exists for this object and we can get a graphical representation of the multiple confidence intervals as shown in Figure 5.5. It appears that there is only evidence for a difference in the B and J genotypes. Note that the particular method implemented in `TukeyHSD` is applicable only to balanced and mildly unbalanced designs (which is the case here). Alternative approaches, applicable to unbalanced designs and more general research questions, will be introduced and discussed in Chapter 14.

### *5.3.3 Water Hardness and Mortality*

The water hardness and mortality data for 61 large towns in England and Wales (see Table 3.3) was analysed in Chapter 3 and here we will extend the analysis by an assessment of the differences of both hardness and mortality

```
R> plot(foster_hsd)
```

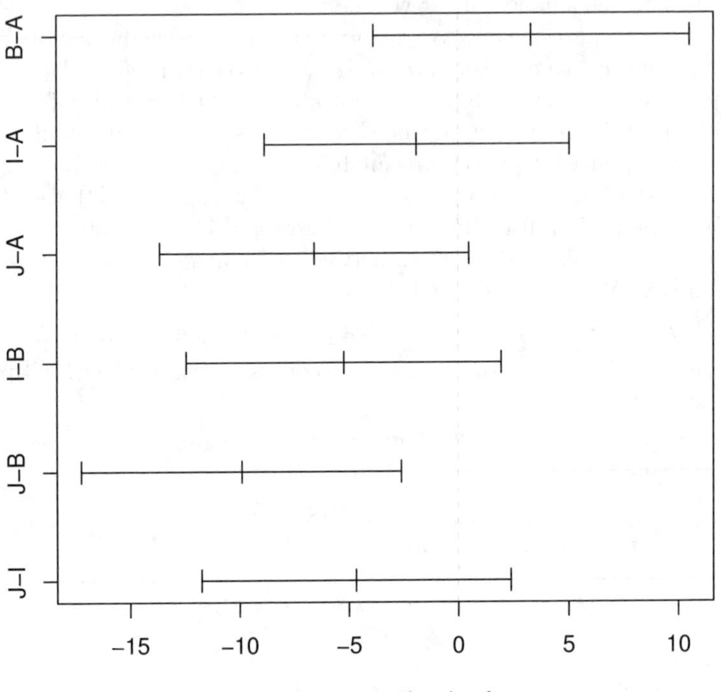


**Figure 5.5** Graphical presentation of multiple comparison results for the `foster` feeding data.

in the North or South. The hypothesis that the two-dimensional mean-vector of water hardness and mortality is the same for cities in the North and the South can be tested by *Hotelling-Lawley* test in a multivariate analysis of variance framework. The R function `manova` can be used to fit such a model and the corresponding `summary` method performs the test specified by the `test` argument

```
R> data("water", package = "HSAUR2")
R> summary(manova(cbind(hardness, mortality) ~ location,
+      data = water), test = "Hotelling-Lawley")

          Df Hotelling approx F num Df den Df    Pr(>F)
location   1    0.9002  26.1062      2     58 8.217e-09
Residuals 59
```

The `cbind` statement in the left hand side of the formula indicates that a *multivariate* response variable is to be modelled. The $p$-value associated with the *Hotelling-Lawley* statistic is very small and there is strong evidence that the mean vectors of the two variables are not the same in the two regions. Looking at the sample means

```
R> tapply(water$hardness, water$location, mean)

   North    South
30.40000 69.76923

R> tapply(water$mortality, water$location, mean)

   North    South
1633.600 1376.808
```

we see large differences in the two regions both in water hardness and mortality, where low mortality is associated with hard water in the South and high mortality with soft water in the North (see Figure 3.8 also).

### 5.3.4 Male Egyptian Skulls

We can begin by looking at a table of mean values for the four measurements within each of the five epochs. The measurements are available in the *data.frame* `skulls` and we can compute the means over all epochs by

```
R> data("skulls", package = "HSAUR2")
R> means <- aggregate(skulls[,c("mb", "bh", "bl", "nh")],
+                     list(epoch = skulls$epoch), mean)
R> means

    epoch       mb       bh       bl       nh
1 c4000BC 131.3667 133.6000 99.16667 50.53333
2 c3300BC 132.3667 132.7000 99.06667 50.23333
3 c1850BC 134.4667 133.8000 96.03333 50.56667
4  c200BC 135.5000 132.3000 94.53333 51.96667
5  cAD150 136.1667 130.3333 93.50000 51.36667
```

It may also be useful to look at these means graphically and this could be done in a variety of ways. Here we construct a scatterplot matrix of the means using the code attached to Figure 5.6.

There appear to be quite large differences between the epoch means, at least on some of the four measurements. We can now test for a difference more formally by using MANOVA with the following R code to apply each of the four possible test criteria mentioned earlier;

```
R> skulls_manova <- manova(cbind(mb, bh, bl, nh) ~ epoch,
+                          data = skulls)
R> summary(skulls_manova, test = "Pillai")

           Df Pillai approx F num Df den Df    Pr(>F)
epoch       4 0.3533   3.5120     16    580 4.675e-06
Residuals 145
```

```
R> pairs(means[,-1],
+      panel = function(x, y) {
+          text(x, y, abbreviate(levels(skulls$epoch)))
+      })
```

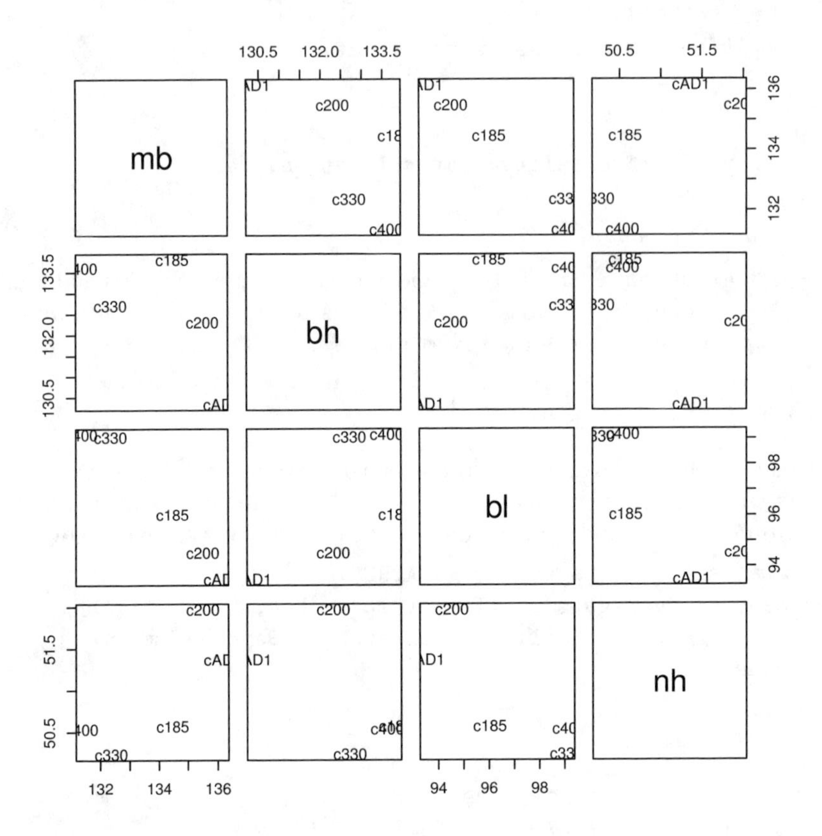


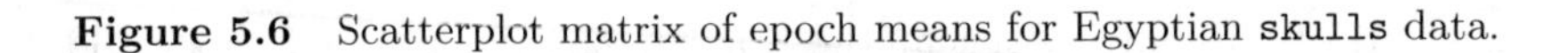

**Figure 5.6** Scatterplot matrix of epoch means for Egyptian skulls data.

```
R> summary(skulls_manova, test = "Wilks")

              Df  Wilks approx F num Df den Df    Pr(>F)
epoch       4.00 0.6636   3.9009  16.00 434.45 7.01e-07
Residuals 145.00

R> summary(skulls_manova, test = "Hotelling-Lawley")

           Df Hotelling approx F num Df den Df     Pr(>F)
epoch       4    0.4818   4.2310     16    562 8.278e-08
Residuals 145

R> summary(skulls_manova, test = "Roy")
```

```
             Df      Roy approx F num Df den Df    Pr(>F)
epoch         4  0.4251  15.4097      4     145 1.588e-10
Residuals   145
```

The $p$-value associated with each four test criteria is very small and there is strong evidence that the skull measurements differ between the five epochs. We might now move on to investigate which epochs differ and on which variables. We can look at the univariate $F$-tests for each of the four variables by using the code

```
R> summary.aov(skulls_manova)
```

```
 Response mb :
             Df  Sum Sq Mean Sq F value    Pr(>F)
epoch         4  502.83  125.71  5.9546 0.0001826
Residuals   145 3061.07   21.11

 Response bh :
             Df Sum Sq Mean Sq F value  Pr(>F)
epoch         4  229.9    57.5  2.4474 0.04897
Residuals   145 3405.3    23.5

 Response bl :
             Df Sum Sq Mean Sq F value    Pr(>F)
epoch         4  803.3   200.8  8.3057 4.636e-06
Residuals   145 3506.0    24.2

 Response nh :
             Df  Sum Sq Mean Sq F value Pr(>F)
epoch         4   61.20   15.30   1.507 0.2032
Residuals   145 1472.13   10.15
```

We see that the results for the maximum breadths (`mb`) and basialiveolar length (`bl`) are highly significant, with those for the other two variables, in particular for nasal heights (`nh`), suggesting little evidence of a difference. To look at the pairwise multivariate tests (any of the four test criteria are equivalent in the case of a one-way layout with two levels only) we can use the `summary` method and `manova` function as follows:

```
R> summary(manova(cbind(mb, bh, bl, nh) ~ epoch, data = skulls,
+                subset = epoch %in% c("c4000BC", "c3300BC")))
```

```
          Df  Pillai approx F num Df den Df Pr(>F)
epoch      1 0.02767  0.39135      4     55  0.814
Residuals 58
```

```
R> summary(manova(cbind(mb, bh, bl, nh) ~ epoch, data = skulls,
+                subset = epoch %in% c("c4000BC", "c1850BC")))
```

```
          Df Pillai approx F num Df den Df  Pr(>F)
epoch      1 0.1876   3.1744      4     55 0.02035
Residuals 58
```

```
R> summary(manova(cbind(mb, bh, bl, nh) ~ epoch, data = skulls,
+                 subset = epoch %in% c("c4000BC", "c200BC")))

          Df Pillai approx F num Df den Df    Pr(>F)
epoch      1 0.3030   5.9766      4     55 0.0004564
Residuals 58
R> summary(manova(cbind(mb, bh, bl, nh) ~ epoch, data = skulls,
+                 subset = epoch %in% c("c4000BC", "cAD150")))

          Df Pillai approx F num Df den Df    Pr(>F)
epoch      1 0.3618   7.7956      4     55 4.736e-05
Residuals 58
```

To keep the overall significance level for the set of all pairwise multivariate tests under some control (and still maintain a reasonable power), Stevens (2001) recommends setting the nominal level $\alpha = 0.15$ and carrying out each test at the $\alpha/m$ level where $m$ s the number of tests performed. The results of the four pairwise tests suggest that as the epochs become further separated in time the four skull measurements become increasingly distinct.

For more details of applying multiple comparisons in the multivariate situation see Stevens (2001).

## 5.4 Summary

Analysis of variance is one of the most widely used of statistical techniques and is easily applied using R as is the extension to multivariate data. An analysis of variance needs to be supplemented by graphical material prior to formal analysis and often to more detailed investigation of group differences using multiple comparison techniques.

## Exercises

Ex. 5.1 Examine the residuals (observed value – fitted value) from fitting a main effects only model to the data in Table 5.1. What conclusions do you draw?

Ex. 5.2 The data in Table 5.4 below arise from a sociological study of Australian Aboriginal and white children reported by Quine (1975). In this study, children of both sexes from four age groups (final grade in primary schools and first, second and third form in secondary school) and from two cultural groups were used. The children in each age group were classified as slow or average learners. The response variable was the number of days absent from school during the school year. (Children who had suffered a serious illness during the years were excluded.) Carry out what you consider to be an appropriate analysis of variance of the data noting that (i) there are unequal numbers of observations in each cell and (ii) the response variable here is a count. Interpret your results with the aid of some suitable tables of means and some informative graphs.

**Table 5.4**: `schooldays` data. Days absent from school.

| `race` | `gender` | `school` | `learner` | `absent` |
|---|---|---|---|---|
| aboriginal | male | F0 | slow | 2 |
| aboriginal | male | F0 | slow | 11 |
| aboriginal | male | F0 | slow | 14 |
| aboriginal | male | F0 | average | 5 |
| aboriginal | male | F0 | average | 5 |
| aboriginal | male | F0 | average | 13 |
| aboriginal | male | F0 | average | 20 |
| aboriginal | male | F0 | average | 22 |
| aboriginal | male | F1 | slow | 6 |
| aboriginal | male | F1 | slow | 6 |
| aboriginal | male | F1 | slow | 15 |
| aboriginal | male | F1 | average | 7 |
| aboriginal | male | F1 | average | 14 |
| aboriginal | male | F2 | slow | 6 |
| aboriginal | male | F2 | slow | 32 |
| ⋮ | ⋮ | ⋮ | ⋮ | ⋮ |

Ex. 5.3 The data in Table 5.5 arise from a large study of risk taking (see Timm, 2002). Students were randomly assigned to three different treatments labelled AA, C and NC. Students were administered two parallel forms of a test called 'low' and 'high'. Carry out a test of the equality of the bivariate means of each treatment population.

**Table 5.5**: `students` data. Treatment and results of two tests in three groups of students.

| treatment | low | high | treatment | low | high |
|---|---|---|---|---|---|
| AA | 8 | 28 | C | 34 | 4 |
| AA | 18 | 28 | C | 34 | 4 |
| AA | 8 | 23 | C | 44 | 7 |
| AA | 12 | 20 | C | 39 | 5 |
| AA | 15 | 30 | C | 20 | 0 |
| AA | 12 | 32 | C | 43 | 11 |
| AA | 18 | 31 | NC | 50 | 5 |
| AA | 29 | 25 | NC | 57 | 51 |
| AA | 6 | 28 | NC | 62 | 52 |
| AA | 7 | 28 | NC | 56 | 52 |
| AA | 6 | 24 | NC | 59 | 40 |
| AA | 14 | 30 | NC | 61 | 68 |
| AA | 11 | 23 | NC | 66 | 49 |
| AA | 12 | 20 | NC | 57 | 49 |
| C | 46 | 13 | NC | 62 | 58 |
| C | 26 | 10 | NC | 47 | 58 |
| C | 47 | 22 | NC | 53 | 40 |
| C | 44 | 14 | | | |

*Source*: From Timm, N. H., *Applied Multivariate Analysis*, Springer, New York, 2002. With kind permission of Springer Science and Business Media.

CHAPTER 6

# Simple and Multiple Linear Regression: How Old is the Universe and Cloud Seeding

## 6.1 Introduction

Freedman et al. (2001) give the relative velocity and the distance of 24 galaxies, according to measurements made using the Hubble Space Telescope – the data are contained in the **gamair** package accompanying Wood (2006), see Table 6.1. Velocities are assessed by measuring the Doppler red shift in the spectrum of light observed from the galaxies concerned, although some correction for 'local' velocity components is required. Distances are measured using the known relationship between the period of Cepheid variable stars and their luminosity. How can these data be used to estimate the age of the universe? Here we shall show how this can be done using simple linear regression.

**Table 6.1**: `hubble` data. Distance and velocity for 24 galaxies.

| galaxy | velocity | distance | galaxy | velocity | distance |
|---|---|---|---|---|---|
| NGC0300 | 133 | 2.00 | NGC3621 | 609 | 6.64 |
| NGC0925 | 664 | 9.16 | NGC4321 | 1433 | 15.21 |
| NGC1326A | 1794 | 16.14 | NGC4414 | 619 | 17.70 |
| NGC1365 | 1594 | 17.95 | NGC4496A | 1424 | 14.86 |
| NGC1425 | 1473 | 21.88 | NGC4548 | 1384 | 16.22 |
| NGC2403 | 278 | 3.22 | NGC4535 | 1444 | 15.78 |
| NGC2541 | 714 | 11.22 | NGC4536 | 1423 | 14.93 |
| NGC2090 | 882 | 11.75 | NGC4639 | 1403 | 21.98 |
| NGC3031 | 80 | 3.63 | NGC4725 | 1103 | 12.36 |
| NGC3198 | 772 | 13.80 | IC4182 | 318 | 4.49 |
| NGC3351 | 642 | 10.00 | NGC5253 | 232 | 3.15 |
| NGC3368 | 768 | 10.52 | NGC7331 | 999 | 14.72 |

*Source*: From Freedman W. L., et al., *The Astrophysical Journal*, 553, 47–72, 2001. With permission.

**Table 6.2:** `clouds` data. Cloud seeding experiments in Florida – see above for explanations of the variables.

| seeding | time | sne | cloudcover | prewetness | echomotion | rainfall |
|---|---|---|---|---|---|---|
| no | 0 | 1.75 | 13.4 | 0.274 | stationary | 12.85 |
| yes | 1 | 2.70 | 37.9 | 1.267 | moving | 5.52 |
| yes | 3 | 4.10 | 3.9 | 0.198 | stationary | 6.29 |
| no | 4 | 2.35 | 5.3 | 0.526 | moving | 6.11 |
| yes | 6 | 4.25 | 7.1 | 0.250 | moving | 2.45 |
| no | 9 | 1.60 | 6.9 | 0.018 | stationary | 3.61 |
| no | 18 | 1.30 | 4.6 | 0.307 | moving | 0.47 |
| no | 25 | 3.35 | 4.9 | 0.194 | moving | 4.56 |
| no | 27 | 2.85 | 12.1 | 0.751 | moving | 6.35 |
| yes | 28 | 2.20 | 5.2 | 0.084 | moving | 5.06 |
| yes | 29 | 4.40 | 4.1 | 0.236 | moving | 2.76 |
| yes | 32 | 3.10 | 2.8 | 0.214 | moving | 4.05 |
| no | 33 | 3.95 | 6.8 | 0.796 | moving | 5.74 |
| yes | 35 | 2.90 | 3.0 | 0.124 | moving | 4.84 |
| yes | 38 | 2.05 | 7.0 | 0.144 | moving | 11.86 |
| no | 39 | 4.00 | 11.3 | 0.398 | moving | 4.45 |
| no | 53 | 3.35 | 4.2 | 0.237 | stationary | 3.66 |
| yes | 55 | 3.70 | 3.3 | 0.960 | moving | 4.22 |
| no | 56 | 3.80 | 2.2 | 0.230 | moving | 1.16 |
| yes | 59 | 3.40 | 6.5 | 0.142 | stationary | 5.45 |
| yes | 65 | 3.15 | 3.1 | 0.073 | moving | 2.02 |
| no | 68 | 3.15 | 2.6 | 0.136 | moving | 0.82 |
| yes | 82 | 4.01 | 8.3 | 0.123 | moving | 1.09 |
| no | 83 | 4.65 | 7.4 | 0.168 | moving | 0.28 |

Weather modification, or cloud seeding, is the treatment of individual clouds or storm systems with various inorganic and organic materials in the hope of achieving an increase in rainfall. Introduction of such material into a cloud that contains supercooled water, that is, liquid water colder than zero degrees of Celsius, has the aim of inducing freezing, with the consequent ice particles growing at the expense of liquid droplets and becoming heavy enough to fall as rain from clouds that otherwise would produce none.

The data shown in Table 6.2 were collected in the summer of 1975 from an experiment to investigate the use of massive amounts of silver iodide (100 to 1000 grams per cloud) in cloud seeding to increase rainfall (Woodley et al., 1977). In the experiment, which was conducted in an area of Florida, 24 days were judged suitable for seeding on the basis that a measured suitability criterion, denoted *S-Ne*, was not less than 1.5. Here $S$ is the 'seedability', the difference between the maximum height of a cloud if seeded and the same cloud if not seeded predicted by a suitable cloud model, and $Ne$ is the number of hours between 1300 and 1600 G.M.T. with 10 centimetre echoes in the target; this quantity biases the decision for experimentation against naturally rainy days. Consequently, optimal days for seeding are those on which seedability is large and the natural rainfall early in the day is small.

On suitable days, a decision was taken at random as to whether to seed or not. For each day the following variables were measured:

`seeding`: a factor indicating whether seeding action occurred (yes or no),

`time`: number of days after the first day of the experiment,

`cloudcover`: the percentage cloud cover in the experimental area, measured using radar,

`prewetness`: the total rainfall in the target area one hour before seeding (in cubic metres $\times 10^7$),

`echomotion`: a factor showing whether the radar echo was moving or stationary,

`rainfall`: the amount of rain in cubic metres $\times 10^7$,

`sne`: suitability criterion, see above.

The objective in analysing these data is to see how rainfall is related to the explanatory variables and, in particular, to determine the effectiveness of seeding. The method to be used is *multiple linear regression*.

## 6.2 Simple Linear Regression

Assume $y_i$ represents the value of what is generally known as the *response variable* on the $i$th individual and that $x_i$ represents the individual's values on what is most often called an *explanatory variable*. The simple linear regression model is

$$y_i = \beta_0 + \beta_1 x_i + \varepsilon_i$$

where $\beta_0$ is the intercept and $\beta_1$ is the slope of the linear relationship assumed between the response and explanatory variables and $\varepsilon_i$ is an error term. (The 'simple' here means that the model contains only a single explanatory variable; we shall deal with the situation where there are several explanatory variables in the next section.) The error terms are assumed to be independent random variables having a normal distribution with mean zero and constant variance $\sigma^2$.

The regression coefficients, $\beta_0$ and $\beta_1$, may be estimated as $\hat{\beta}_0$ and $\hat{\beta}_1$ using *least squares estimation*, in which the sum of squared differences between the observed values of the response variable $y_i$ and the values 'predicted' by the regression equation $\hat{y}_i = \hat{\beta}_0 + \hat{\beta}_1 x_i$ is minimised, leading to the estimates;

$$\hat{\beta}_1 = \frac{\sum_{i=1}^{n} (y_i - \bar{y})(x_i - \bar{x})}{\sum_{i=1}^{n} (x_i - \bar{x})^2}$$

$$\hat{\beta}_0 = \bar{y} - \hat{\beta}_1 \bar{x}$$

where $\bar{y}$ and $\bar{x}$ are the means of the response and explanatory variable, respectively.

The predicted values of the response variable $y$ from the model are $\hat{y}_i = \hat{\beta}_0 + \hat{\beta}_1 x_i$. The variance $\sigma^2$ of the error terms is estimated as

$$\hat{\sigma}^2 = \frac{1}{n-2} \sum_{i=1}^{n} (y_i - \hat{y}_i)^2.$$

The estimated variance of the estimate of the slope parameter is

$$\mathsf{Var}(\hat{\beta}_1) = \frac{\hat{\sigma}^2}{\sum_{i=1}^{n} (x_i - \bar{x})^2},$$

whereas the estimated variance of a predicted value $y_{\text{pred}}$ at a given value of $x$, say $x_0$ is

$$\mathsf{Var}(y_{\text{pred}}) = \hat{\sigma}^2 \sqrt{\frac{1}{n} + 1 + \frac{(x_0 - \bar{x})^2}{\sum_{i=1}^{n} (x_i - \bar{x})^2}}.$$

In some applications of simple linear regression a model without an intercept is required (when the data is such that the line must go through the origin), i.e., a model of the form

$$y_i = \beta_1 x_i + \varepsilon_i.$$

In this case application of least squares gives the following estimator for $\beta_1$

$$\hat{\beta}_1 = \frac{\sum_{i=1}^{n} x_i y_i}{\sum_{i=1}^{n} x_i^2}. \tag{6.1}$$

## 6.3 Multiple Linear Regression

Assume $y_i$ represents the value of the response variable on the $i$th individual, and that $x_{i1}, x_{i2}, \dots, x_{iq}$ represents the individual's values on $q$ explanatory variables, with $i = 1, \dots, n$. The multiple linear regression model is given by

$$y_i = \beta_0 + \beta_1 x_{i1} + \cdots + \beta_q x_{iq} + \varepsilon_i.$$

The error terms $\varepsilon_i$, $i = 1, \dots, n$, are assumed to be independent random variables having a normal distribution with mean zero and constant variance $\sigma^2$. Consequently, the distribution of the random response variable, $y$, is also normal with expected value given by the linear combination of the explanatory variables

$$\mathsf{E}(y|x_1, \dots, x_q) = \beta_0 + \beta_1 x_1 + \dots + \beta_q x_q$$

and with variance $\sigma^2$.

The parameters of the model $\beta_k$, $k = 1, \dots, q$, are known as regression coefficients with $\beta_0$ corresponding to the overall mean. The regression coefficients represent the expected change in the response variable associated with a unit change in the corresponding explanatory variable, when the remaining explanatory variables are held constant. The *linear* in multiple linear regression applies to the regression parameters, not to the response or explanatory variables. Consequently, models in which, for example, the logarithm of a response variable is modelled in terms of quadratic functions of some of the explanatory variables would be included in this class of models.

The multiple linear regression model can be written most conveniently for all $n$ individuals by using matrices and vectors as $\mathbf{y} = \mathbf{X}\beta + \varepsilon$ where $\mathbf{y}^\top = (y_1, \dots, y_n)$ is the vector of response variables, $\beta^\top = (\beta_0, \beta_1, \dots, \beta_q)$ is the vector of regression coefficients, and $\varepsilon^\top = (\varepsilon_1, \dots, \varepsilon_n)$ are the error terms. The *design* or *model matrix* $\mathbf{X}$ consists of the $q$ continuously measured explanatory variables and a column of ones corresponding to the *intercept* term

$$\mathbf{X} = \begin{pmatrix} 1 & x_{11} & x_{12} & \dots & x_{1q} \\ 1 & x_{21} & x_{22} & \dots & x_{2q} \\ \vdots & \vdots & \vdots & \ddots & \vdots \\ 1 & x_{n1} & x_{n2} & \dots & x_{nq} \end{pmatrix}.$$

In case one or more of the explanatory variables are nominal or ordinal variables, they are represented by a zero-one dummy coding. Assume that $x_1$ is a factor at $m$ levels, the submatrix of $\mathbf{X}$ corresponding to $x_1$ is a $n \times m$ matrix of zeros and ones, where the $j$th element in the $i$th row is one when $x_{i1}$ is at the $j$th level.

Assuming that the cross-product $\mathbf{X}^\top\mathbf{X}$ is non-singular, i.e., can be inverted, then the least squares estimator of the parameter vector $\beta$ is unique and can be calculated by $\hat{\beta} = (\mathbf{X}^\top\mathbf{X})^{-1}\mathbf{X}^\top\mathbf{y}$. The expectation and covariance of this estimator $\hat{\beta}$ are given by $\mathsf{E}(\hat{\beta}) = \beta$ and $\mathsf{Var}(\hat{\beta}) = \sigma^2(\mathbf{X}^\top\mathbf{X})^{-1}$. The diagonal elements of the covariance matrix $\mathsf{Var}(\hat{\beta})$ give the variances of $\hat{\beta}_j, j = 0, \dots, q$, whereas the off diagonal elements give the covariances between pairs of $\hat{\beta}_j$ and $\hat{\beta}_k$. The square roots of the diagonal elements of the covariance matrix are thus the standard errors of the estimates $\hat{\beta}_j$.

If the cross-product $\mathbf{X}^\top\mathbf{X}$ is singular we need to reformulate the model to $\mathbf{y} = \mathbf{X}\mathbf{C}\beta^\star + \varepsilon$ such that $\mathbf{X}^\star = \mathbf{X}\mathbf{C}$ has full rank. The matrix $\mathbf{C}$ is called the *contrast matrix* in S and R and the result of the model fit is an estimate $\hat{\beta}^\star$.

By default, a contrast matrix derived from *treatment contrasts* is used. For the theoretical details we refer to Searle (1971), the implementation of contrasts in S and R is discussed by Chambers and Hastie (1992) and Venables and Ripley (2002).

The regression analysis can be assessed using the following analysis of variance table (Table 6.3):

**Table 6.3**: Analysis of variance table for the multiple linear regression model.

| Source of variation | Sum of squares | Degrees of freedom |
|---|---|---|
| Regression | $\sum_{i=1}^{n} (\hat{y}_i - \bar{y})^2$ | $q$ |
| Residual | $\sum_{i=1}^{n} (\hat{y}_i - y_i)^2$ | $n - q - 1$ |
| Total | $\sum_{i=1}^{n} (y_i - \bar{y})^2$ | $n - 1$ |

where $\hat{y}_i$ is the predicted value of the response variable for the $i$th individual $\hat{y}_i = \hat{\beta}_0 + \hat{\beta}_1 x_{i1} + \cdots + \hat{\beta}_q x_{q1}$ and $\bar{y} = \sum_{i=1}^{n} y_i/n$ is the mean of the response variable.

The mean square ratio

$$F = \frac{\sum_{i=1}^{n} (\hat{y}_i - \bar{y})^2/q}{\sum_{i=1}^{n} (\hat{y}_i - y_i)^2/(n - q - 1)}$$

provides an $F$-test of the general hypothesis

$$H_0 : \beta_1 = \cdots = \beta_q = 0.$$

Under $H_0$, the test statistic $F$ has an $F$-distribution with $q$ and $n - q - 1$ degrees of freedom. An estimate of the variance $\sigma^2$ is

$$\hat{\sigma}^2 = \frac{1}{n - q - 1} \sum_{i=1}^{n} (y_i - \hat{y}_i)^2.$$

The correlation between the observed values $y_i$ and the fitted values $\hat{y}_i$ is known as the *multiple correlation coefficient*. Individual regression coefficients can be assessed by using the ratio $t$-statistics $t_j = \hat{\beta}_j/\sqrt{\mathsf{Var}(\hat{\beta})_{jj}}$, although these ratios should be used only as rough guides to the 'significance' of the coefficients. The problem of selecting the 'best' subset of variables to be included in a model is one of the most delicate ones in statistics and we refer to Miller (2002) for the theoretical details and practical limitations (and see Exercise 6.4).

### 6.3.1 Regression Diagnostics

The possible influence of outliers and the checking of assumptions made in fitting the multiple regression model, i.e., constant variance and normality of error terms, can both be undertaken using a variety of diagnostic tools, of which the simplest and most well known are the estimated residuals, i.e., the differences between the observed values of the response and the fitted values of the response. In essence these residuals estimate the error terms in the simple and multiple linear regression model. So, after estimation, the next stage in the analysis should be an examination of such residuals from fitting the chosen model to check on the normality and constant variance assumptions and to identify outliers. The most useful plots of these residuals are:

- A plot of residuals against each explanatory variable in the model. The presence of a non-linear relationship, for example, may suggest that a higher-order term, in the explanatory variable should be considered.
- A plot of residuals against fitted values. If the variance of the residuals appears to increase with predicted value, a transformation of the response variable may be in order.
- A normal probability plot of the residuals. After all the systematic variation has been removed from the data, the residuals should look like a sample from a standard normal distribution. A plot of the ordered residuals against the expected order statistics from a normal distribution provides a graphical check of this assumption.

## 6.4 Analysis Using R

### 6.4.1 Estimating the Age of the Universe

Prior to applying a simple regression to the data it will be useful to look at a plot to assess their major features. The R code given in Figure 6.1 produces a scatterplot of velocity and distance. The diagram shows a clear, strong relationship between velocity and distance. The next step is to fit a simple linear regression model to the data, but in this case the nature of the data requires a model without intercept because if distance is zero so is relative speed. So the model to be fitted to these data is

$$\text{velocity} = \beta_1 \text{distance} + \varepsilon.$$

This is essentially what astronomers call Hubble's Law and $\beta_1$ is known as Hubble's constant; $\beta_1^{-1}$ gives an approximate age of the universe.

To fit this model we are estimating $\beta_1$ using formula (6.1). Although this operation is rather easy

```
R> sum(hubble$distance * hubble$velocity) /
+      sum(hubble$distance^2)
```

```
[1] 76.58117
```

it is more convenient to apply R's linear modelling function

```
R> plot(velocity ~ distance, data = hubble)
```

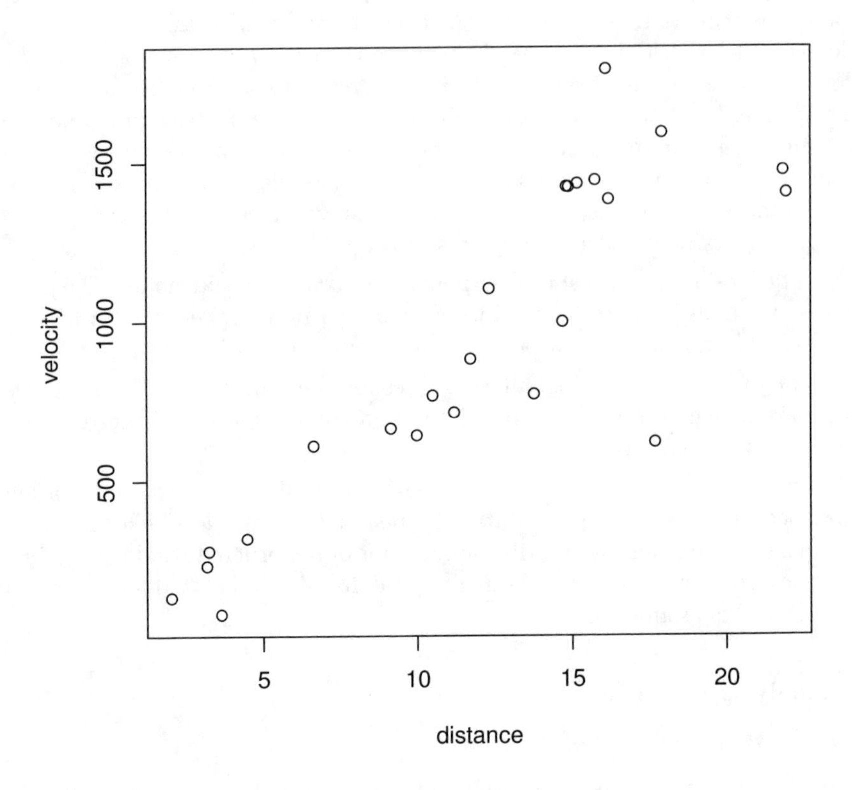


**Figure 6.1** Scatterplot of velocity and distance.

```
R> hmod <- lm(velocity ~ distance - 1, data = hubble)
```

Note that the model formula specifies a model without intercept. We can now extract the estimated model coefficients via

```
R> coef(hmod)
distance
76.58117
```

and add this estimated regression line to the scatterplot; the result is shown in Figure 6.2. In addition, we produce a scatterplot of the residuals $y_i - \hat{y}_i$ against fitted values $\hat{y}_i$ to assess the quality of the model fit. It seems that for higher distance values the variance of velocity increases; however, we are interested in only the estimated parameter $\hat{\beta}_1$ which remains valid under variance heterogeneity (in contrast to $t$-tests and associated $p$-values).

Now we can use the estimated value of $\beta_1$ to find an approximate value

```
R> layout(matrix(1:2, ncol = 2))
R> plot(velocity ~ distance, data = hubble)
R> abline(hmod)
R> plot(hmod, which = 1)
```

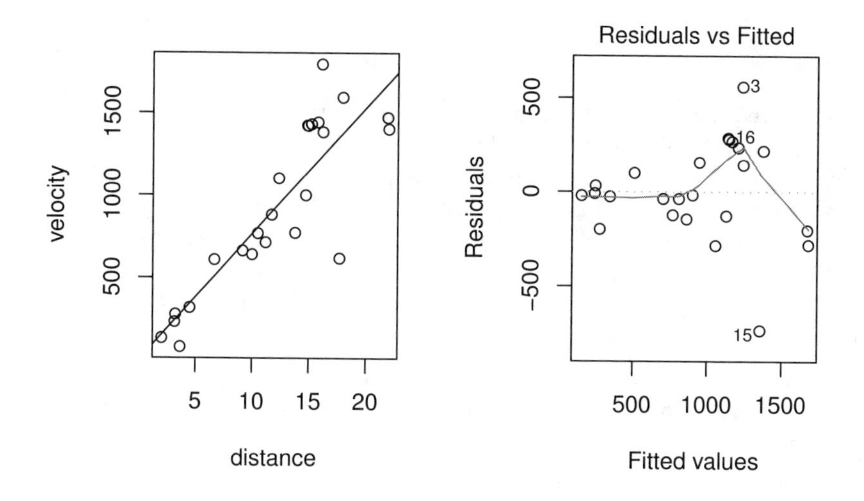


**Figure 6.2** Scatterplot of velocity and distance with estimated regression line (left) and plot of residuals against fitted values (right).

for the age of the universe. The Hubble constant itself has units of km $\times$ sec$^{-1}$ $\times$ Mpc$^{-1}$. A mega-parsec (Mpc) is $3.09 \times 10^{19}$km, so we need to divide the estimated value of $\beta_1$ by this amount in order to obtain Hubble's constant with units of sec$^{-1}$. The approximate age of the universe in seconds will then be the inverse of this calculation. Carrying out the necessary computations

```
R> Mpc <- 3.09 * 10^19
R> ysec <- 60^2 * 24 * 365.25
R> Mpcyear <- Mpc / ysec
R> 1 / (coef(hmod) / Mpcyear)

   distance
12785935335
```

gives an estimated age of roughly 12.8 billion years.

### *6.4.2 Cloud Seeding*

Again, a graphical display highlighting the most important aspects of the data will be helpful. Here we will construct boxplots of the rainfall in each category of the dichotomous explanatory variables and scatterplots of rainfall against each of the continuous explanatory variables.

Both the boxplots (Figure 6.3) and the scatterplots (Figure 6.4) show some evidence of outliers. The row names of the extreme observations in the `clouds` *data.frame* can be identified via

```
R> rownames(clouds)[clouds$rainfall %in% c(bxpseeding$out,
+                                              bxpecho$out)]
```

```
[1] "1"  "15"
```

where `bxpseeding` and `bxpecho` are variables created by `boxplot` in Figure 6.3. Now we shall not remove these observations but bear in mind during the modelling process that they may cause problems.

In this example it is sensible to assume that the effect that some of the other explanatory variables is modified by seeding and therefore consider a model that includes seeding as covariate and, furthermore, allows interaction terms for `seeding` with each of the covariates except `time`. This model can be described by the *formula*

```
R> clouds_formula <- rainfall ~ seeding +
+      seeding:(sne + cloudcover + prewetness + echomotion) +
+      time
```

and the design matrix $\mathbf{X}^{\star}$ can be computed via

```
R> Xstar <- model.matrix(clouds_formula, data = clouds)
```

By default, treatment contrasts have been applied to the dummy codings of the factors `seeding` and `echomotion` as can be seen from the inspection of the `contrasts` attribute of the model matrix

```
R> attr(Xstar, "contrasts")
```

```
$seeding
[1] "contr.treatment"

$echomotion
[1] "contr.treatment"
```

The default contrasts can be changed via the `contrasts.arg` argument to `model.matrix` or the `contrasts` argument to the fitting function, for example `lm` or `aov` as shown in Chapter 5.

However, such internals are hidden and performed by high-level model-fitting functions such as `lm` which will be used to fit the linear model defined by the *formula* `clouds_formula`:

```
R> clouds_lm <- lm(clouds_formula, data = clouds)
R> class(clouds_lm)
```

```
[1] "lm"
```

The results of the model fitting is an object of class *lm* for which a `summary` method showing the conventional regression analysis output is available. The

```
R> data("clouds", package = "HSAUR2")
R> layout(matrix(1:2, nrow = 2))
R> bxpseeding <- boxplot(rainfall ~ seeding, data = clouds,
+      ylab = "Rainfall", xlab = "Seeding")
R> bxpecho <- boxplot(rainfall ~ echomotion, data = clouds,
+      ylab = "Rainfall", xlab = "Echo Motion")
```

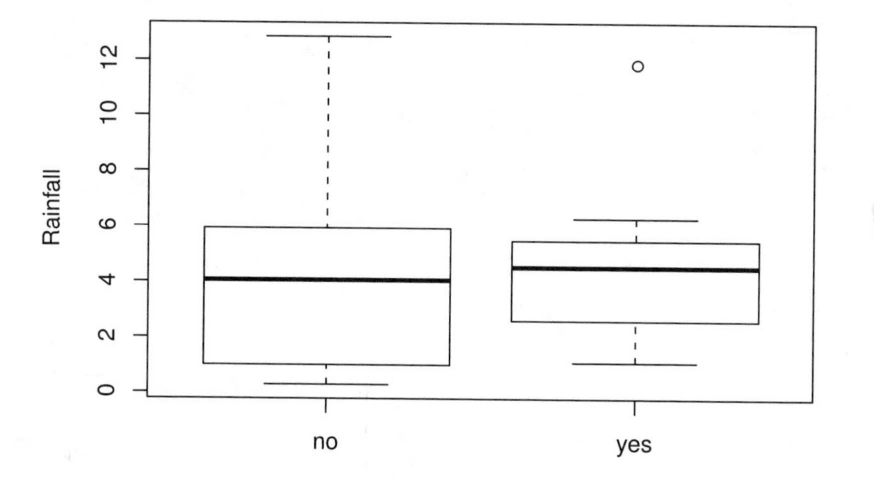

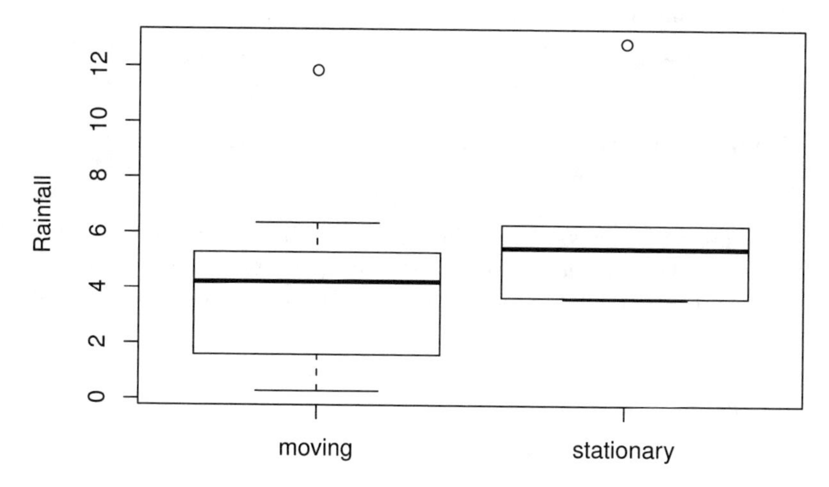


**Figure 6.3** Boxplots of `rainfall`.

```
R> layout(matrix(1:4, nrow = 2))
R> plot(rainfall ~ time, data = clouds)
R> plot(rainfall ~ cloudcover, data = clouds)
R> plot(rainfall ~ sne, data = clouds, xlab="S-Ne criterion")
R> plot(rainfall ~ prewetness, data = clouds)
```

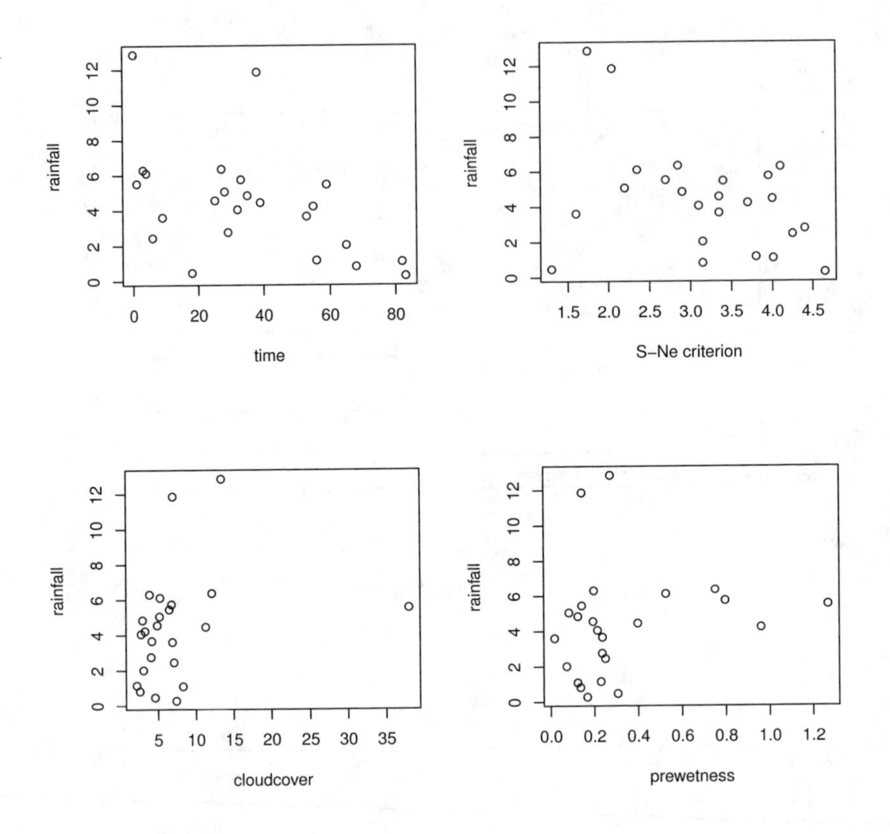


**Figure 6.4** Scatterplots of `rainfall` against the continuous covariates.

output in Figure 6.5 shows the estimates $\hat{\beta}^\star$ with corresponding standard errors and $t$-statistics as well as the $F$-statistic with associated $p$-value.

Many methods are available for extracting components of the fitted model. The estimates $\hat{\beta}^\star$ can be assessed via

```
R> betastar <- coef(clouds_lm)
R> betastar
```

```
  (Intercept)
  -0.34624093
   seedingyes
```

```
R> summary(clouds_lm)

Call:
lm(formula = clouds_formula, data = clouds)

Residuals:
    Min      1Q  Median      3Q     Max 
-2.5259 -1.1486 -0.2704  1.0401  4.3913 

Coefficients:
                                Estimate Std. Error t value
(Intercept)                     -0.34624    2.78773  -0.124
seedingyes                      15.68293    4.44627   3.527
time                            -0.04497    0.02505  -1.795
seedingno:sne                    0.41981    0.84453   0.497
seedingyes:sne                  -2.77738    0.92837  -2.992
seedingno:cloudcover             0.38786    0.21786   1.780
seedingyes:cloudcover           -0.09839    0.11029  -0.892
seedingno:prewetness             4.10834    3.60101   1.141
seedingyes:prewetness            1.55127    2.69287   0.576
seedingno:echomotionstationary   3.15281    1.93253   1.631
seedingyes:echomotionstationary  2.59060    1.81726   1.426
                                Pr(>|t|)
(Intercept)                      0.90306
seedingyes                       0.00372
time                             0.09590
seedingno:sne                    0.62742
seedingyes:sne                   0.01040
seedingno:cloudcover             0.09839
seedingyes:cloudcover            0.38854
seedingno:prewetness             0.27450
seedingyes:prewetness            0.57441
seedingno:echomotionstationary   0.12677
seedingyes:echomotionstationary  0.17757

Residual standard error: 2.205 on 13 degrees of freedom
Multiple R-squared: 0.7158,      Adjusted R-squared: 0.4972
F-statistic: 3.274 on 10 and 13 DF,  p-value: 0.02431
```

**Figure 6.5** R output of the linear model fit for the clouds data.

```
    15.68293481 
           time 
    -0.04497427 
  seedingno:sne 
     0.41981393 
 seedingyes:sne 
    -2.77737613 
```

```
           seedingno:cloudcover
                     0.38786207
          seedingyes:cloudcover
                    -0.09839285
           seedingno:prewetness
                     4.10834188
          seedingyes:prewetness
                     1.55127493
 seedingno:echomotionstationary
                     3.15281358
seedingyes:echomotionstationary
                     2.59059513
```

and the corresponding covariance matrix $\mathsf{Cov}(\hat{\beta}^{\star})$ is available from the `vcov` method

```
R> Vbetastar <- vcov(clouds_lm)
```

where the square roots of the diagonal elements are the standard errors as shown in Figure 6.5

```
R> sqrt(diag(Vbetastar))
                    (Intercept)
                     2.78773403
                     seedingyes
                     4.44626606
                           time
                     0.02505286
                   seedingno:sne
                     0.84452994
                  seedingyes:sne
                     0.92837010
           seedingno:cloudcover
                     0.21785501
          seedingyes:cloudcover
                     0.11028981
           seedingno:prewetness
                     3.60100694
          seedingyes:prewetness
                     2.69287308
 seedingno:echomotionstationary
                     1.93252592
seedingyes:echomotionstationary
                     1.81725973
```

The results of the linear model fit, as shown in Figure 6.5, suggests that rainfall can be increased by cloud seeding. Moreover, the model indicates that higher values of the S-Ne criterion lead to less rainfall, but only on days when cloud seeding happened, i.e., the interaction of seeding with S-Ne significantly affects rainfall. A suitable graph will help in the interpretation of this result. We can plot the relationship between rainfall and S-Ne for seeding and non-seeding days using the R code shown with Figure 6.6.

```
R> psymb <- as.numeric(clouds$seeding)
R> plot(rainfall ~ sne, data = clouds, pch = psymb,
+       xlab = "S-Ne criterion")
R> abline(lm(rainfall ~ sne, data = clouds,
+             subset = seeding == "no"))
R> abline(lm(rainfall ~ sne, data = clouds,
+             subset = seeding == "yes"), lty = 2)
R> legend("topright", legend = c("No seeding", "Seeding"),
+         pch = 1:2, lty = 1:2, bty = "n")
```

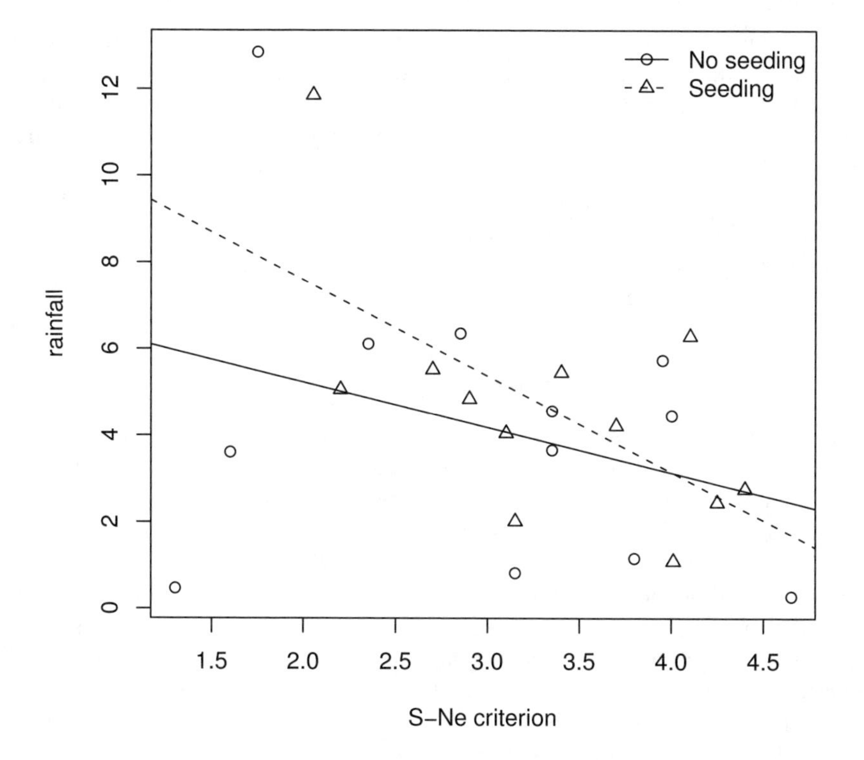


**Figure 6.6** Regression relationship between S-Ne criterion and rainfall with and without seeding.

The plot suggests that for smaller S-Ne values, seeding produces greater rainfall than no seeding, whereas for larger values of S-Ne it tends to produce less. The cross-over occurs at an S-Ne value of approximately four which suggests that seeding is best carried out when S-Ne is less than four. But the number of observations is small and we should perhaps now consider the influence of any outlying observations on these results.

In order to investigate the quality of the model fit, we need access to the residuals and the fitted values. The residuals can be found by the `residuals` method and the fitted values of the response from the `fitted` (or `predict`) method

```
R> clouds_resid <- residuals(clouds_lm)
R> clouds_fitted <- fitted(clouds_lm)
```

Now the residuals and the fitted values can be used to construct diagnostic plots; for example the residual plot in Figure 6.7 where each observation is labelled by its number. Observations 1 and 15 give rather large residual values and the data should perhaps be reanalysed after these two observations are removed. The normal probability plot of the residuals shown in Figure 6.8 shows a reasonable agreement between theoretical and sample quantiles, however, observations 1 and 15 are extreme again.

A further diagnostic that is often very useful is an index plot of the Cook's distances for each observation. This statistic is defined as

$$D_k = \frac{1}{(q+1)\hat{\sigma}^2} \sum_{i=1}^{n} (\hat{y}_{i(k)} - y_i)^2$$

where $\hat{y}_{i(k)}$ is the fitted value of the $i$th observation when the $k$th observation is omitted from the model. The values of $D_k$ assess the impact of the $k$th observation on the estimated regression coefficients. Values of $D_k$ greater than one are suggestive that the corresponding observation has undue influence on the estimated regression coefficients (see Cook and Weisberg, 1982).

An index plot of the Cook's distances for each observation (and many other plots including those constructed above from using the basic functions) can be found from applying the `plot` method to the object that results from the application of the `lm` function. Figure 6.9 suggests that observations 2 and 18 have undue influence on the estimated regression coefficients, but the two outliers identified previously do not. Again it may be useful to look at the results after these two observations have been removed (see Exercise 6.2).

## 6.5 Summary

Multiple regression is used to assess the relationship between a set of explanatory variables and a response variable (with simple linear regression, there is a single exploratory variable). The response variable is assumed to be normally distributed with a mean that is a linear function of the explanatory variables and a variance that is independent of the explanatory variables. An important

```
R> plot(clouds_fitted, clouds_resid, xlab = "Fitted values",
+       ylab = "Residuals", type = "n",
+       ylim = max(abs(clouds_resid)) * c(-1, 1))
R> abline(h = 0, lty = 2)
R> text(clouds_fitted, clouds_resid, labels = rownames(clouds))
```

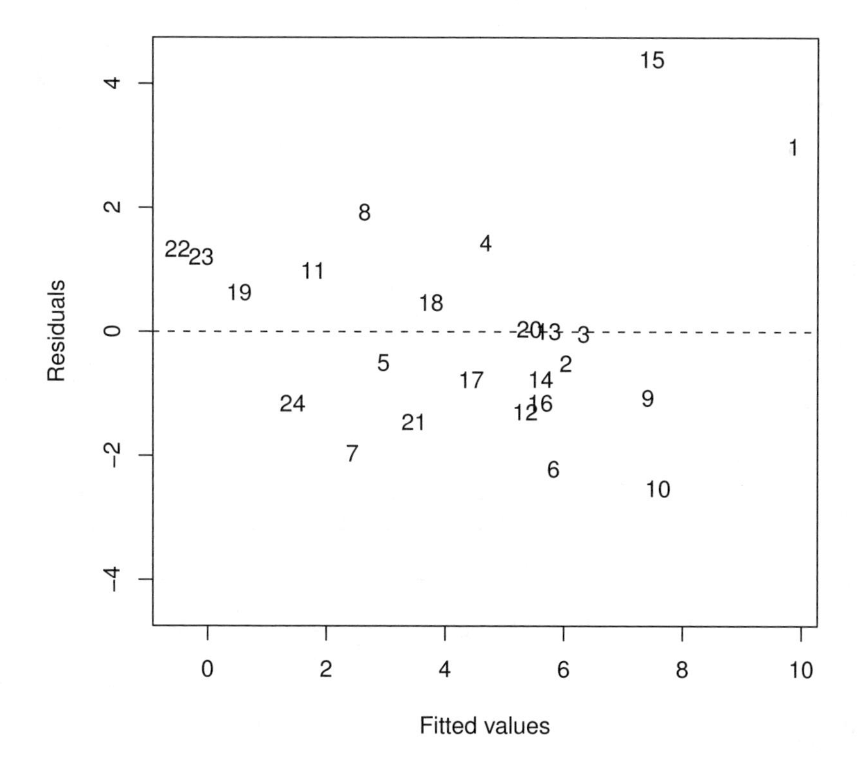


**Figure 6.7** Plot of residuals against fitted values for clouds seeding data.

part of any regression analysis involves the graphical examination of residuals and other diagnostic statistics to help identify departures from assumptions.

## Exercises

Ex. 6.1 The simple residuals calculated as the difference between an observed and predicted value have a distribution that is scale dependent since the variance of each is a function of both $\sigma^2$ and the diagonal elements of the

```
R> qqnorm(clouds_resid, ylab = "Residuals")
R> qqline(clouds_resid)
```

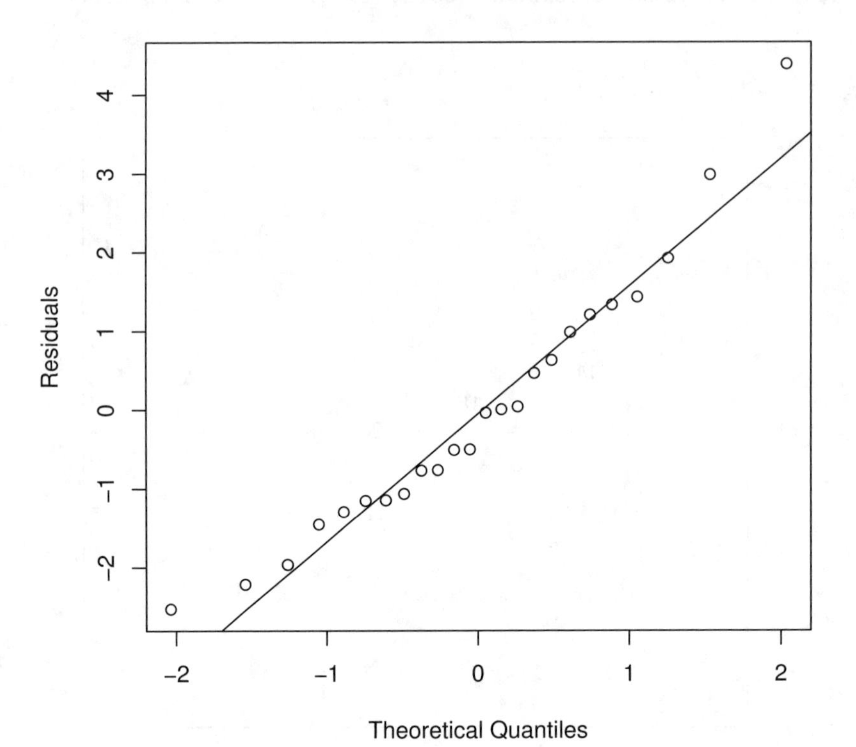


**Figure 6.8** Normal probability plot of residuals from cloud seeding model `clouds_lm`.

*hat matrix* $\mathbf{H}$ given by

$$\mathbf{H} = \mathbf{X}(\mathbf{X}^\top \mathbf{X})^{-1}\mathbf{X}^\top.$$

Consequently it is often more useful to work with the standardised version of the residuals that does not depend on either of these quantities. These standardised residuals are calculated as

$$r_i = \frac{y_i - \hat{y}_i}{\hat{\sigma}\sqrt{1 - h_{ii}}}$$

where $\hat{\sigma}^2$ is the estimator of $\sigma^2$ and $h_{ii}$ is the $i$th diagonal element of $\mathbf{H}$. Write an R function to calculate these residuals and use it to obtain some

```
R> plot(clouds_lm)
```

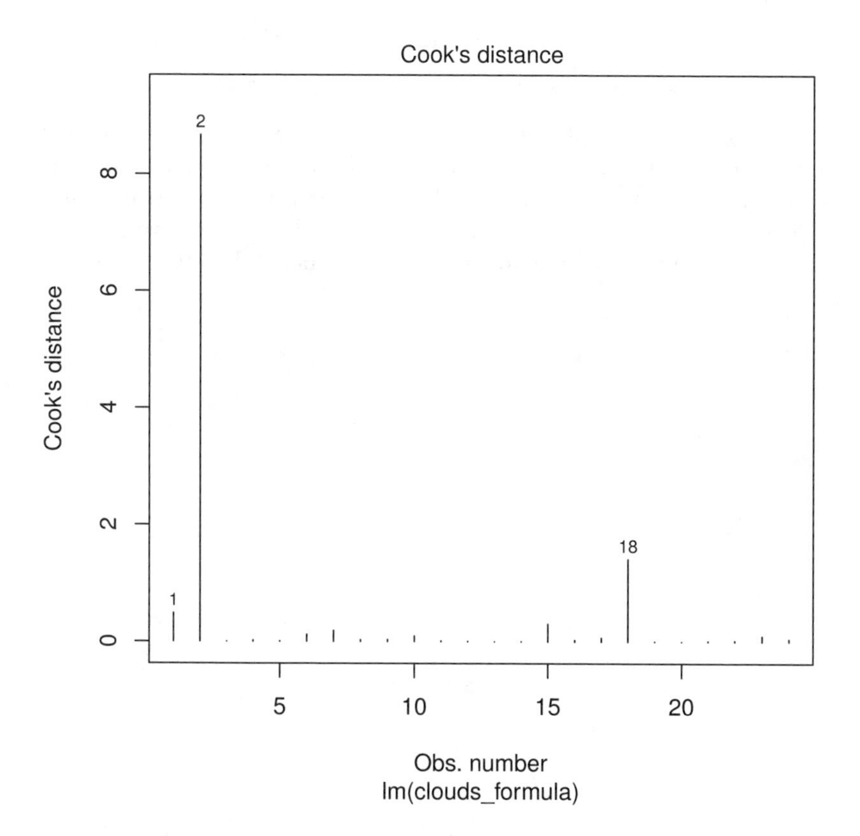


**Figure 6.9** Index plot of Cook's distances for cloud seeding data.

diagnostic plots similar to those mentioned in the text. (The elements of the hat matrix can be obtained from the `lm.influence` function.)

Ex. 6.2 Investigate refitting the cloud seeding data after removing any observations which may give cause for concern.

Ex. 6.3 Show how the analysis of variance table for the data in Table 5.1 of the previous chapter can be constructed from the results of applying an appropriate multiple linear regression to the data.

Ex. 6.4 Investigate the use of the `leaps` function from package **leaps** (Lumley and Miller, 2009) for selecting the 'best' set of variables predicting rainfall in the cloud seeding data.

Ex. 6.5 Remove the observations for galaxies having leverage greater than 0.08 and refit the zero intercept model. What is the estimated age of the universe from this model?

Ex. 6.6 Fit a quadratic regression model, i.e, a model of the form

$$\text{velocity} = \beta_1 \times \text{distance} + \beta_2 \times \text{distance}^2 + \varepsilon,$$

to the `hubble` data and plot the fitted curve and the simple linear regression fit on a scatterplot of the data. Which model do you consider most sensible considering the nature of the data? (The 'quadratic model' here is still regarded as a linear regression model since the term *linear* relates to the parameters of the model not to the powers of the explanatory variable.)

CHAPTER 7

# Logistic Regression and Generalised Linear Models: Blood Screening, Women's Role in Society, Colonic Polyps, and Driving and Back Pain

## 7.1 Introduction

The erythrocyte sedimentation rate (ESR) is the rate at which red blood cells (erythrocytes) settle out of suspension in blood plasma, when measured under standard conditions. If the ESR increases when the level of certain proteins in the blood plasma rise in association with conditions such as rheumatic diseases, chronic infections and malignant diseases, its determination might be useful in screening blood samples taken from people suspected of suffering from one of the conditions mentioned. The absolute value of the ESR is not of great importance; rather, less than 20mm/hr indicates a 'healthy' individual. To assess whether the ESR is a useful diagnostic tool, Collett and Jemain (1985) collected the data shown in Table 7.1. The question of interest is whether there is any association between the probability of an ESR reading greater than 20mm/hr and the levels of the two plasma proteins. If there is not then the determination of ESR would not be useful for diagnostic purposes.

**Table 7.1**: `plasma` data. Blood plasma data.

| fibrinogen | globulin | ESR | fibrinogen | globulin | ESR |
|---|---|---|---|---|---|
| 2.52 | 38 | ESR < 20 | 2.88 | 30 | ESR < 20 |
| 2.56 | 31 | ESR < 20 | 2.65 | 46 | ESR < 20 |
| 2.19 | 33 | ESR < 20 | 2.28 | 36 | ESR < 20 |
| 2.18 | 31 | ESR < 20 | 2.67 | 39 | ESR < 20 |
| 3.41 | 37 | ESR < 20 | 2.29 | 31 | ESR < 20 |
| 2.46 | 36 | ESR < 20 | 2.15 | 31 | ESR < 20 |
| 3.22 | 38 | ESR < 20 | 2.54 | 28 | ESR < 20 |
| 2.21 | 37 | ESR < 20 | 3.34 | 30 | ESR < 20 |
| 3.15 | 39 | ESR < 20 | 2.99 | 36 | ESR < 20 |
| 2.60 | 41 | ESR < 20 | 3.32 | 35 | ESR < 20 |
| 2.29 | 36 | ESR < 20 | 5.06 | 37 | ESR > 20 |
| 2.35 | 29 | ESR < 20 | 3.34 | 32 | ESR > 20 |
| 3.15 | 36 | ESR < 20 | 2.38 | 37 | ESR > 20 |
| 2.68 | 34 | ESR < 20 | 3.53 | 46 | ESR > 20 |

**Table 7.1**: `plasma` data (continued).

| fibrinogen | globulin | ESR | fibrinogen | globulin | ESR |
|---|---|---|---|---|---|
| 2.60 | 38 | ESR < 20 | 2.09 | 44 | ESR > 20 |
| 2.23 | 37 | ESR < 20 | 3.93 | 32 | ESR > 20 |

*Source*: From Collett, D., Jemain, A., *Sains Malay.*, 4, 493–511, 1985. With permission.

In a survey carried out in 1974/1975 each respondent was asked if he or she agreed or disagreed with the statement "Women should take care of running their homes and leave running the country up to men". The responses are summarised in Table 7.2 (from Haberman, 1973) and also given in Collett (2003). The questions of interest here are whether the responses of men and women differ and how years of education affect the response.

**Table 7.2**: `womensrole` data. Women's role in society data.

| education | gender | agree | disagree |
|---|---|---|---|
| 0 | Male | 4 | 2 |
| 1 | Male | 2 | 0 |
| 2 | Male | 4 | 0 |
| 3 | Male | 6 | 3 |
| 4 | Male | 5 | 5 |
| 5 | Male | 13 | 7 |
| 6 | Male | 25 | 9 |
| 7 | Male | 27 | 15 |
| 8 | Male | 75 | 49 |
| 9 | Male | 29 | 29 |
| 10 | Male | 32 | 45 |
| 11 | Male | 36 | 59 |
| 12 | Male | 115 | 245 |
| 13 | Male | 31 | 70 |
| 14 | Male | 28 | 79 |
| 15 | Male | 9 | 23 |
| 16 | Male | 15 | 110 |
| 17 | Male | 3 | 29 |
| 18 | Male | 1 | 28 |
| 19 | Male | 2 | 13 |
| 20 | Male | 3 | 20 |
| 0 | Female | 4 | 2 |
| 1 | Female | 1 | 0 |
| 2 | Female | 0 | 0 |
| 3 | Female | 6 | 1 |
| 4 | Female | 10 | 0 |
| 5 | Female | 14 | 7 |

**Table 7.2**: `womensrole` data (continued).

| education | gender | agree | disagree |
|---|---|---|---|
| 6 | Female | 17 | 5 |
| 7 | Female | 26 | 16 |
| 8 | Female | 91 | 36 |
| 9 | Female | 30 | 35 |
| 10 | Female | 55 | 67 |
| 11 | Female | 50 | 62 |
| 12 | Female | 190 | 403 |
| 13 | Female | 17 | 92 |
| 14 | Female | 18 | 81 |
| 15 | Female | 7 | 34 |
| 16 | Female | 13 | 115 |
| 17 | Female | 3 | 28 |
| 18 | Female | 0 | 21 |
| 19 | Female | 1 | 2 |
| 20 | Female | 2 | 4 |

*Source*: From Haberman, S. J., *Biometrics*, 29, 205–220, 1973. With permission.

Giardiello et al. (1993) and Piantadosi (1997) describe the results of a placebo-controlled trial of a non-steroidal anti-inflammatory drug in the treatment of familial andenomatous polyposis (FAP). The trial was halted after a planned interim analysis had suggested compelling evidence in favour of the treatment. The data shown in Table 7.3 give the number of colonic polyps after a 12-month treatment period. The question of interest is whether the number of polyps is related to treatment and/or age of patients.

**Table 7.3**: `polyps` data. Number of polyps for two treatment arms.

| number | treat | age | number | treat | age |
|---|---|---|---|---|---|
| 63 | placebo | 20 | 3 | drug | 23 |
| 2 | drug | 16 | 28 | placebo | 22 |
| 28 | placebo | 18 | 10 | placebo | 30 |
| 17 | drug | 22 | 40 | placebo | 27 |
| 61 | placebo | 13 | 33 | drug | 23 |
| 1 | drug | 23 | 46 | placebo | 22 |
| 7 | placebo | 34 | 50 | placebo | 34 |
| 15 | placebo | 50 | 3 | drug | 23 |
| 44 | placebo | 19 | 1 | drug | 22 |
| 25 | drug | 17 | 4 | drug | 42 |

**Table 7.4** `backpain` data. Number of drivers (D) and non-drivers ($\bar{\text{D}}$), suburban (S) and city inhabitants ($\bar{\text{S}}$) either suffering from a herniated disc (cases) or not (controls).

| | | | Controls | | | | |
|---|---|---|---|---|---|---|---|
| | | | $\bar{\text{D}}$ | | D | | |
| | | | $\bar{\text{S}}$ | S | $\bar{\text{S}}$ | S | Total |
| Cases | $\bar{\text{D}}$ | $\bar{\text{S}}$ | 9 | 0 | 10 | 7 | 26 |
| | | S | 2 | 2 | 1 | 1 | 6 |
| | D | $\bar{\text{S}}$ | 14 | 1 | 20 | 29 | 64 |
| | | S | 22 | 4 | 32 | 63 | 121 |
| | Total | | 47 | 7 | 63 | 100 | 217 |

The last of the data sets to be considered in this chapter is shown in Table 7.4. These data arise from a study reported in Kelsey and Hardy (1975) which was designed to investigate whether driving a car is a risk factor for low back pain resulting from acute herniated lumbar intervertebral discs (AHLID). A *case-control study* was used with cases selected from people who had recently had X-rays taken of the lower back and had been diagnosed as having AHLID. The controls were taken from patients admitted to the same hospital as a case with a condition unrelated to the spine. Further matching was made on age and gender and a total of 217 matched pairs were recruited, consisting of 89 female pairs and 128 male pairs. As a further potential risk factor, the variable `suburban` indicates whether each member of the pair lives in the suburbs or in the city.

## 7.2 Logistic Regression and Generalised Linear Models

### *7.2.1 Logistic Regression*

One way of writing the multiple regression model described in the previous chapter is as $y \sim \mathcal{N}(\mu, \sigma^2)$ where $\mu = \beta_0 + \beta_1 x_1 + \cdots + \beta_q x_q$. This makes it clear that this model is suitable for continuous response variables with, conditional on the values of the explanatory variables, a normal distribution with constant variance. So clearly the model would not be suitable for applying to the erythrocyte sedimentation rate in Table 7.1, since the response variable is binary. If we were to model the expected value of this type of response, i.e., the probability of it taking the value one, say $\pi$, directly as a linear function of explanatory variables, it could lead to fitted values of the response probability outside the range $[0, 1]$, which would clearly not be sensible. And if we write the value of the binary response as $y = \pi(x_1, x_2, \dots, x_q) + \varepsilon$ it soon becomes clear that the assumption of normality for $\varepsilon$ is also wrong. In fact here $\varepsilon$ may assume only one of two possible values. If $y = 1$, then $\varepsilon = 1 - \pi(x_1, x_2, \dots, x_q)$

with probability $\pi(x_1, x_2, \ldots, x_q)$ and if $y = 0$ then $\varepsilon = \pi(x_1, x_2, \ldots, x_q)$ with probability $1 - \pi(x_1, x_2, \ldots, x_q)$. So $\varepsilon$ has a distribution with mean zero and variance equal to $\pi(x_1, x_2, \ldots, x_q)(1 - \pi(x_1, x_2, \ldots, x_q))$, i.e., the conditional distribution of our binary response variable follows a binomial distribution with probability given by the conditional mean, $\pi(x_1, x_2, \ldots, x_q)$.

So instead of modelling the expected value of the response directly as a linear function of explanatory variables, a suitable transformation is modelled. In this case the most suitable transformation is the *logistic* or *logit* function of $\pi$ leading to the model

$$\text{logit}(\pi) = \log\left(\frac{\pi}{1-\pi}\right) = \beta_0 + \beta_1 x_1 + \cdots + \beta_q x_q. \tag{7.1}$$

The logit of a probability is simply the log of the odds of the response taking the value one. Equation (7.1) can be rewritten as

$$\pi(x_1, x_2, \ldots, x_q) = \frac{\exp(\beta_0 + \beta_1 x_1 + \cdots + \beta_q x_q)}{1 + \exp(\beta_0 + \beta_1 x_1 + \cdots + \beta_q x_q)}. \tag{7.2}$$

The logit function can take any real value, but the associated probability always lies in the required $[0, 1]$ interval. In a logistic regression model, the parameter $\beta_j$ associated with explanatory variable $x_j$ is such that $\exp(\beta_j)$ is the odds that the response variable takes the value one when $x_j$ increases by one, conditional on the other explanatory variables remaining constant. The parameters of the logistic regression model (the vector of regression coefficients $\beta$) are estimated by maximum likelihood; details are given in Collett (2003).

### *7.2.2 The Generalised Linear Model*

The analysis of variance models considered in Chapter 5 and the multiple regression model described in Chapter 6 are, essentially, completely equivalent. Both involve a linear combination of a set of explanatory variables (dummy variables in the case of analysis of variance) as a model for the observed response variable. And both include residual terms assumed to have a normal distribution. The equivalence of analysis of variance and multiple regression is spelt out in more detail in Everitt (2001).

The logistic regression model described in this chapter also has similarities to the analysis of variance and multiple regression models. Again a linear combination of explanatory variables is involved, although here the expected value of the binary response is not modelled directly but via a logistic transformation. In fact all three techniques can be unified in the *generalised linear model* (GLM), first introduced in a landmark paper by Nelder and Wedderburn (1972). The GLM enables a wide range of seemingly disparate problems of statistical modelling and inference to be set in an elegant unifying framework of great power and flexibility. A comprehensive technical account of the model is given in McCullagh and Nelder (1989). Here we describe GLMs only briefly. Essentially GLMs consist of three main features:

1. An *error distribution* giving the distribution of the response around its mean. For analysis of variance and multiple regression this will be the normal; for logistic regression it is the binomial. Each of these (and others used in other situations to be described later) come from the same, *exponential family* of probability distributions, and it is this family that is used in generalised linear modelling (see Everitt and Pickles, 2000).

2. A *link function*, $g$, that shows how the linear function of the explanatory variables is related to the expected value of the response:

$$g(\mu) = \beta_0 + \beta_1 x_1 + \cdots + \beta_q x_q.$$

   For analysis of variance and multiple regression the link function is simply the identity function; in logistic regression it is the logit function.

3. The *variance function* that captures how the variance of the response variable depends on the mean. We will return to this aspect of GLMs later in the chapter.

Estimation of the parameters in a GLM is usually achieved through a maximum likelihood approach – see McCullagh and Nelder (1989) for details. Having estimated a GLM for a data set, the question of the quality of its fit arises. Clearly the investigator needs to be satisfied that the chosen model describes the data adequately, before drawing conclusions about the parameter estimates themselves. In practise, most interest will lie in comparing the fit of competing models, particularly in the context of selecting subsets of explanatory variables that describe the data in a parsimonious manner. In GLMs a measure of fit is provided by a quantity known as the *deviance* which measures how closely the model-based fitted values of the response approximate the observed value. Comparing the deviance values for two models gives a likelihood ratio test of the two models that can be compared by using a statistic having a $\chi^2$-distribution with degrees of freedom equal to the difference in the number of parameters estimated under each model. More details are given in Cook (1998).

## 7.3 Analysis Using R

### *7.3.1 ESR and Plasma Proteins*

We begin by looking at the ESR data from Table 7.1. As always it is good practise to begin with some simple graphical examination of the data before undertaking any formal modelling. Here we will look at conditional density plots of the response variable given the two explanatory variables; such plots describe how the conditional distribution of the categorical variable ESR changes as the numerical variables fibrinogen and gamma globulin change. The required R code to construct these plots is shown with Figure 7.1. It appears that higher levels of each protein are associated with ESR values above 20 mm/hr.

We can now fit a logistic regression model to the data using the `glm` func-

```
R> data("plasma", package = "HSAUR2")
R> layout(matrix(1:2, ncol = 2))
R> cdplot(ESR ~ fibrinogen, data = plasma)
R> cdplot(ESR ~ globulin, data = plasma)
```

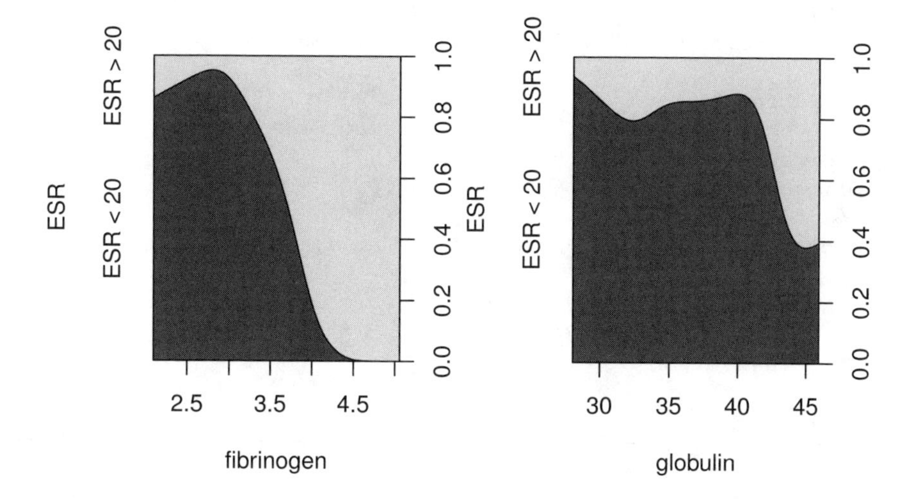


**Figure 7.1** Conditional density plots of the erythrocyte sedimentation rate (ESR) given fibrinogen and globulin.

tion. We start with a model that includes only a single explanatory variable, `fibrinogen`. The code to fit the model is

```
R> plasma_glm_1 <- glm(ESR ~ fibrinogen, data = plasma,
+                       family = binomial())
```

The formula implicitly defines a parameter for the global mean (the intercept term) as discussed in Chapter 5 and Chapter 6. The distribution of the response is defined by the `family` argument, a binomial distribution in our case. (The default link function when the binomial family is requested is the logistic function.)

A description of the fitted model can be obtained from the summary method applied to the fitted model. The output is shown in Figure 7.2.

From the results in Figure 7.2 we see that the regression coefficient for fibrinogen is significant at the 5% level. An increase of one unit in this variable increases the log-odds in favour of an ESR value greater than 20 by an estimated 1.83 with 95% confidence interval

```
R> confint(plasma_glm_1, parm = "fibrinogen")

    2.5 %    97.5 %
0.3387619 3.9984921
```

```
R> summary(plasma_glm_1)

Call:
glm(formula = ESR ~ fibrinogen, family = binomial(),
    data = plasma)

Deviance Residuals:
    Min       1Q   Median       3Q      Max
-0.9298  -0.5399  -0.4382  -0.3356   2.4794

Coefficients:
            Estimate Std. Error z value Pr(>|z|)
(Intercept)  -6.8451     2.7703  -2.471   0.0135
fibrinogen    1.8271     0.9009   2.028   0.0425

(Dispersion parameter for binomial family taken to be 1)

    Null deviance: 30.885  on 31  degrees of freedom
Residual deviance: 24.840  on 30  degrees of freedom
AIC: 28.840

Number of Fisher Scoring iterations: 5
```

**Figure 7.2** R output of the summary method for the logistic regression model fitted to ESR and fibrigonen.

These values are more helpful if converted to the corresponding values for the odds themselves by exponentiating the estimate

```
R> exp(coef(plasma_glm_1)["fibrinogen"])

fibrinogen
  6.215715
```

and the confidence interval

```
R> exp(confint(plasma_glm_1, parm = "fibrinogen"))

    2.5 %    97.5 %
 1.403209 54.515884
```

The confidence interval is very wide because there are few observations overall and very few where the ESR value is greater than 20. Nevertheless it seems likely that increased values of fibrinogen lead to a greater probability of an ESR value greater than 20.

We can now fit a logistic regression model that includes both explanatory variables using the code

```
R> plasma_glm_2 <- glm(ESR ~ fibrinogen +  globulin,
+      data = plasma, family = binomial())
```

and the output of the summary method is shown in Figure 7.3.

```
R> summary(plasma_glm_2)

Call:
glm(formula = ESR ~ fibrinogen + globulin,
  family = binomial(), data = plasma)

Deviance Residuals:
    Min       1Q   Median       3Q      Max
-0.9683  -0.6122  -0.3458  -0.2116   2.2636

Coefficients:
            Estimate Std. Error z value Pr(>|z|)
(Intercept) -12.7921     5.7963  -2.207   0.0273
fibrinogen    1.9104     0.9710   1.967   0.0491
globulin      0.1558     0.1195   1.303   0.1925

(Dispersion parameter for binomial family taken to be 1)

    Null deviance: 30.885  on 31  degrees of freedom
Residual deviance: 22.971  on 29  degrees of freedom
AIC: 28.971

Number of Fisher Scoring iterations: 5
```

**Figure 7.3** R output of the summary method for the logistic regression model fitted to ESR and both globulin and fibrinogen.

The coefficient for gamma globulin is not significantly different from zero. Subtracting the residual deviance of the second model from the corresponding value for the first model we get a value of 1.87. Tested using a $\chi^2$-distribution with a single degree of freedom this is not significant at the 5% level and so we conclude that gamma globulin is not associated with ESR level. In R, the task of comparing the two nested models can be performed using the anova function

```
R> anova(plasma_glm_1, plasma_glm_2, test = "Chisq")

Analysis of Deviance Table

Model 1: ESR ~ fibrinogen
Model 2: ESR ~ fibrinogen + globulin
  Resid. Df Resid. Dev Df Deviance P(>|Chi|)
1        30    24.8404
2        29    22.9711  1   1.8692    0.1716
```

Nevertheless we shall use the predicted values from the second model and plot them against the values of *both* explanatory variables using a *bubbleplot* to illustrate the use of the symbols function. The estimated conditional proba-

```
R> plot(globulin ~ fibrinogen, data = plasma, xlim = c(2, 6),
+       ylim = c(25, 55), pch = ".")
R> symbols(plasma$fibrinogen, plasma$globulin, circles = prob,
+         add = TRUE)
```

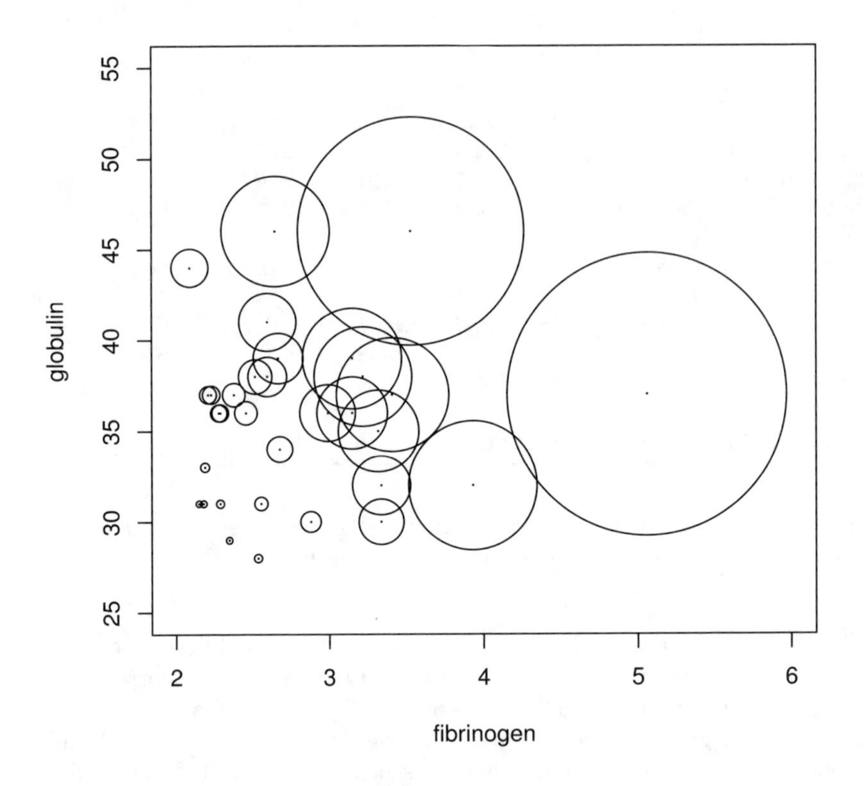


**Figure 7.4** Bubbleplot of fitted values for a logistic regression model fitted to the `plasma` data.

bility of a ESR value larger 20 for all observations can be computed, following formula (7.2), by

```
R> prob <- predict(plasma_glm_2, type = "response")
```

and now we can assign a larger circle to observations with larger probability as shown in Figure 7.4. The plot clearly shows the increasing probability of an ESR value above 20 (larger circles) as the values of fibrinogen, and to a lesser extent, gamma globulin, increase.

### 7.3.2 Women's Role in Society

Originally the data in Table 7.2 would have been in a completely equivalent form to the data in Table 7.1 data, but here the individual observations have been grouped into counts of numbers of agreements and disagreements for the two explanatory variables, `gender` and `education`. To fit a logistic regression model to such grouped data using the `glm` function we need to specify the number of agreements and disagreements as a two-column matrix on the left hand side of the model formula. We first fit a model that includes the two explanatory variables using the code

```
R> data("womensrole", package = "HSAUR2")
R> fm1 <- cbind(agree, disagree) ~ gender + education
R> womensrole_glm_1 <- glm(fm1, data = womensrole,
+                          family = binomial())
```

```
R> summary(womensrole_glm_1)

Call:
glm(formula = fm1, family = binomial(), data = womensrole)

Deviance Residuals:
     Min        1Q    Median        3Q       Max
-2.72544  -0.86302  -0.06525   0.84340   3.13315

Coefficients:
             Estimate Std. Error z value Pr(>|z|)
(Intercept)   2.50937    0.18389  13.646   <2e-16
genderFemale -0.01145    0.08415  -0.136    0.892
education    -0.27062    0.01541 -17.560   <2e-16

(Dispersion parameter for binomial family taken to be 1)

    Null deviance: 451.722  on 40  degrees of freedom
Residual deviance:  64.007  on 38  degrees of freedom
AIC: 208.07

Number of Fisher Scoring iterations: 4
```

**Figure 7.5** R output of the `summary` method for the logistic regression model fitted to the `womensrole` data.

From the `summary` output in Figure 7.5 it appears that education has a highly significant part to play in predicting whether a respondent will agree with the statement read to them, but the respondent's gender is apparently unimportant. As years of education increase the probability of agreeing with the statement declines. We now are going to construct a plot comparing the observed proportions of agreeing with those fitted by our fitted model. Because we will reuse this plot for another fitted object later on, we define a function which plots years of education against some fitted probabilities, e.g.,

```
R> role.fitted1 <- predict(womensrole_glm_1, type = "response")
```

and labels each observation with the person's gender:

```
1  R> myplot <- function(role.fitted)  {
2  +      f <- womensrole$gender == "Female"
3  +      plot(womensrole$education, role.fitted, type = "n",
4  +           ylab = "Probability of agreeing",
5  +           xlab = "Education", ylim = c(0,1))
6  +      lines(womensrole$education[!f], role.fitted[!f], lty = 1)
7  +      lines(womensrole$education[f], role.fitted[f], lty = 2)
8  +      lgtxt <- c("Fitted (Males)", "Fitted (Females)")
9  +      legend("topright", lgtxt, lty = 1:2, bty = "n")
10 +      y <-  womensrole$agree / (womensrole$agree +
11 +                                womensrole$disagree)
12 +      text(womensrole$education, y, ifelse(f, "\\VE", "\\MA"),
13 +           family = "HersheySerif", cex = 1.25)
14 + }
```

In lines 3–5 of function `myplot`, an empty scatterplot of education and fitted probabilities (`type = "n"`) is set up, basically to set the scene for the following plotting actions. Then, two lines are drawn (using function `lines` in lines 6 and 7), one for males (with line type 1) and one for females (with line type 2, i.e., a dashed line), where the logical vector `f` describes both genders. In line 9 a legend is added. Finally, in lines 12 and 13 we plot 'observed' values, i.e., the frequencies of agreeing in each of the groups (`y` as computed in lines 10 and 11) and use the Venus and Mars symbols to indicate gender.

The two curves for males and females in Figure 7.6 are almost the same reflecting the non-significant value of the regression coefficient for gender in `womensrole_glm_1`. But the observed values plotted on Figure 7.6 suggest that there might be an interaction of education and gender, a possibility that can be investigated by applying a further logistic regression model using

```
R> fm2 <- cbind(agree,disagree) ~ gender * education
R> womensrole_glm_2 <- glm(fm2, data = womensrole,
+                          family = binomial())
```

The `gender` and `education` interaction term is seen to be highly significant, as can be seen from the `summary` output in Figure 7.7.

Interpreting this interaction effect is made simpler if we again plot fitted and observed values using the same code as previously after getting fitted values from `womensrole_glm_2`. The plot is shown in Figure 7.8. We see that for fewer years of education women have a higher probability of agreeing with the statement than men, but when the years of education exceed about ten then this situation reverses.

A range of residuals and other diagnostics is available for use in association with logistic regression to check whether particular components of the model

```
R> myplot(role.fitted1)
```

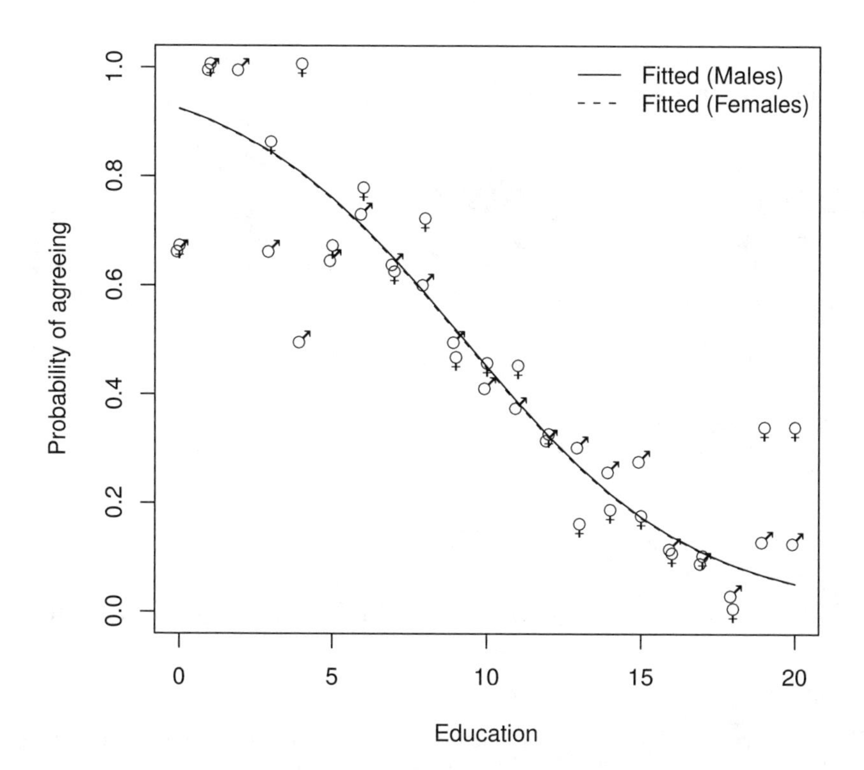


**Figure 7.6** Fitted (from `womensrole_glm_1`) and observed probabilities of agreeing for the `womensrole` data.

are adequate. A comprehensive account of these is given in Collett (2003); here we shall demonstrate only the use of what is known as the *deviance residual*. This is the signed square root of the contribution of the $i$th observation to the overall deviance. Explicitly it is given by

$$d_i = \text{sign}(y_i - \hat{y}_i) \left( 2y_i \log \left( \frac{y_i}{\hat{y}_i} \right) + 2(n_i - y_i) \log \left( \frac{n_i - y_i}{n_i - \hat{y}_i} \right) \right)^{1/2} \quad (7.3)$$

where sign is the function that makes $d_i$ positive when $y_i \geq \hat{y}_i$ and negative else. In (7.3) $y_i$ is the observed number of ones for the $i$th observation (the number of people who agree for each combination of covariates in our example), and $\hat{y}_i$ is its fitted value from the model. The residual provides information about how well the model fits each particular observation.

```
R> summary(womensrole_glm_2)

Call:
glm(formula = fm2, family = binomial(), data = womensrole)

Deviance Residuals:
     Min        1Q    Median        3Q       Max
-2.39097  -0.88062   0.01532   0.72783   2.45262

Coefficients:
                          Estimate Std. Error z value Pr(>|z|)
(Intercept)                2.09820    0.23550   8.910  < 2e-16
genderFemale               0.90474    0.36007   2.513  0.01198
education                 -0.23403    0.02019 -11.592  < 2e-16
genderFemale:education -0.08138    0.03109  -2.617  0.00886

(Dispersion parameter for binomial family taken to be 1)

    Null deviance: 451.722  on 40  degrees of freedom
Residual deviance:  57.103  on 37  degrees of freedom
AIC: 203.16

Number of Fisher Scoring iterations: 4
```

**Figure 7.7** R output of the `summary` method for the logistic regression model fitted to the `womensrole` data.

We can obtain a plot of deviance residuals plotted against fitted values using the following code above Figure 7.9. The residuals fall into a horizontal band between $-2$ and 2. This pattern does not suggest a poor fit for any particular observation or subset of observations.

### *7.3.3 Colonic Polyps*

The data on colonic polyps in Table 7.3 involves *count* data. We could try to model this using multiple regression but there are two problems. The first is that a response that is a count can take only positive values, and secondly such a variable is unlikely to have a normal distribution. Instead we will apply a GLM with a log link function, ensuring that fitted values are positive, and a Poisson error distribution, i.e.,

$$\mathsf{P}(y) = \frac{e^{-\lambda}\lambda^y}{y!}.$$

This type of GLM is often known as *Poisson regression*. We can apply the model using

```
R> role.fitted2 <- predict(womensrole_glm_2, type = "response")
R> myplot(role.fitted2)
```

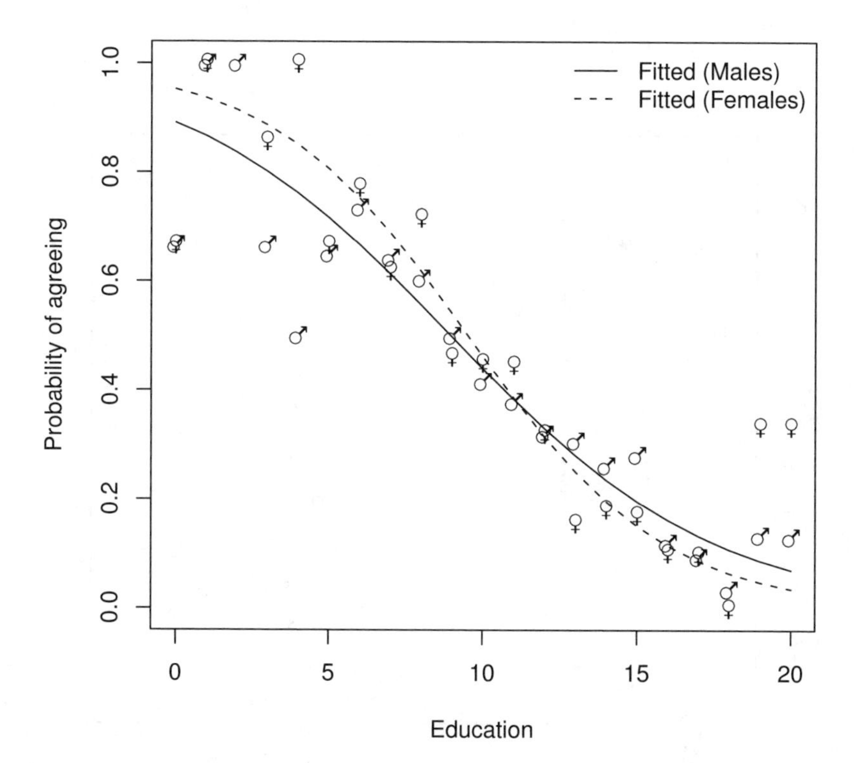


**Figure 7.8** Fitted (from `womensrole_glm_2`) and observed probabilities of agreeing for the `womensrole` data.

```
R> data("polyps", package = "HSAUR2")
R> polyps_glm_1 <- glm(number ~ treat + age, data = polyps,
+                      family = poisson())
```

(The default link function when the Poisson family is requested is the log function.)

From Figure 7.10 we see that the regression coefficients for both age and treatment are highly significant. But there is a problem with the model, but before we can deal with it we need a short digression to describe in more detail the third component of GLMs mentioned in the previous section, namely their variance functions, $V(\mu)$.

```
R> res <- residuals(womensrole_glm_2, type = "deviance")
R> plot(predict(womensrole_glm_2), res,
+       xlab="Fitted values", ylab = "Residuals",
+       ylim = max(abs(res)) * c(-1,1))
R> abline(h = 0, lty = 2)
```

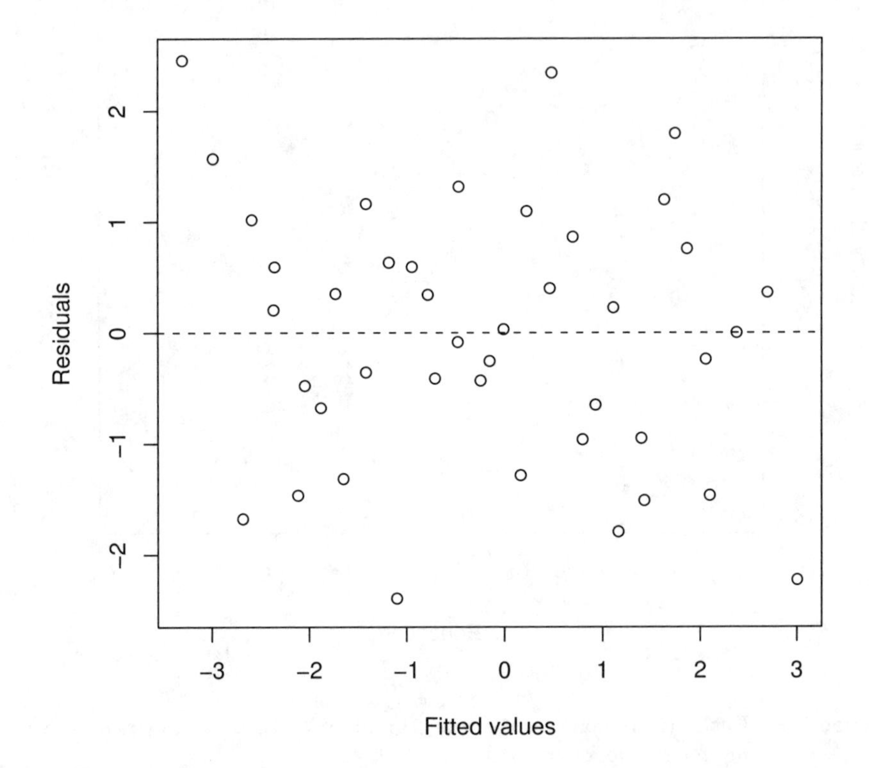


**Figure 7.9** Plot of deviance residuals from logistic regression model fitted to the `womensrole` data.

The variance function of a GLM captures how the variance of a response variable depends upon its mean. The general form of the relationship is

$$\text{Var(response)} = \phi V(\mu)$$

where $\phi$ is constant and $V(\mu)$ specifies how the variance depends on the mean. For the error distributions considered previously this general form becomes:

**Normal:** $V(\mu) = 1, \phi = \sigma^2$; here the variance does not depend on the mean.

**Binomial:** $V(\mu) = \mu(1 - \mu), \phi = 1$.

```
R> summary(polyps_glm_1)

Call:
glm(formula = number ~ treat + age, family = poisson(),
  data = polyps)

Deviance Residuals:
    Min        1Q    Median        3Q       Max
-4.2212   -3.0536   -0.1802    1.4459    5.8301

Coefficients:
             Estimate Std. Error z value Pr(>|z|)
(Intercept)  4.529024   0.146872   30.84   < 2e-16
treatdrug   -1.359083   0.117643  -11.55   < 2e-16
age         -0.038830   0.005955   -6.52 7.02e-11

(Dispersion parameter for poisson family taken to be 1)

    Null deviance: 378.66  on 19  degrees of freedom
Residual deviance: 179.54  on 17  degrees of freedom
AIC: 273.88

Number of Fisher Scoring iterations: 5
```

**Figure 7.10** R output of the **summary** method for the Poisson regression model fitted to the **polyps** data.

**Poisson:** $V(\mu) = \mu, \phi = 1$.

In the case of a Poisson variable we see that the mean and variance are equal, and in the case of a binomial variable where the mean is the probability of the variable taking the value one, $\pi$, the variance is $\pi(1 - \pi)$.

Both the Poisson and binomial distributions have variance functions that are completely determined by the mean. There is no free parameter for the variance since, in applications of the generalised linear model with binomial or Poisson error distributions the dispersion parameter, $\phi$, is defined to be one (see previous results for logistic and Poisson regression). But in some applications this becomes too restrictive to fully account for the empirical variance in the data; in such cases it is common to describe the phenomenon as *overdispersion*. For example, if the response variable is the proportion of family members who have been ill in the past year, observed in a large number of families, then the individual binary observations that make up the observed proportions are likely to be correlated rather than independent. The non-independence can lead to a variance that is greater (less) than on the assumption of binomial variability. And observed counts often exhibit larger variance than would be expected from the Poisson assumption, a fact noted over 80 years ago by Greenwood and Yule (1920).

When fitting generalised models with binomial or Poisson error distributions, overdispersion can often be spotted by comparing the residual deviance with its degrees of freedom. For a well-fitting model the two quantities should be approximately equal. If the deviance is far greater than the degrees of freedom overdispersion may be indicated. This is the case for the results in Figure 7.10. So what can we do?

We can deal with overdispersion by using a procedure known as *quasi-likelihood*, which allows the estimation of model parameters without fully knowing the error distribution of the response variable. McCullagh and Nelder (1989) give full details of the quasi-likelihood approach. In many respects it simply allows for the estimation of $\phi$ from the data rather than defining it to be unity for the binomial and Poisson distributions. We can apply quasi-likelihood estimation to the colonic polyps data using the following R code

```
R> polyps_glm_2 <- glm(number ~ treat + age, data = polyps,
+                      family = quasipoisson())
R> summary(polyps_glm_2)
```

```
Call:
glm(formula = number ~ treat + age,
  family = quasipoisson(), data = polyps)

Deviance Residuals:
    Min        1Q    Median        3Q       Max
-4.2212  -3.0536   -0.1802    1.4459    5.8301

Coefficients:
            Estimate Std. Error t value Pr(>|t|)
(Intercept)  4.52902    0.48106   9.415 3.72e-08
treatdrug   -1.35908    0.38533  -3.527  0.00259
age         -0.03883    0.01951  -1.991  0.06284

(Dispersion parameter for quasipoisson family taken to be 10.73)

    Null deviance: 378.66  on 19  degrees of freedom
Residual deviance: 179.54  on 17  degrees of freedom
AIC: NA

Number of Fisher Scoring iterations: 5
```

The regression coefficients for both explanatory variables remain significant but their estimated standard errors are now much greater than the values given in Figure 7.10. A possible reason for overdispersion in these data is that polyps do not occur independently of one another, but instead may ‘cluster’ together.

### 7.3.4 Driving and Back Pain

A frequently used design in medicine is the matched case-control study in which each patient suffering from a particular condition of interest included in the study is matched to one or more people without the condition. The most commonly used matching variables are age, ethnic group, mental status etc. A design with $m$ controls per case is known as a $1 : m$ matched study. In many cases $m$ will be one, and it is the $1 : 1$ matched study that we shall concentrate on here where we analyse the data on low back pain given in Table 7.4. To begin we shall describe the form of the logistic model appropriate for case-control studies in the simplest case where there is only one binary explanatory variable.

With matched pairs data the form of the logistic model involves the probability, $\varphi$, that in matched pair number $i$, for a given value of the explanatory variable the member of the pair is a case. Specifically the model is

$$\text{logit}(\varphi_i) = \alpha_i + \beta x.$$

The odds that a subject with $x = 1$ is a case equals $\exp(\beta)$ times the odds that a subject with $x = 0$ is a case.

The model generalises to the situation where there are $q$ explanatory variables as

$$\text{logit}(\varphi_i) = \alpha_i + \beta_1 x_1 + \beta_2 x_2 + \dots \beta_q x_q.$$

Typically one $x$ is an explanatory variable of real interest, such as past exposure to a risk factor, with the others being used as a form of statistical control in addition to the variables already controlled by virtue of using them to form matched pairs. This is the case in our back pain example where it is the effect of car driving on lower back pain that is of most interest.

The problem with the model above is that the number of parameters increases at the same rate as the sample size with the consequence that maximum likelihood estimation is no longer viable. We can overcome this problem if we regard the parameters $\alpha_i$ as of little interest and so are willing to forgo their estimation. If we do, we can then create a *conditional likelihood function* that will yield maximum likelihood estimators of the coefficients, $\beta_1, \dots, \beta_q$, that are consistent and asymptotically normally distributed. The mathematics behind this are described in Collett (2003).

The model can be fitted using the `clogit` function from package **survival**; the results are shown in Figure 7.11.

```
R> library("survival")
R> backpain_glm <- clogit(I(status == "case") ~
+      driver + suburban + strata(ID), data = backpain)
```

The response has to be a logical (`TRUE` for cases) and the `strata` command specifies the matched pairs.

The estimate of the odds ratio of a herniated disc occurring in a driver relative to a nondriver is 1.93 with a 95% confidence interval of (1.09, 3.44).

```
R> print(backpain_glm)

Call:
clogit(I(status == "case") ~ driver + suburban + strata(ID),
    data = backpain)


             coef exp(coef) se(coef)     z     p
driveryes   0.658      1.93    0.294 2.24 0.025
suburbanyes 0.255      1.29    0.226 1.13 0.260

Likelihood ratio test=9.55  on 2 df, p=0.00846  n= 434
```

**Figure 7.11** R output of the `print` method for the conditional logistic regression model fitted to the `backpain` data.

Conditional on residence we can say that the risk of a herniated disc occurring in a driver is about twice that of a nondriver. There is no evidence that where a person lives affects the risk of lower back pain.

## 7.4 Summary

Generalised linear models provide a very powerful and flexible framework for the application of regression models to a variety of non-normal response variables, for example, logistic regression to binary responses and Poisson regression to count data.

## Exercises

Ex. 7.1 Construct a perspective plot of the fitted values from a logistic regression model fitted to the `plasma` data in which both fibrinogen and gamma globulin are included as explanatory variables.

Ex. 7.2 Collett (2003) argues that two outliers need to be removed from the `plasma` data. Try to identify those two unusual observations by means of a scatterplot.

Ex. 7.3 The data shown in Table 7.5 arise from 31 male patients who have been treated for superficial bladder cancer (see Seeber, 1998), and give the number of recurrent tumours during a particular time after the removal of the primary tumour, along with the size of the original tumour (whether smaller or larger than 3 cm). Use Poisson regression to estimate the effect of size of tumour on the number of recurrent tumours.

**Table 7.5**: `bladdercancer` data. Number of recurrent tumours for bladder cancer patients.

| time | tumorsize | number | time | tumorsize | number |
|---|---|---|---|---|---|
| 2 | <=3cm | 1 | 13 | <=3cm | 2 |
| 3 | <=3cm | 1 | 15 | <=3cm | 2 |
| 6 | <=3cm | 1 | 18 | <=3cm | 2 |
| 8 | <=3cm | 1 | 23 | <=3cm | 2 |
| 9 | <=3cm | 1 | 20 | <=3cm | 3 |
| 10 | <=3cm | 1 | 24 | <=3cm | 4 |
| 11 | <=3cm | 1 | 1 | >3cm | 1 |
| 13 | <=3cm | 1 | 5 | >3cm | 1 |
| 14 | <=3cm | 1 | 17 | >3cm | 1 |
| 16 | <=3cm | 1 | 18 | >3cm | 1 |
| 21 | <=3cm | 1 | 25 | >3cm | 1 |
| 22 | <=3cm | 1 | 18 | >3cm | 2 |
| 24 | <=3cm | 1 | 25 | >3cm | 2 |
| 26 | <=3cm | 1 | 4 | >3cm | 3 |
| 27 | <=3cm | 1 | 19 | >3cm | 4 |
| 7 | <=3cm | 2 | | | |

*Source*: From Seeber, G. U. H., in *Encyclopedia of Biostatistics*, John Wiley & Sons, Chichester, UK, 1998. With permission.

Ex. 7.4 The data in Table 7.6 show the survival times from diagnosis of patients suffering from leukemia and the values of two explanatory variables, the white blood cell count (`wbc`) and the presence or absence of a morphological characteristic of the white blood cells (`ag`) (the data are available in package **MASS**, Venables and Ripley, 2002). Define a binary outcome variable according to whether or not patients lived for at least 24 weeks after diagnosis and then fit a logistic regression model to the data. It may be advisable to transform the very large white blood counts to avoid regression coefficients very close to 0 (and odds ratios very close to 1). And a model that contains only the two explanatory variables may not be adequate for these data. Construct some graphics useful in the interpretation of the final model you fit.

**Table 7.6**: `leuk` data (package **MASS**). Survival times of patients suffering from leukemia.

| wbc | ag | time | wbc | ag | time |
|---|---|---|---|---|---|
| 2300 | present | 65 | 4400 | absent | 56 |
| 750 | present | 156 | 3000 | absent | 65 |
| 4300 | present | 100 | 4000 | absent | 17 |
| 2600 | present | 134 | 1500 | absent | 7 |
| 6000 | present | 16 | 9000 | absent | 16 |
| 10500 | present | 108 | 5300 | absent | 22 |
| 10000 | present | 121 | 10000 | absent | 3 |
| 17000 | present | 4 | 19000 | absent | 4 |
| 5400 | present | 39 | 27000 | absent | 2 |
| 7000 | present | 143 | 28000 | absent | 3 |
| 9400 | present | 56 | 31000 | absent | 8 |
| 32000 | present | 26 | 26000 | absent | 4 |
| 35000 | present | 22 | 21000 | absent | 3 |
| 100000 | present | 1 | 79000 | absent | 30 |
| 100000 | present | 1 | 100000 | absent | 4 |
| 52000 | present | 5 | 100000 | absent | 43 |
| 100000 | present | 65 | | | |

CHAPTER 8

# Density Estimation: Erupting Geysers and Star Clusters

## 8.1 Introduction

Geysers are natural fountains that shoot up into the air, at more or less regular intervals, a column of heated water and steam. Old Faithful is one such geyser and is the most popular attraction of Yellowstone National Park, although it is not the largest or grandest geyser in the park. Old Faithful can vary in height from 100–180 feet with an average near 130–140 feet. Eruptions normally last between 1.5 to 5 minutes.

From August 1 to August 15, 1985, Old Faithful was observed and the waiting times between successive eruptions noted. There were 300 eruptions observed, so 299 waiting times were (in minutes) recorded and those shown in Table 8.1.

**Table 8.1**: `faithful` data (package **datasets**). Old Faithful geyser waiting times between two eruptions.

| `waiting` | `waiting` | `waiting` | `waiting` | `waiting` |
|---:|---:|---:|---:|---:|
| 79 | 83 | 75 | 76 | 50 |
| 54 | 71 | 59 | 63 | 82 |
| 74 | 64 | 89 | 88 | 54 |
| 62 | 77 | 79 | 52 | 75 |
| 85 | 81 | 59 | 93 | 78 |
| 55 | 59 | 81 | 49 | 79 |
| 88 | 84 | 50 | 57 | 78 |
| 85 | 48 | 85 | 77 | 78 |
| 51 | 82 | 59 | 68 | 70 |
| 85 | 60 | 87 | 81 | 79 |
| 54 | 92 | 53 | 81 | 70 |
| 84 | 78 | 69 | 73 | 54 |
| 78 | 78 | 77 | 50 | 86 |
| 47 | 65 | 56 | 85 | 50 |
| 83 | 73 | 88 | 74 | 90 |
| 52 | 82 | 81 | 55 | 54 |
| 62 | 56 | 45 | 77 | 54 |
| 84 | 79 | 82 | 83 | 77 |
| 52 | 71 | 55 | 83 | 79 |

**Table 8.1**: `faithful` data (continued).

| `waiting` | `waiting` | `waiting` | `waiting` | `waiting` |
|---|---|---|---|---|
| 79 | 62 | 90 | 51 | 64 |
| 51 | 76 | 45 | 78 | 75 |
| 47 | 60 | 83 | 84 | 47 |
| 78 | 78 | 56 | 46 | 86 |
| 69 | 76 | 89 | 83 | 63 |
| 74 | 83 | 46 | 55 | 85 |
| 83 | 75 | 82 | 81 | 82 |
| 55 | 82 | 51 | 57 | 57 |
| 76 | 70 | 86 | 76 | 82 |
| 78 | 65 | 53 | 84 | 67 |
| 79 | 73 | 79 | 77 | 74 |
| 73 | 88 | 81 | 81 | 54 |
| 77 | 76 | 60 | 87 | 83 |
| 66 | 80 | 82 | 77 | 73 |
| 80 | 48 | 77 | 51 | 73 |
| 74 | 86 | 76 | 78 | 88 |
| 52 | 60 | 59 | 60 | 80 |
| 48 | 90 | 80 | 82 | 71 |
| 80 | 50 | 49 | 91 | 83 |
| 59 | 78 | 96 | 53 | 56 |
| 90 | 63 | 53 | 78 | 79 |
| 80 | 72 | 77 | 46 | 78 |
| 58 | 84 | 77 | 77 | 84 |
| 84 | 75 | 65 | 84 | 58 |
| 58 | 51 | 81 | 49 | 83 |
| 73 | 82 | 71 | 83 | 43 |
| 83 | 62 | 70 | 71 | 60 |
| 64 | 88 | 81 | 80 | 75 |
| 53 | 49 | 93 | 49 | 81 |
| 82 | 83 | 53 | 75 | 46 |
| 59 | 81 | 89 | 64 | 90 |
| 75 | 47 | 45 | 76 | 46 |
| 90 | 84 | 86 | 53 | 74 |
| 54 | 52 | 58 | 94 | |
| 80 | 86 | 78 | 55 | |
| 54 | 81 | 66 | 76 | |

The Hertzsprung-Russell (H-R) diagram forms the basis of the theory of stellar evolution. The diagram is essentially a plot of the energy output of stars plotted against their surface temperature. Data from the H-R diagram of Star Cluster CYG OB1, calibrated according to Vanisma and De Greve (1972) are shown in Table 8.2 (from Hand et al., 1994).

**Table 8.2**: `CYGOB1` data. Energy output and surface temperature of Star Cluster CYG OB1.

| logst | logli | logst | logli | logst | logli |
|---|---|---|---|---|---|
| 4.37 | 5.23 | 4.23 | 3.94 | 4.45 | 5.22 |
| 4.56 | 5.74 | 4.42 | 4.18 | 3.49 | 6.29 |
| 4.26 | 4.93 | 4.23 | 4.18 | 4.23 | 4.34 |
| 4.56 | 5.74 | 3.49 | 5.89 | 4.62 | 5.62 |
| 4.30 | 5.19 | 4.29 | 4.38 | 4.53 | 5.10 |
| 4.46 | 5.46 | 4.29 | 4.22 | 4.45 | 5.22 |
| 3.84 | 4.65 | 4.42 | 4.42 | 4.53 | 5.18 |
| 4.57 | 5.27 | 4.49 | 4.85 | 4.43 | 5.57 |
| 4.26 | 5.57 | 4.38 | 5.02 | 4.38 | 4.62 |
| 4.37 | 5.12 | 4.42 | 4.66 | 4.45 | 5.06 |
| 3.49 | 5.73 | 4.29 | 4.66 | 4.50 | 5.34 |
| 4.43 | 5.45 | 4.38 | 4.90 | 4.45 | 5.34 |
| 4.48 | 5.42 | 4.22 | 4.39 | 4.55 | 5.54 |
| 4.01 | 4.05 | 3.48 | 6.05 | 4.45 | 4.98 |
| 4.29 | 4.26 | 4.38 | 4.42 | 4.42 | 4.50 |
| 4.42 | 4.58 | 4.56 | 5.10 | | |

## 8.2 Density Estimation

The goal of density estimation is to approximate the probability density function of a random variable (univariate or multivariate) given a sample of observations of the variable. Univariate histograms are a simple example of a density estimate; they are often used for two purposes, counting and displaying the distribution of a variable, but according to Wilkinson (1992), they are effective for neither. For bivariate data, two-dimensional histograms can be constructed, but for small and moderate sized data sets that is not of any real use for estimating the bivariate density function, simply because most of the 'boxes' in the histogram will contain too few observations, or if the number of boxes is reduced the resulting histogram will be too coarse a representation of the density function.

The density estimates provided by one- and two-dimensional histograms can be improved on in a number of ways. If, of course, we are willing to assume a particular form for the variable's distribution, for example, Gaussian, density estimation would be reduced to estimating the parameters of the assumed distribution. More commonly, however, we wish to allow the data to speak for themselves and so one of a variety of non-parametric estimation procedures that are now available might be used. Density estimation is covered in detail in several books, including Silverman (1986), Scott (1992), Wand and Jones (1995) and Simonoff (1996). One of the most popular classes of procedures is the kernel density estimators, which we now briefly describe for univariate and bivariate data.

### *8.2.1 Kernel Density Estimators*

From the definition of a probability density, if the random $X$ has a density $f$,

$$f(x) = \lim_{h \to 0} \frac{1}{2h} \mathsf{P}(x - h < X < x + h). \tag{8.1}$$

For any given $h$ a naïve estimator of $\mathsf{P}(x - h < X < x + h)$ is the proportion of the observations $x_1, x_2, \dots, x_n$ falling in the interval $(x - h, x + h)$, that is

$$\hat{f}(x) = \frac{1}{2hn} \sum_{i=1}^{n} I(x_i \in (x - h, x + h)), \tag{8.2}$$

i.e., the number of $x_1, \dots, x_n$ falling in the interval $(x - h, x + h)$ divided by $2hn$. If we introduce a weight function $W$ given by

$$W(x) = \begin{cases} \frac{1}{2} & |x| < 1 \\ 0 & \text{else} \end{cases}$$

then the naïve estimator can be rewritten as

$$\hat{f}(x) = \frac{1}{n} \sum_{i=1}^{n} \frac{1}{h} W\left(\frac{x - x_i}{h}\right). \tag{8.3}$$

Unfortunately this estimator is not a continuous function and is not particularly satisfactory for practical density estimation. It does however lead naturally to the kernel estimator defined by

$$\hat{f}(x) = \frac{1}{hn} \sum_{i=1}^{n} K\left(\frac{x - x_i}{h}\right) \tag{8.4}$$

where $K$ is known as the *kernel function* and $h$ as the *bandwidth* or *smoothing parameter*. The kernel function must satisfy the condition

$$\int_{-\infty}^{\infty} K(x)dx = 1.$$

Usually, but not always, the kernel function will be a symmetric density function, for example, the normal. Three commonly used kernel functions are

**rectangular:**

$$K(x) = \begin{cases} \frac{1}{2} & |x| < 1 \\ 0 & \text{else} \end{cases}$$

**triangular:**

$$K(x) = \begin{cases} 1 - |x| & |x| < 1 \\ 0 & \text{else} \end{cases}$$

**Gaussian:**

$$K(x) = \frac{1}{\sqrt{2\pi}} e^{-\frac{1}{2}x^2}$$

The three kernel functions are implemented in R as shown in lines 1–3 of Figure 8.1. For some grid x, the kernel functions are plotted using the R statements in lines 5–11 (Figure 8.1).

The kernel estimator $\hat{f}$ is a sum of 'bumps' placed at the observations. The kernel function determines the shape of the bumps while the window width $h$ determines their width. Figure 8.2 (redrawn from a similar plot in Silverman, 1986) shows the individual bumps $n^{-1}h^{-1}K((x - x_i)/h)$, as well as the estimate $\hat{f}$ obtained by adding them up for an artificial set of data points

```
R> x <- c(0, 1, 1.1, 1.5, 1.9, 2.8, 2.9, 3.5)
R> n <- length(x)
```

For a grid

```
R> xgrid <- seq(from = min(x) - 1, to = max(x) + 1, by = 0.01)
```

on the real line, we can compute the contribution of each measurement in x, with $h = 0.4$, by the Gaussian kernel (defined in Figure 8.1, line 3) as follows;

```
R> h <- 0.4
R> bumps <- sapply(x, function(a) gauss((xgrid - a)/h)/(n * h))
```

A plot of the individual bumps and their sum, the kernel density estimate $\hat{f}$, is shown in Figure 8.2.

The kernel density estimator considered as a sum of 'bumps' centred at the observations has a simple extension to two dimensions (and similarly for more than two dimensions). The bivariate estimator for data $(x_1, y_1)$, $(x_2, y_2)$, ..., $(x_n, y_n)$ is defined as

$$\hat{f}(x, y) = \frac{1}{nh_x h_y} \sum_{i=1}^{n} K\left(\frac{x - x_i}{h_x}, \frac{y - y_i}{h_y}\right). \tag{8.5}$$

In this estimator each coordinate direction has its own smoothing parameter $h_x$ and $h_y$. An alternative is to scale the data equally for both dimensions and use a single smoothing parameter.

```
R> rec <- function(x) (abs(x) < 1) * 0.5
R> tri <- function(x) (abs(x) < 1) * (1 - abs(x))
R> gauss <- function(x) 1/sqrt(2*pi) * exp(-(x^2)/2)
R> x <- seq(from = -3, to = 3, by = 0.001)
R> plot(x, rec(x), type = "l", ylim = c(0,1), lty = 1,
+      ylab = expression(K(x)))
R> lines(x, tri(x), lty = 2)
R> lines(x, gauss(x), lty = 3)
R> legend(-3, 0.8, legend = c("Rectangular", "Triangular",
+        "Gaussian"), lty = 1:3, title = "kernel functions",
+        bty = "n")
```

**Figure 8.1** Three commonly used kernel functions.

```
R> plot(xgrid, rowSums(bumps), ylab = expression(hat(f)(x)),
+        type = "l", xlab = "x", lwd = 2)
R> rug(x, lwd = 2)
R> out <- apply(bumps, 2, function(b) lines(xgrid, b))
```

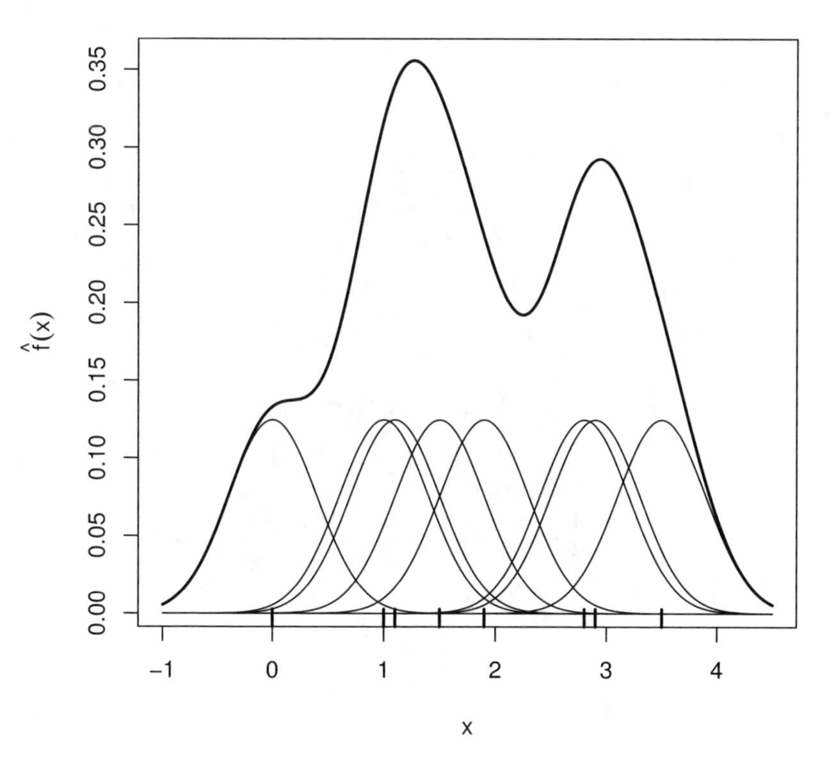


**Figure 8.2** Kernel estimate showing the contributions of Gaussian kernels evaluated for the individual observations with bandwidth $h = 0.4$.

For bivariate density estimation a commonly used kernel function is the standard bivariate normal density

$$K(x, y) = \frac{1}{2\pi} e^{-\frac{1}{2}(x^2+y^2)}.$$

Another possibility is the bivariate Epanechnikov kernel given by

$$K(x, y) = \begin{cases} \frac{2}{\pi}(1 - x^2 - y^2) & x^2 + y^2 < 1 \\ 0 & \text{else} \end{cases}$$

```
R> epa <- function(x, y)
+      ((x^2 + y^2) < 1) * 2/pi * (1 - x^2 - y^2)
R> x <- seq(from = -1.1, to = 1.1, by = 0.05)
R> epavals <- sapply(x, function(a) epa(a, x))
R> persp(x = x, y = x, z = epavals, xlab = "x", ylab = "y",
+        zlab = expression(K(x, y)), theta = -35, axes = TRUE,
+        box = TRUE)
```

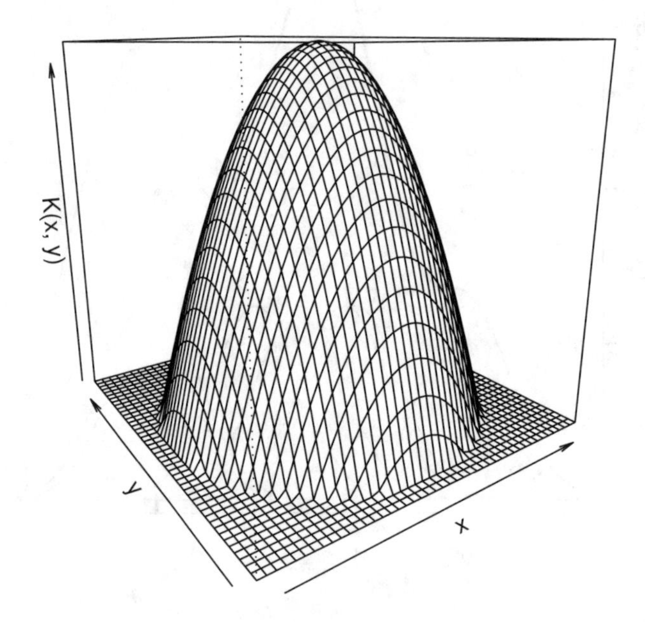


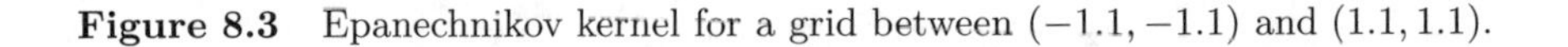

**Figure 8.3** Epanechnikov kernel for a grid between $(-1.1, -1.1)$ and $(1.1, 1.1)$.

which is implemented and depicted in Figure 8.3, here by using the `persp` function for plotting in three dimensions.

According to Venables and Ripley (2002) the bandwidth should be chosen to be proportional to $n^{-1/5}$; unfortunately the constant of proportionality depends on the unknown density. The tricky problem of bandwidth estimation is considered in detail in Silverman (1986).

## 8.3 Analysis Using R

The R function `density` can be used to calculate kernel density estimators with a variety of kernels (`window` argument). We can illustrate the function's use by applying it to the geyser data to calculate three density estimates of the data and plot each on a histogram of the data, using the code displayed with Figure 8.4. The `hist` function places an ordinary histogram of the geyser data in each of the three plotting regions (lines 4, 10, 17). Then, the `density` function with three different kernels (lines 8, 14, 21, with a Gaussian kernel being the default in line 8) is plotted in addition. The `rug` statement simply places the observations in vertical bars onto the x-axis. All three density estimates show that the waiting times between eruptions have a distinctly bimodal form, which we will investigate further in Subsection 8.3.1.

For the bivariate star data in Table 8.2 we can estimate the bivariate density using the `bkde2D` function from package **KernSmooth** (Wand and Ripley, 2009). The resulting estimate can then be displayed as a contour plot (using `contour`) or as a perspective plot (using `persp`). The resulting contour plot is shown in Figure 8.5, and the perspective plot in 8.6. Both clearly show the presence of two separated classes of stars.

### *8.3.1 A Parametric Density Estimate for the Old Faithful Data*

In the previous section we considered the non-parametric kernel density estimators for the Old Faithful data. The estimators showed the clear bimodality of the data and in this section this will be investigated further by fitting a parametric model based on a two-component normal mixture model. Such models are members of the class of finite mixture distributions described in great detail in McLachlan and Peel (2000). The two-component normal mixture distribution was first considered by Karl Pearson over 100 years ago (Pearson, 1894) and is given explicitly by

$$f(x) = p\phi(x, \mu_1, \sigma_1^2) + (1 - p)\phi(x, \mu_2, \sigma_2^2)$$

where $\phi(x, \mu, \sigma^2)$ denotes a normal density with mean $\mu$ and variance $\sigma^2$.

This distribution has five parameters to estimate, the mixing proportion, $p$, and the mean and variance of each component normal distribution. Pearson heroically attempted this by the method of moments, which required solving a polynomial equation of the $9^{\text{th}}$ degree. Nowadays the preferred estimation approach is maximum likelihood. The following R code contains a function to calculate the relevant log-likelihood and then uses the optimiser `optim` to find values of the five parameters that minimise the negative log-likelihood.

```
R> logL <- function(param, x) {
+      d1 <- dnorm(x, mean = param[2], sd = param[3])
+      d2 <- dnorm(x, mean = param[4], sd = param[5])
+      -sum(log(param[1] * d1 + (1 - param[1]) * d2))
+  }
```

```
R> data("faithful", package = "datasets")
R> x <- faithful$waiting
R> layout(matrix(1:3, ncol = 3))
R> hist(x, xlab = "Waiting times (in min.)", ylab = "Frequency",
+       probability = TRUE, main = "Gaussian kernel",
+       border = "gray")
R> lines(density(x, width = 12), lwd = 2)
R> rug(x)
R> hist(x, xlab = "Waiting times (in min.)", ylab = "Frequency",
+       probability = TRUE, main = "Rectangular kernel",
+       border = "gray")
R> lines(density(x, width = 12, window = "rectangular"), lwd = 2)
R> rug(x)
R> hist(x, xlab = "Waiting times (in min.)", ylab = "Frequency",
+       probability = TRUE, main = "Triangular kernel",
+       border = "gray")
R> lines(density(x, width = 12, window = "triangular"), lwd = 2)
R> rug(x)
```

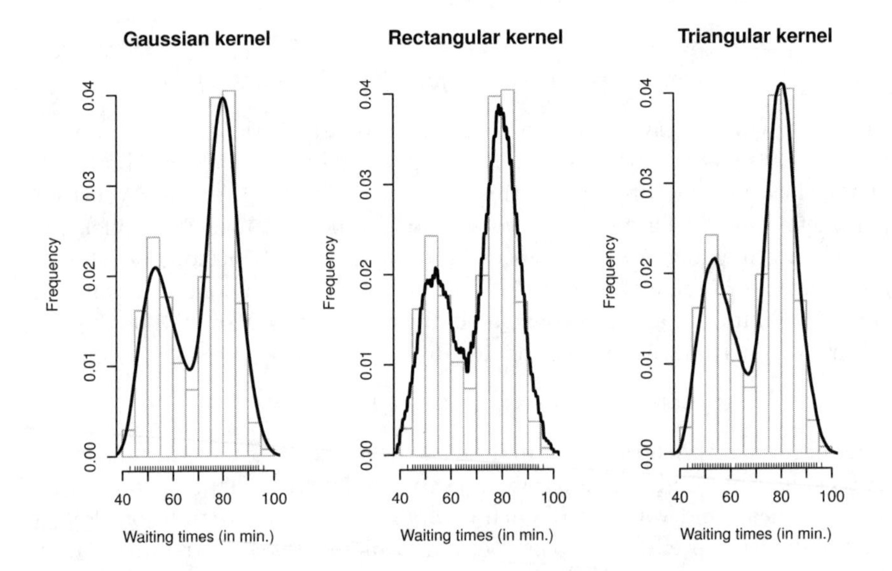


**Figure 8.4** Density estimates of the geyser eruption data imposed on a histogram of the data.

```
R> library("KernSmooth")
R> data("CYGOB1", package = "HSAUR2")
R> CYGOB1d <- bkde2D(CYGOB1, bandwidth = sapply(CYGOB1, dpik))
R> contour(x = CYGOB1d$x1, y = CYGOB1d$x2, z = CYGOB1d$fhat,
+          xlab = "log surface temperature",
+          ylab = "log light intensity")
```

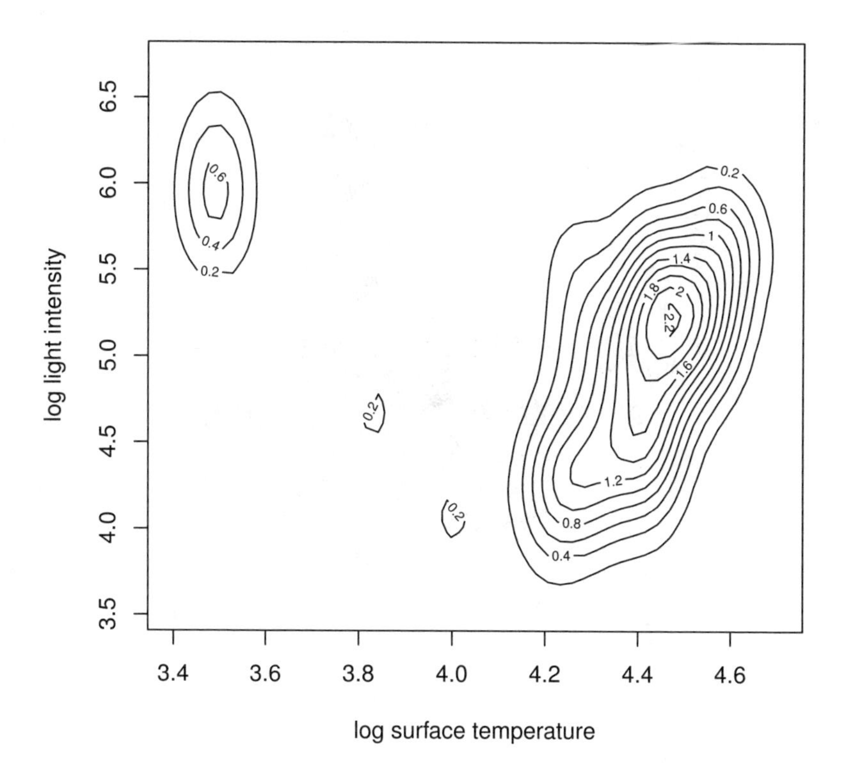


**Figure 8.5** A contour plot of the bivariate density estimate of the CYGOB1 data, i.e., a two-dimensional graphical display for a three-dimensional problem.

```
R> persp(x = CYGOB1d$x1, y = CYGOB1d$x2, z = CYGOB1d$fhat,
+        xlab = "log surface temperature",
+        ylab = "log light intensity",
+        zlab = "estimated density",
+        theta = -35, axes = TRUE, box = TRUE)
```

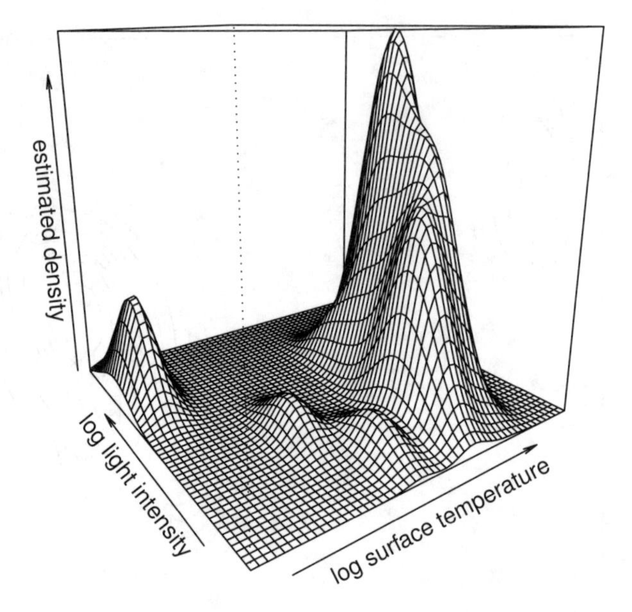


**Figure 8.6** The bivariate density estimate of the CYGOB1 data, here shown in a three-dimensional fashion using the persp function.

```
R> startparam <- c(p = 0.5, mu1 = 50, sd1 = 3, mu2 = 80, sd2 = 3)
R> opp <- optim(startparam, logL, x = faithful$waiting,
+               method = "L-BFGS-B",
+               lower = c(0.01, rep(1, 4)),
+               upper = c(0.99, rep(200, 4)))
R> opp

$par
        p       mu1       sd1       mu2       sd2
```

```
 0.360891 54.612125  5.872379 80.093414  5.867288

$value
[1] 1034.002

$counts
function gradient
      55       55

$convergence
[1] 0
```

Of course, optimising the appropriate likelihood 'by hand' is not very convenient. In fact, (at least) two packages offer high-level functionality for estimating mixture models. The first one is package **mclust** (Fraley et al., 2009) implementing the methodology described in Fraley and Raftery (2002). Here, a Bayesian information criterion (BIC) is applied to choose the form of the mixture model:

```
R> library("mclust")
R> mc <- Mclust(faithful$waiting)
R> mc

 best model: equal variance with 2 components
```

and the estimated means are

```
R> mc$parameters$mean

       1        2
54.61911 80.09384
```

with estimated standard deviation (found to be equal within both groups)

```
R> sqrt(mc$parameters$variance$sigmasq)

[1] 5.86848
```

The proportion is $\hat{p} = 0.36$. The second package is called **flexmix** whose functionality is described by Leisch (2004). A mixture of two normals can be fitted using

```
R> library("flexmix")
R> fl <- flexmix(waiting ~ 1, data = faithful, k = 2)
```

with $\hat{p} = 0.36$ and estimated parameters

```
R> parameters(fl, component = 1)

                    Comp.1
coef.(Intercept) 54.628701
sigma             5.895234

R> parameters(fl, component = 2)

                    Comp.2
coef.(Intercept) 80.098582
sigma             5.871749
```

```
R> opar <- as.list(opp$par)
R> rx <- seq(from = 40, to = 110, by = 0.1)
R> d1 <- dnorm(rx, mean = opar$mu1, sd = opar$sd1)
R> d2 <- dnorm(rx, mean = opar$mu2, sd = opar$sd2)
R> f <- opar$p * d1 + (1 - opar$p) * d2
R> hist(x, probability = TRUE, xlab = "Waiting times (in min.)",
+      border = "gray", xlim = range(rx), ylim = c(0, 0.06),
+      main = "")
R> lines(rx, f, lwd = 2)
R> lines(rx, dnorm(rx, mean = mean(x), sd = sd(x)), lty = 2,
+       lwd = 2)
R> legend(50, 0.06, lty = 1:2, bty = "n",
+        legend = c("Fitted two-component mixture density",
+                   "Fitted single normal density"))
```

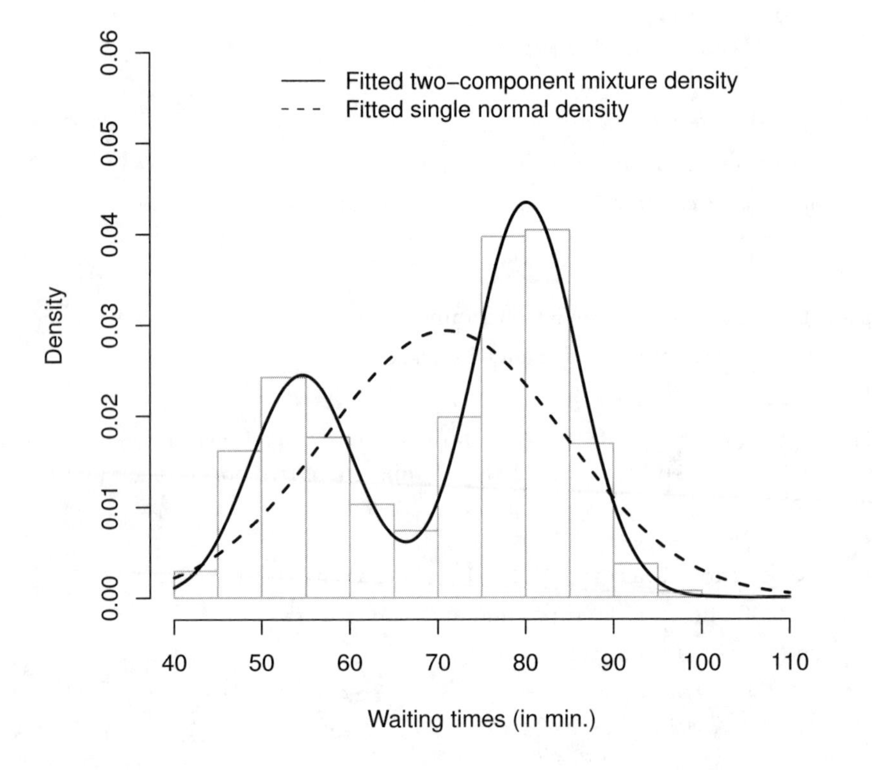


**Figure 8.7** Fitted normal density and two-component normal mixture for geyser eruption data.

The results are identical for all practical purposes and we can plot the fitted mixture and a single fitted normal into a histogram of the data using the R code which produces Figure 8.7. The `dnorm` function can be used to evaluate the normal density with given mean and standard deviation, here as estimated for the two-components of our mixture model, which are then collapsed into our density estimate `f`. Clearly the two-component mixture is a far better fit than a single normal distribution for these data.

We can get standard errors for the five parameter estimates by using a bootstrap approach (see Efron and Tibshirani, 1993). The original data are slightly perturbed by drawing $n$ out of $n$ observations *with replacement* and those artificial replications of the original data are called *bootstrap samples*. Now, we can fit the mixture for each bootstrap sample and assess the variability of the estimates, for example using confidence intervals. Some suitable R code based on the `Mclust` function follows. First, we define a function that, for a bootstrap sample `indx`, fits a two-component mixture model and returns $\hat{p}$ and the estimated means (note that we need to make sure that we always get an estimate of $p$, not $1 - p$):

```
R> library("boot")
R> fit <- function(x, indx) {
+      a <- Mclust(x[indx], minG = 2, maxG = 2)$parameters
+      if (a$pro[1] < 0.5)
+          return(c(p = a$pro[1], mu1 = a$mean[1],
+                                 mu2 = a$mean[2]))
+      return(c(p = 1 - a$pro[1], mu1 = a$mean[2],
+                                 mu2 = a$mean[1]))
+ }
```

The function `fit` can now be fed into the `boot` function (Canty and Ripley, 2009) for bootstrapping (here 1000 bootstrap samples are drawn)

```
R> bootpara <- boot(faithful$waiting, fit, R = 1000)
```

We assess the variability of our estimates $\hat{p}$ by means of adjusted bootstrap percentile (BCa) confidence intervals, which for $\hat{p}$ can be obtained from

```
R> boot.ci(bootpara, type = "bca", index = 1)

BOOTSTRAP CONFIDENCE INTERVAL CALCULATIONS
Based on 1000 bootstrap replicates

CALL :
boot.ci(boot.out = bootpara, type = "bca", index = 1)

Intervals :
Level       BCa
95%   ( 0.3041,  0.4233 )
Calculations and Intervals on Original Scale
```

We see that there is a reasonable variability in the mixture model; however, the means in the two components are rather stable, as can be seen from

```
R> boot.ci(bootpara, type = "bca", index = 2)

BOOTSTRAP CONFIDENCE INTERVAL CALCULATIONS
Based on 1000 bootstrap replicates

CALL :
boot.ci(boot.out = bootpara, type = "bca", index = 2)

Intervals :
Level       BCa
95%   (53.42, 56.07 )
Calculations and Intervals on Original Scale
```

for $\hat{\mu}_1$ and for $\hat{\mu}_2$ from

```
R> boot.ci(bootpara, type = "bca", index = 3)

BOOTSTRAP CONFIDENCE INTERVAL CALCULATIONS
Based on 1000 bootstrap replicates

CALL :
boot.ci(boot.out = bootpara, type = "bca", index = 3)

Intervals :
Level       BCa
95%   (79.05, 81.01 )
Calculations and Intervals on Original Scale
```

Finally, we show a graphical representation of both the bootstrap distribution of the mean estimates *and* the corresponding confidence intervals. For convenience, we define a function for plotting, namely

```
R> bootplot <- function(b, index, main = "") {
+      dens <- density(b$t[,index])
+      ci <- boot.ci(b, type = "bca", index = index)$bca[4:5]
+      est <- b$t0[index]
+      plot(dens, main = main)
+      y <- max(dens$y) / 10
+      segments(ci[1], y, ci[2], y, lty = 2)
+      points(ci[1], y, pch = "(")
+      points(ci[2], y, pch = ")")
+      points(est, y, pch = 19)
+  }
```

The element `t` of an object created by `boot` contains the bootstrap replications of our estimates, i.e., the values computed by `fit` for each of the 1000 bootstrap samples of the geyser data. First, we plot a simple density estimate and then construct a line representing the confidence interval. We apply this function to the bootstrap distributions of our estimates $\hat{\mu}_1$ and $\hat{\mu}_2$ in Figure 8.8.

```
R> layout(matrix(1:2, ncol = 2))
R> bootplot(bootpara, 2, main = expression(mu[1]))
R> bootplot(bootpara, 3, main = expression(mu[2]))
```

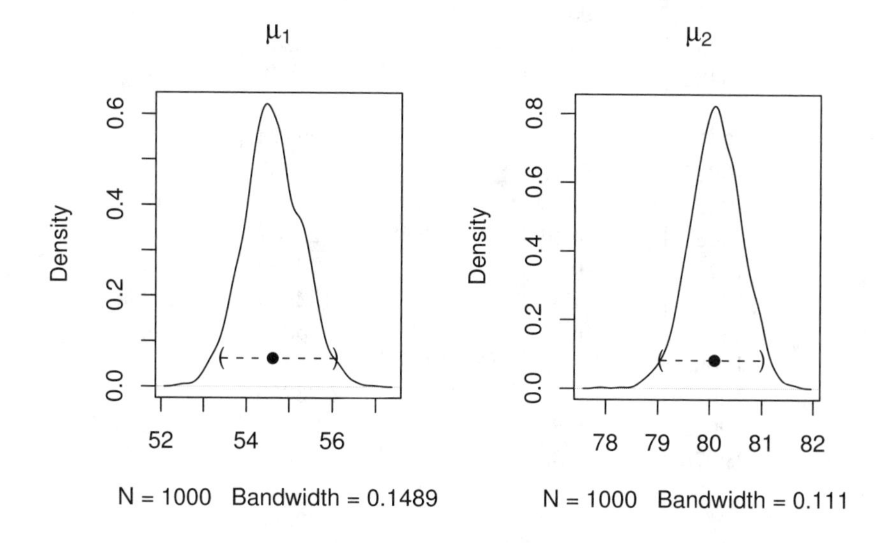


**Figure 8.8** Bootstrap distribution and confidence intervals for the mean estimates of a two-component mixture for the geyser data.

## 8.4 Summary

Histograms and scatterplots are frequently used to give graphical representations of univariate and bivariate data. But both can often be improved and made more helpful by adding some form of density estimate. For scatterplots in particular, adding a contour plot of the estimated bivariate density can be particularly useful in aiding in the identification of clusters, gaps and outliers.

## Exercises

Ex. 8.1 The data shown in Table 8.3 are the velocities of 82 galaxies from six well-separated conic sections of space (Postman et al., 1986, Roeder, 1990). The data are intended to shed light on whether or not the observable universe contains superclusters of galaxies surrounded by large voids. The evidence for the existence of superclusters would be the multimodality of the distribution of velocities. Construct a histogram of the data and add a variety of kernel estimates of the density function. What do you conclude about the possible existence of superclusters of galaxies?

**Table 8.3**: `galaxies` data (package **MASS**). Velocities of 82 galaxies.

| galaxies | galaxies | galaxies | galaxies | galaxies |
|---|---|---|---|---|
| 9172 | 19349 | 20196 | 22209 | 23706 |
| 9350 | 19440 | 20215 | 22242 | 23711 |
| 9483 | 19473 | 20221 | 22249 | 24129 |
| 9558 | 19529 | 20415 | 22314 | 24285 |
| 9775 | 19541 | 20629 | 22374 | 24289 |
| 10227 | 19547 | 20795 | 22495 | 24366 |
| 10406 | 19663 | 20821 | 22746 | 24717 |
| 16084 | 19846 | 20846 | 22747 | 24990 |
| 16170 | 19856 | 20875 | 22888 | 25633 |
| 18419 | 19863 | 20986 | 22914 | 26690 |
| 18552 | 19914 | 21137 | 23206 | 26995 |
| 18600 | 19918 | 21492 | 23241 | 32065 |
| 18927 | 19973 | 21701 | 23263 | 32789 |
| 19052 | 19989 | 21814 | 23484 | 34279 |
| 19070 | 20166 | 21921 | 23538 | |
| 19330 | 20175 | 21960 | 23542 | |
| 19343 | 20179 | 22185 | 23666 | |

*Source*: From Roeder, K., *J. Am. Stat. Assoc.*, 85, 617–624, 1990. Reprinted with permission from *The Journal of the American Statistical Association.* 

Ex. 8.2 The data in Table 8.4 give the birth and death rates for 69 countries (from Hartigan, 1975). Produce a scatterplot of the data that shows a contour plot of the estimated bivariate density. Does the plot give you any interesting insights into the possible structure of the data?

**Table 8.4**: `birthdeathrates` data. Birth and death rates for 69 countries.

| birth | death | birth | death | birth | death |
|---|---|---|---|---|---|
| 36.4 | 14.6 | 26.2 | 4.3 | 18.2 | 12.2 |
| 37.3 | 8.0 | 34.8 | 7.9 | 16.4 | 8.2 |
| 42.1 | 15.3 | 23.4 | 5.1 | 16.9 | 9.5 |
| 55.8 | 25.6 | 24.8 | 7.8 | 17.6 | 19.8 |
| 56.1 | 33.1 | 49.9 | 8.5 | 18.1 | 9.2 |
| 41.8 | 15.8 | 33.0 | 8.4 | 18.2 | 11.7 |
| 46.1 | 18.7 | 47.7 | 17.3 | 18.0 | 12.5 |
| 41.7 | 10.1 | 46.6 | 9.7 | 17.4 | 7.8 |
| 41.4 | 19.7 | 45.1 | 10.5 | 13.1 | 9.9 |
| 35.8 | 8.5 | 42.9 | 7.1 | 22.3 | 11.9 |
| 34.0 | 11.0 | 40.1 | 8.0 | 19.0 | 10.2 |
| 36.3 | 6.1 | 21.7 | 9.6 | 20.9 | 8.0 |
| 32.1 | 5.5 | 21.8 | 8.1 | 17.5 | 10.0 |
| 20.9 | 8.8 | 17.4 | 5.8 | 19.0 | 7.5 |
| 27.7 | 10.2 | 45.0 | 13.5 | 23.5 | 10.8 |
| 20.5 | 3.9 | 33.6 | 11.8 | 15.7 | 8.3 |
| 25.0 | 6.2 | 44.0 | 11.7 | 21.5 | 9.1 |
| 17.3 | 7.0 | 44.2 | 13.5 | 14.8 | 10.1 |
| 46.3 | 6.4 | 27.7 | 8.2 | 18.9 | 9.6 |
| 14.8 | 5.7 | 22.5 | 7.8 | 21.2 | 7.2 |
| 33.5 | 6.4 | 42.8 | 6.7 | 21.4 | 8.9 |
| 39.2 | 11.2 | 18.8 | 12.8 | 21.6 | 8.7 |
| 28.4 | 7.1 | 17.1 | 12.7 | 25.5 | 8.8 |

*Source*: From Hartigan, J. A., *Clustering Algorithms*, Wiley, New York, 1975. With permission.

Ex. 8.3 A sex difference in the age of onset of schizophrenia was noted by Kraepelin (1919). Subsequent epidemiological studies of the disorder have consistently shown an earlier onset in men than in women. One model that has been suggested to explain this observed difference is known as the *subtype model* which postulates two types of schizophrenia, one characterised by early onset, typical symptoms and poor premorbid competence, and the other by late onset, atypical symptoms and good premorbid competence. The early onset type is assumed to be largely a disorder of men and the late onset largely a disorder of women. By fitting finite mixtures of normal densities separately to the onset data for men and women given in Table 8.5 see if you can produce some evidence for or against the subtype model.

**Table 8.5**: `schizophrenia` data. Age on onset of schizophrenia for both sexes.

| age | gender | age | gender | age | gender | age | gender |
|---|---|---|---|---|---|---|---|
| 20 | female | 20 | female | 22 | male | 27 | male |
| 30 | female | 43 | female | 19 | male | 18 | male |
| 21 | female | 39 | female | 16 | male | 43 | male |
| 23 | female | 40 | female | 16 | male | 20 | male |
| 30 | female | 26 | female | 18 | male | 17 | male |
| 25 | female | 50 | female | 16 | male | 21 | male |
| 13 | female | 17 | female | 33 | male | 5 | male |
| 19 | female | 17 | female | 22 | male | 27 | male |
| 16 | female | 23 | female | 23 | male | 25 | male |
| 25 | female | 44 | female | 10 | male | 18 | male |
| 20 | female | 30 | female | 14 | male | 24 | male |
| 25 | female | 35 | female | 15 | male | 33 | male |
| 27 | female | 20 | female | 20 | male | 32 | male |
| 43 | female | 41 | female | 11 | male | 29 | male |
| 6 | female | 18 | female | 25 | male | 34 | male |
| 21 | female | 39 | female | 9 | male | 20 | male |
| 15 | female | 27 | female | 22 | male | 21 | male |
| 26 | female | 28 | female | 25 | male | 31 | male |
| 23 | female | 30 | female | 20 | male | 22 | male |
| 21 | female | 34 | female | 19 | male | 15 | male |
| 23 | female | 33 | female | 22 | male | 27 | male |
| 23 | female | 30 | female | 23 | male | 26 | male |
| 34 | female | 29 | female | 24 | male | 23 | male |
| 14 | female | 46 | female | 29 | male | 47 | male |
| 17 | female | 36 | female | 24 | male | 17 | male |
| 18 | female | 58 | female | 22 | male | 21 | male |
| 21 | female | 28 | female | 26 | male | 16 | male |
| 16 | female | 30 | female | 20 | male | 21 | male |
| 35 | female | 28 | female | 25 | male | 19 | male |
| 32 | female | 37 | female | 17 | male | 31 | male |
| 48 | female | 31 | female | 25 | male | 34 | male |
| 53 | female | 29 | female | 28 | male | 23 | male |
| 51 | female | 32 | female | 22 | male | 23 | male |
| 48 | female | 48 | female | 22 | male | 20 | male |
| 29 | female | 49 | female | 23 | male | 21 | male |
| 25 | female | 30 | female | 35 | male | 18 | male |
| 44 | female | 21 | male | 16 | male | 26 | male |
| 23 | female | 18 | male | 29 | male | 30 | male |
| 36 | female | 23 | male | 33 | male | 17 | male |
| 58 | female | 21 | male | 15 | male | 21 | male |
| 28 | female | 27 | male | 29 | male | 19 | male |

**Table 8.5**: `schizophrenia` data (continued).

| age | gender | age | gender | age | gender | age | gender |
|---|---|---|---|---|---|---|---|
| 51 | female | 24 | male | 20 | male | 22 | male |
| 40 | female | 20 | male | 29 | male | 52 | male |
| 43 | female | 12 | male | 24 | male | 19 | male |
| 21 | female | 15 | male | 39 | male | 24 | male |
| 48 | female | 19 | male | 10 | male | 19 | male |
| 17 | female | 21 | male | 20 | male | 19 | male |
| 23 | female | 22 | male | 23 | male | 33 | male |
| 28 | female | 19 | male | 15 | male | 32 | male |
| 44 | female | 24 | male | 18 | male | 29 | male |
| 28 | female | 9 | male | 20 | male | 58 | male |
| 21 | female | 19 | male | 21 | male | 39 | male |
| 31 | female | 18 | male | 30 | male | 42 | male |
| 22 | female | 17 | male | 21 | male | 32 | male |
| 56 | female | 23 | male | 18 | male | 32 | male |
| 60 | female | 17 | male | 19 | male | 46 | male |
| 15 | female | 23 | male | 15 | male | 38 | male |
| 21 | female | 19 | male | 19 | male | 44 | male |
| 30 | female | 37 | male | 18 | male | 35 | male |
| 26 | female | 26 | male | 25 | male | 45 | male |
| 28 | female | 22 | male | 17 | male | 41 | male |
| 23 | female | 24 | male | 15 | male | 31 | male |
| 21 | female | 19 | male | 42 | male | | |

CHAPTER 9

# Recursive Partitioning: Predicting Body Fat and Glaucoma Diagnosis

## 9.1 Introduction

Worldwide, overweight and obesity are considered to be major health problems because of their strong association with a higher risk of diseases of the metabolic syndrome, including diabetes mellitus and cardiovascular disease, as well as with certain forms of cancer. Obesity is frequently evaluated by using simple indicators such as body mass index, waist circumference, or waist-to-hip ratio. Specificity and adequacy of these indicators are still controversial, mainly because they do not allow a precise assessment of body composition. Body fat, especially visceral fat, is suggested to be a better predictor of diseases of the metabolic syndrome. Garcia et al. (2005) report on the development of a multiple linear regression model for body fat content by means of $p = 9$ common anthropometric measurements which were obtained for $n = 71$ healthy German women. In addition, the women's body composition was measured by Dual Energy X-Ray Absorptiometry (DXA). This reference method is very accurate in measuring body fat but finds little applicability in practical environments, mainly because of high costs and the methodological efforts needed. Therefore, a simple regression model for predicting DXA measurements of body fat is of special interest for the practitioner. The following variables are available (the measurements are given in Table 9.1):

`DEXfat`: body fat measured by DXA, the response variable,

`age`: age of the subject in years,

`waistcirc`: waist circumference,

`hipcirc`: hip circumference,

`elbowbreadth`: breadth of the elbow, and

`kneebreadth`: breadth of the knee.

**Table 9.1**: `bodyfat` data (package **mboost**). Body fat prediction by skinfold thickness, circumferences, and bone breadths.

| DEXfat | age | waistcirc | hipcirc | elbowbreadth | kneebreadth |
|---|---|---|---|---|---|
| 41.68 | 57 | 100.0 | 112.0 | 7.1 | 9.4 |
| 43.29 | 65 | 99.5 | 116.5 | 6.5 | 8.9 |

**Table 9.1**: `bodyfat` data (continued).

| DEXfat | age | waistcirc | hipcirc | elbowbreadth | kneebreadth |
|---:|---:|---:|---:|---:|---:|
| 35.41 | 59 | 96.0 | 108.5 | 6.2 | 8.9 |
| 22.79 | 58 | 72.0 | 96.5 | 6.1 | 9.2 |
| 36.42 | 60 | 89.5 | 100.5 | 7.1 | 10.0 |
| 24.13 | 61 | 83.5 | 97.0 | 6.5 | 8.8 |
| 29.83 | 56 | 81.0 | 103.0 | 6.9 | 8.9 |
| 35.96 | 60 | 89.0 | 105.0 | 6.2 | 8.5 |
| 23.69 | 58 | 80.0 | 97.0 | 6.4 | 8.8 |
| 22.71 | 62 | 79.0 | 93.0 | 7.0 | 8.8 |
| 23.42 | 63 | 79.0 | 99.0 | 6.2 | 8.6 |
| 23.24 | 62 | 72.0 | 94.0 | 6.7 | 8.7 |
| 26.25 | 64 | 81.5 | 95.0 | 6.2 | 8.2 |
| 21.94 | 60 | 65.0 | 90.0 | 5.7 | 8.2 |
| 30.13 | 61 | 79.0 | 107.5 | 5.8 | 8.6 |
| 36.31 | 66 | 98.5 | 109.0 | 6.9 | 9.6 |
| 27.72 | 63 | 79.5 | 101.5 | 7.0 | 9.4 |
| 46.99 | 57 | 117.0 | 116.0 | 7.1 | 10.7 |
| 42.01 | 49 | 100.5 | 112.0 | 6.9 | 9.4 |
| 18.63 | 65 | 82.0 | 91.0 | 6.6 | 8.8 |
| 38.65 | 58 | 101.0 | 107.5 | 6.4 | 8.6 |
| 21.20 | 63 | 80.0 | 96.0 | 6.9 | 8.6 |
| 35.40 | 60 | 89.0 | 101.0 | 6.2 | 9.2 |
| 29.63 | 59 | 89.5 | 99.5 | 6.0 | 8.1 |
| 25.16 | 32 | 73.0 | 99.0 | 7.2 | 8.6 |
| 31.75 | 42 | 87.0 | 102.0 | 6.9 | 10.8 |
| 40.58 | 49 | 90.2 | 110.3 | 7.1 | 9.5 |
| 21.69 | 63 | 80.5 | 97.0 | 5.8 | 8.8 |
| 46.60 | 57 | 102.0 | 124.0 | 6.6 | 11.2 |
| 27.62 | 44 | 86.0 | 102.0 | 6.3 | 8.3 |
| 41.30 | 61 | 102.0 | 122.5 | 6.3 | 10.8 |
| 42.76 | 62 | 103.0 | 125.0 | 7.3 | 11.1 |
| 28.84 | 24 | 81.0 | 100.0 | 6.6 | 9.7 |
| 36.88 | 54 | 85.5 | 113.0 | 6.2 | 9.6 |
| 25.09 | 65 | 75.3 | 101.2 | 5.2 | 9.3 |
| 29.73 | 67 | 81.0 | 104.3 | 5.7 | 8.1 |
| 28.92 | 45 | 85.0 | 106.0 | 6.7 | 10.0 |
| 43.80 | 51 | 102.2 | 118.5 | 6.8 | 10.6 |
| 26.74 | 49 | 78.0 | 99.0 | 6.2 | 9.8 |
| 33.79 | 52 | 93.3 | 109.0 | 6.8 | 9.8 |
| 62.02 | 66 | 106.5 | 126.0 | 6.4 | 11.4 |
| 40.01 | 63 | 102.0 | 117.0 | 6.6 | 10.6 |
| 42.72 | 42 | 111.0 | 109.0 | 6.7 | 9.9 |
| 32.49 | 50 | 102.0 | 108.0 | 6.2 | 9.8 |

**Table 9.1**: `bodyfat` data (continued).

| DEXfat | age | waistcirc | hipcirc | elbowbreadth | kneebreadth |
|---|---|---|---|---|---|
| 45.92 | 63 | 116.8 | 132.0 | 6.1 | 9.8 |
| 42.23 | 62 | 112.0 | 127.0 | 7.2 | 11.0 |
| 47.48 | 42 | 115.0 | 128.5 | 6.6 | 10.0 |
| 60.72 | 41 | 115.0 | 125.0 | 7.3 | 11.8 |
| 32.74 | 67 | 89.8 | 109.0 | 6.3 | 9.6 |
| 27.04 | 67 | 82.2 | 103.6 | 7.2 | 9.2 |
| 21.07 | 43 | 75.0 | 99.3 | 6.0 | 8.4 |
| 37.49 | 54 | 98.0 | 109.5 | 7.0 | 10.0 |
| 38.08 | 49 | 105.0 | 116.3 | 7.0 | 9.5 |
| 40.83 | 25 | 89.5 | 122.0 | 6.5 | 10.0 |
| 18.51 | 26 | 87.8 | 94.0 | 6.6 | 9.0 |
| 26.36 | 33 | 79.2 | 107.7 | 6.5 | 9.0 |
| 20.08 | 36 | 80.0 | 95.0 | 6.4 | 9.0 |
| 43.71 | 38 | 105.5 | 122.5 | 6.6 | 10.0 |
| 31.61 | 26 | 95.0 | 109.0 | 6.7 | 9.5 |
| 28.98 | 52 | 81.5 | 102.3 | 6.4 | 9.2 |
| 18.62 | 29 | 71.0 | 92.0 | 6.4 | 8.5 |
| 18.64 | 31 | 68.0 | 93.0 | 5.7 | 7.2 |
| 13.70 | 19 | 68.0 | 88.0 | 6.5 | 8.2 |
| 14.88 | 35 | 68.5 | 94.5 | 6.5 | 8.8 |
| 16.46 | 27 | 75.0 | 95.0 | 6.4 | 9.1 |
| 11.21 | 40 | 66.6 | 92.2 | 6.1 | 8.5 |
| 11.21 | 53 | 66.6 | 92.2 | 6.1 | 8.5 |
| 14.18 | 31 | 69.7 | 93.2 | 6.2 | 8.1 |
| 20.84 | 27 | 66.5 | 100.0 | 6.5 | 8.5 |
| 19.00 | 52 | 76.5 | 103.0 | 7.4 | 8.5 |
| 18.07 | 59 | 71.0 | 88.3 | 5.7 | 8.9 |

A second set of data that will also be used in this chapter involves the investigation reported in Mardin et al. (2003) of whether laser scanner images of the eye background can be used to classify a patient's eye as suffering from glaucoma or not. Glaucoma is a neuro-degenerative disease of the optic nerve and is one of the major reasons for blindness in elderly people. For 196 people, 98 patients suffering glaucoma and 98 controls which have been matched by age and gender, 62 numeric variables derived from the laser scanning images are available. The data are available as `GlaucomaM` from package **ipred** (Peters et al., 2002). The variables describe the morphology of the optic nerve head, i.e., measures of volumes and areas in certain regions of the eye background. Those regions have been manually outlined by a physician. Our aim is to construct a prediction model which is able to decide whether an eye is affected by glaucomateous changes based on the laser image data.

Both sets of data described above could be analysed using the regression models described in Chapter 6 and Chapter 7, i.e., regression models for numeric and binary response variables based on a linear combination of the covariates. But here we shall employ an alternative approach known as *recursive partitioning*, where the resulting models are usually called *regression or classification trees*. This method was originally invented to deal with possible non-linear relationships between covariates and response. The basic idea is to partition the covariate space and to compute simple statistics of the dependent variable, like the mean or median, inside each cell.

## 9.2 Recursive Partitioning

There exist many algorithms for the construction of classification or regression trees but the majority of algorithms follow a simple general rule: First partition the observations by univariate splits in a recursive way and second fit a constant model in each cell of the resulting partition. An overview of this field of regression models is given by Murthy (1998).

In more details, for the first step, one selects a covariate $x_j$ from the $q$ available covariates $x_1, \dots, x_q$ and estimates a split point which separates the response values $y_i$ into two groups. For an ordered covariate $x_j$ a split point is a number $\xi$ dividing the observations into two groups. The first group consists of all observations with $x_j \leq \xi$ and the second group contains the observations satisfying $x_j > \xi$. For a nominal covariate $x_j$, the two groups are defined by a set of levels $A$ where either $x_j \in A$ or $x_j \notin A$.

Once the splits $\xi$ or $A$ for some selected covariate $x_j$ have been estimated, one applies the procedure sketched above for all observations in the first group and, recursively, splits this set of observations further. The same happens for all observations in the second group. The recursion is stopped when some stopping criterion is fulfilled.

The available algorithms mostly differ with respect to three points: how the covariate is selected in each step, how the split point is estimated and which stopping criterion is applied. One of the most popular algorithms is described in the *Classification and Regression Trees* book by Breiman et al. (1984) and is available in R by the functions in package **rpart** (Therneau and Atkinson, 1997, Therneau et al., 2009). This algorithm first examines all possible splits for all covariates and chooses the split which leads to two groups that are 'purer' than the current group with respect to the values of the response variable $y$. There are many possible measures of impurity available, for regression problems with nominal response the *Gini* criterion is the default in **rpart**, alternatives and a more detailed description of tree based methods can be found in Ripley (1996).

The question when the recursion needs to stop is all but trivial. In fact, trees with too many leaves will suffer from overfitting and small trees will miss important aspects of the problem. Commonly, this problem is addressed by so-called *pruning* methods. As the name suggests, one first grows a very large tree using a trivial stopping criterion as the number of observations in a leaf, say, and then prunes branches that are not necessary.

Once that a tree has been grown, a simple summary statistic is computed for each leaf. The mean or median can be used for continuous responses and for nominal responses the proportions of the classes is commonly used. The prediction of a new observation is simply the corresponding summary statistic of the leaf to which this observation belongs.

However, even the right-sized tree consists of binary splits which are, of course, hard decisions. When the underlying relationship between covariate and response is smooth, such a split point estimate will be affected by high variability. This problem is addressed by so called *ensemble methods*. Here, multiple trees are grown on perturbed instances of the data set and their predictions are averaged. The simplest representative of such a procedure is called *bagging* (Breiman, 1996) and works as follows. We draw $B$ bootstrap samples from the original data set, i.e., we draw $n$ out of $n$ observations with replacement from our $n$ original observations. For each of those bootstrap samples we grow a very large tree. When we are interested in the prediction for a new observation, we pass this observation through all $B$ trees and average their predictions. It has been shown that the goodness of the predictions for future cases can be improved dramatically by this or similar simple procedures. More details can be found in Bühlmann (2004).

## 9.3 Analysis Using R

### 9.3.1 Predicting Body Fat Content

The `rpart` function from **rpart** can be used to grow a regression tree. The response variable and the covariates are defined by a model formula in the same way as for `lm`, say. By default, a large initial tree is grown, we restrict the number of observations required to establish a potential binary split to at least ten:

```
R> library("rpart")
R> data("bodyfat", package = "mboost")
R> bodyfat_rpart <- rpart(DEXfat ~ age + waistcirc + hipcirc +
+      elbowbreadth + kneebreadth, data = bodyfat,
+      control = rpart.control(minsplit = 10))
```

A `print` method for *rpart* objects is available; however, a graphical representation (here utilising functionality offered from package **partykit**, Hothorn and Zeileis, 2009) shown in Figure 9.1 is more convenient. Observations that satisfy the condition shown for each node go to the left and observations that don't are element of the right branch in each node. As expected, higher values for waist- and hip circumferences and wider knees correspond to higher values of body fat content. The rightmost terminal node consists of only three rather extreme observations.

```
R> library("partykit")
R> plot(as.party(bodyfat_rpart), tp_args = list(id = FALSE))
```

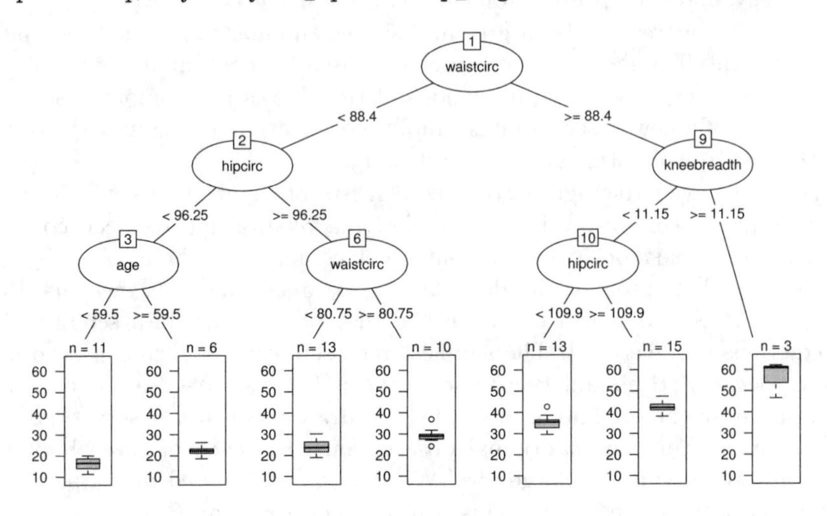


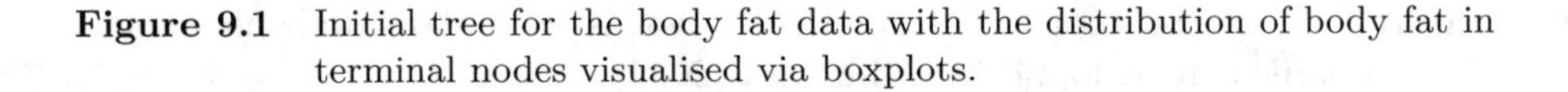

**Figure 9.1** Initial tree for the body fat data with the distribution of body fat in terminal nodes visualised via boxplots.

To determine if the tree is appropriate or if some of the branches need to be subjected to pruning we can use the cptable element of the *rpart* object:

```
R> print(bodyfat_rpart$cptable)
```

```
          CP nsplit  rel error    xerror       xstd
1 0.66289544      0 1.00000000 1.0270918 0.16840424
2 0.09376252      1 0.33710456 0.4273989 0.09430024
3 0.07703606      2 0.24334204 0.4449342 0.08686150
4 0.04507506      3 0.16630598 0.3535449 0.06957080
5 0.01844561      4 0.12123092 0.2642626 0.05974575
6 0.01818982      5 0.10278532 0.2855892 0.06221393
7 0.01000000      6 0.08459549 0.2785367 0.06242559
```

```
R> opt <- which.min(bodyfat_rpart$cptable[,"xerror"])
```

The xerror column contains of estimates of cross-validated prediction error for different numbers of splits (nsplit). The best tree has four splits. Now we can prune back the large initial tree using

```
R> cp <- bodyfat_rpart$cptable[opt, "CP"]
R> bodyfat_prune <- prune(bodyfat_rpart, cp = cp)
```

The result is shown in Figure 9.2. Note that the inner nodes three and six have been removed from the tree. Still, the rightmost terminal node might give very unreliable extreme predictions.

```
R> plot(as.party(bodyfat_prune), tp_args = list(id = FALSE))
```

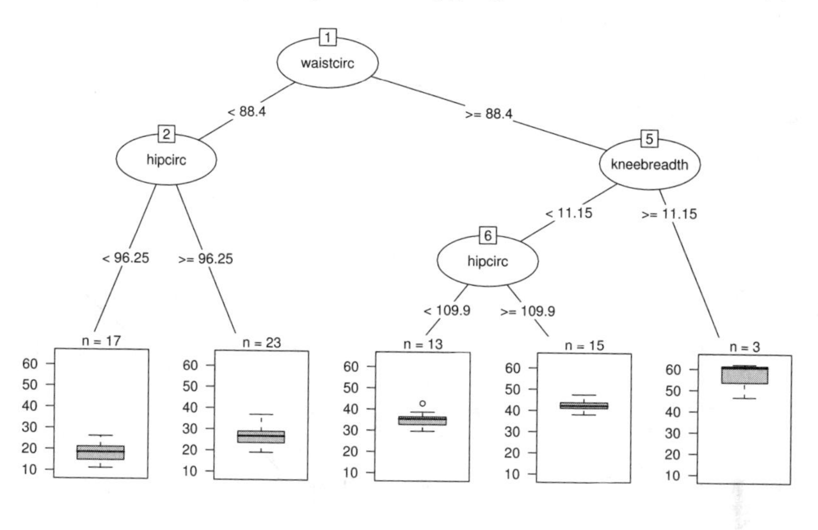


**Figure 9.2** Pruned regression tree for body fat data.

Given this model, one can predict the (unknown, in real circumstances) body fat content based on the covariate measurements. Here, using the known values of the response variable, we compare the model predictions with the actually measured body fat as shown in Figure 9.3. The three observations with large body fat measurements in the rightmost terminal node can be identified easily.

### *9.3.2 Glaucoma Diagnosis*

We start with a large initial tree and prune back branches according to the cross-validation criterion. The default is to use 10 runs of 10-fold cross-validation and we choose 100 runs of 10-fold cross-validation for reasons to be explained later.

```
R> data("GlaucomaM", package = "ipred")
R> glaucoma_rpart <- rpart(Class ~ ., data = GlaucomaM,
+       control = rpart.control(xval = 100))
R> glaucoma_rpart$cptable
```

```
          CP nsplit rel error    xerror       xstd
1 0.65306122      0 1.0000000 1.5306122 0.06054391
2 0.07142857      1 0.3469388 0.3877551 0.05647630
3 0.01360544      2 0.2755102 0.3775510 0.05590431
4 0.01000000      5 0.2346939 0.4489796 0.05960655
```

```
R> DEXfat_pred <- predict(bodyfat_prune, newdata = bodyfat)
R> xlim <- range(bodyfat$DEXfat)
R> plot(DEXfat_pred ~ DEXfat, data = bodyfat, xlab = "Observed",
+       ylab = "Predicted", ylim = xlim, xlim = xlim)
R> abline(a = 0, b = 1)
```

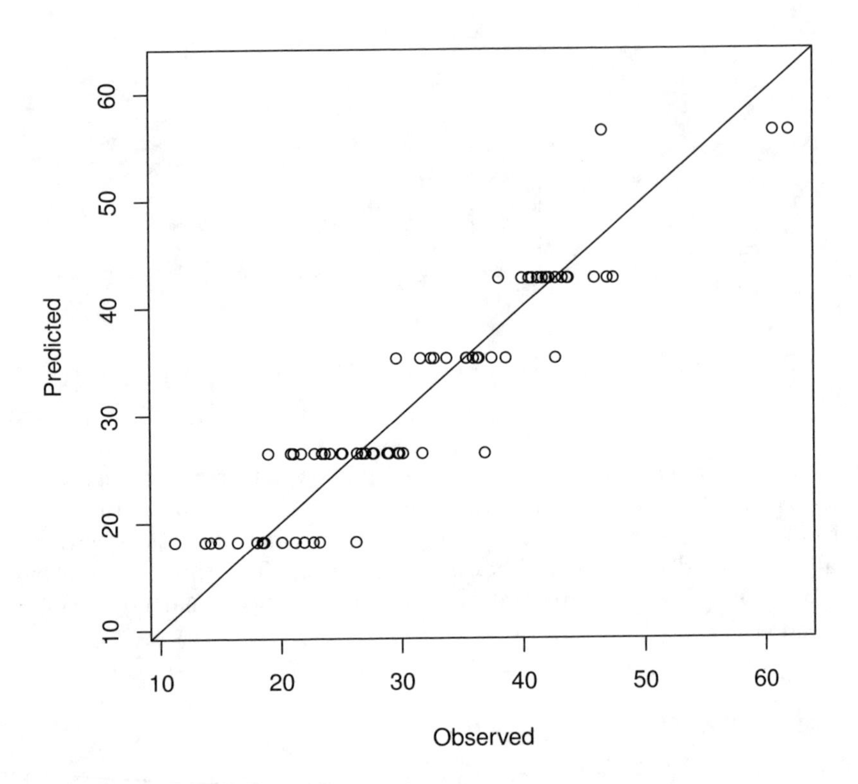


**Figure 9.3** Observed and predicted DXA measurements.

```
R> opt <- which.min(glaucoma_rpart$cptable[,"xerror"])
R> cp <- glaucoma_rpart$cptable[opt, "CP"]
R> glaucoma_prune <- prune(glaucoma_rpart, cp = cp)
```

The pruned tree consists of three leaves only (Figure 9.4); the class distribution in each leaf is depicted using a barplot. For most eyes, the decision about the disease is based on the variable varg, a measurement of the volume of the optic nerve above some reference plane. A volume larger than 0.209 $mm^3$ indicates that the eye is healthy, and damage of the optic nerve head asso-

```
R> plot(as.party(glaucoma_prune), tp_args = list(id = FALSE))
```

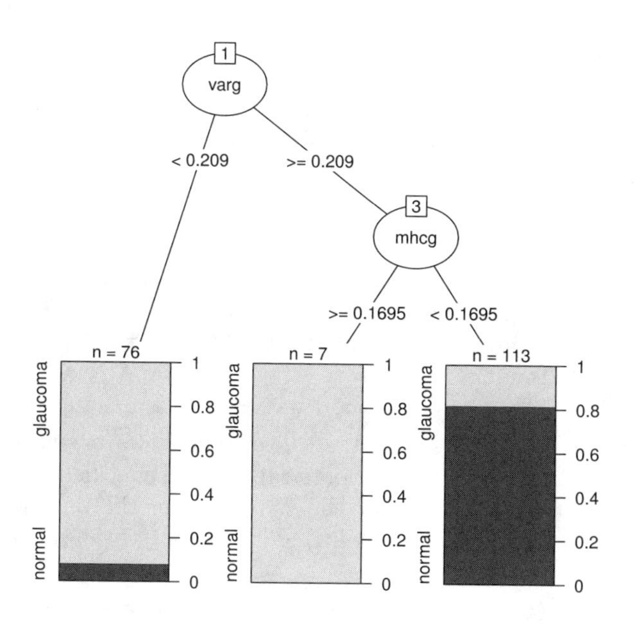


**Figure 9.4** Pruned classification tree of the glaucoma data with class distribution in the leaves.

ciated with loss of optic nerves (`varg` smaller than 0.209 mm$^3$) indicates a glaucomateous change.

As we discussed earlier, the choice of the appropriatly sized tree is not a trivial problem. For the glaucoma data, the above choice of three leaves is very unstable across multiple runs of cross-validation. As an illustration of this problem we repeat the very same analysis as shown above and record the optimal number of splits as suggested by the cross-validation runs.

```
R> nsplitopt <- vector(mode = "integer", length = 25)
R> for (i in 1:length(nsplitopt)) {
+      cp <- rpart(Class ~ ., data = GlaucomaM)$cptable
+      nsplitopt[i] <- cp[which.min(cp[,"xerror"]), "nsplit"]
+ }
R> table(nsplitopt)
```

```
nsplitopt
 1  2  5
14  7  4
```

Although for 14 runs of cross-validation a simple tree with one split only is suggested, larger trees would have been favoured in 11 of the cases. This short analysis shows that we should not trust the tree in Figure 9.4 too much.

One way out of this dilemma is the aggregation of multiple trees via bagging. In R, the bagging idea can be implemented by three or four lines of code. Case count or weight vectors representing the bootstrap samples can be drawn from the multinominal distribution with parameters $n$ and $p_1 = 1/n, \dots, p_n = 1/n$ via the `rmultinom` function. For each weight vector, one large tree is constructed without pruning and the *rpart* objects are stored in a list, here called `trees`:

```
R> trees <- vector(mode = "list", length = 25)
R> n <- nrow(GlaucomaM)
R> bootsamples <- rmultinom(length(trees), n, rep(1, n)/n)
R> mod <- rpart(Class ~ ., data = GlaucomaM,
+                  control = rpart.control(xval = 0))
R> for (i in 1:length(trees))
+      trees[[i]] <- update(mod, weights = bootsamples[,i])
```

The `update` function re-evaluates the call of `mod`, however, with the weights being altered, i.e., fits a tree to a bootstrap sample specified by the weights. It is interesting to have a look at the structures of the multiple trees. For example, the variable selected for splitting in the root of the tree is not unique as can be seen by

```
R> table(sapply(trees, function(x) as.character(x$frame$var[1])))
phcg varg vari vars
   1   14    9    1
```

Although `varg` is selected most of the time, other variables such as `vari` occur as well – a further indication that the tree in Figure 9.4 is questionable and that hard decisions are not appropriate for the glaucoma data.

In order to make use of the ensemble of trees in the list `trees` we estimate the conditional probability of suffering from glaucoma given the covariates for each observation in the original data set by

```
R> classprob <- matrix(0, nrow = n, ncol = length(trees))
R> for (i in 1:length(trees)) {
+      classprob[,i] <- predict(trees[[i]],
+                               newdata = GlaucomaM)[,1]
+      classprob[bootsamples[,i] > 0,i] <- NA
+  }
```

Thus, for each observation we get 25 estimates. However, each observation has been used for growing one of the trees with probability 0.632 and thus was not used with probability 0.368. Consequently, the estimate from a tree where an observation was not used for growing is better for judging the quality of the predictions and we label the other estimates with `NA`.

Now, we can average the estimates and we vote for glaucoma when the average of the estimates of the conditional glaucoma probability exceeds 0.5. The comparison between the observed and the predicted classes does not suffer from overfitting since the predictions are computed from those trees for which each single observation was *not* used for growing.

```
R> avg <- rowMeans(classprob, na.rm = TRUE)
R> predictions <- factor(ifelse(avg > 0.5, "glaucoma",
+                                          "normal"))
R> predtab <- table(predictions, GlaucomaM$Class)
R> predtab

predictions glaucoma normal
   glaucoma       77     16
   normal         21     82
```

Thus, an honest estimate of the probability of a glaucoma prediction when the patient is actually suffering from glaucoma is

```
R> round(predtab[1,1] / colSums(predtab)[1] * 100)

glaucoma
      79
```

per cent. For

```
R> round(predtab[2,2] / colSums(predtab)[2] * 100)

normal
    84
```

per cent of normal eyes, the ensemble does not predict a glaucomateous damage.

Although we are mainly interested in a predictor, i.e., a *black box* machine for predicting glaucoma is our main focus, the nature of the black box might be interesting as well. From the classification tree analysis shown above we expect to see a relationship between the volume above the reference plane (`varg`) and the estimated conditional probability of suffering from glaucoma. A graphical approach is sufficient here and we simply plot the observed values of `varg` against the averages of the estimated glaucoma probability (such plots have been used by Breiman, 2001b, Garczarek and Weihs, 2003, for example). In addition, we construct such a plot for another covariate as well, namely `vari`, the volume above the reference plane measured in the inferior part of the optic nerve head only. Figure 9.5 shows that the initial split of $0.209mm^3$ for `varg` (see Figure 9.4) corresponds to the ensemble predictions rather well.

The bagging procedure is a special case of a more general approach called *random forest* (Breiman, 2001a). The package **randomForest** (Breiman et al., 2009) can be used to compute such ensembles via

```
R> library("randomForest")
R> rf <- randomForest(Class ~ ., data = GlaucomaM)
```

and we obtain out-of-bag estimates for the prediction error via

```
R> table(predict(rf), GlaucomaM$Class)

           glaucoma normal
  glaucoma       80     12
  normal         18     86
```

```
R> library("lattice")
R> gdata <- data.frame(avg = rep(avg, 2),
+      class = rep(as.numeric(GlaucomaM$Class), 2),
+      obs = c(GlaucomaM[["varg"]], GlaucomaM[["vari"]]),
+      var = factor(c(rep("varg", nrow(GlaucomaM)),
+                     rep("vari", nrow(GlaucomaM)))))
R> panelf <- function(x, y) {
+            panel.xyplot(x, y, pch = gdata$class)
+            panel.abline(h = 0.5, lty = 2)
+        }
R> print(xyplot(avg ~ obs | var, data = gdata,
+         panel = panelf,
+         scales = "free", xlab = "",
+         ylab = "Estimated Class Probability Glaucoma"))
```

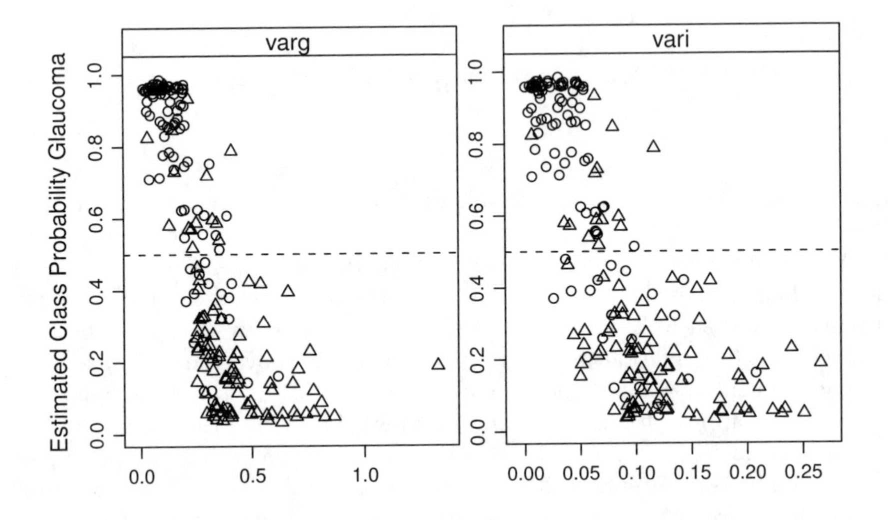


**Figure 9.5** Estimated class probabilities depending on two important variables. The 0.5 cut-off for the estimated glaucoma probability is depicted as a horizontal line. Glaucomateous eyes are plotted as circles and normal eyes are triangles.

```
R> plot(bodyfat_ctree)
```

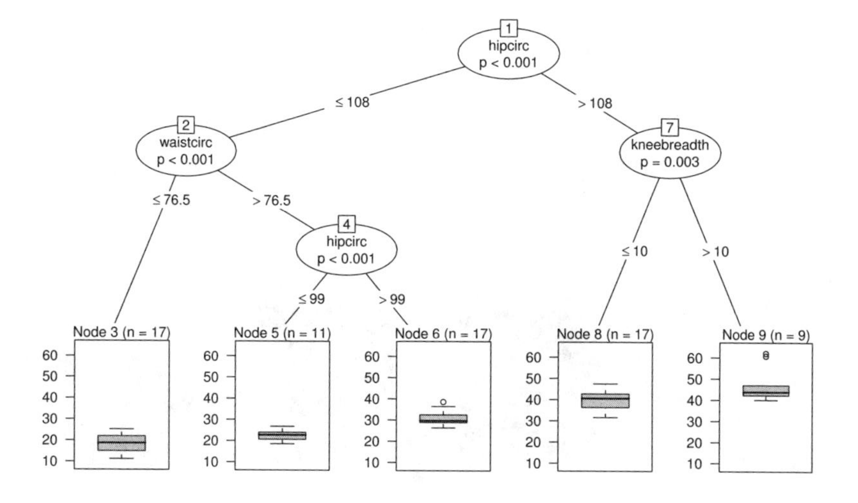


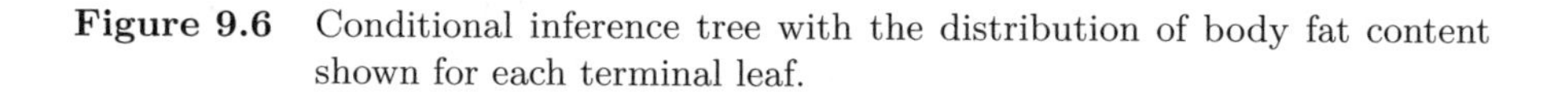

**Figure 9.6** Conditional inference tree with the distribution of body fat content shown for each terminal leaf.

### *9.3.3 Trees Revisited*

Another approach to recursive partitioning, making a connection to classical statistical test problems such as those discussed in Chapter 4, is implemented in the **party** package (Hothorn et al., 2006b, 2009c). In each node of those trees, a significance test on independence between any of the covariates and the response is performed and a split is established when the $p$-value, possibly adjusted for multiple comparisons, is smaller than a pre-specified nominal level $\alpha$. This approach has the advantage that one does not need to prune back large initial trees since we have a statistically motivated stopping criterion – the $p$-value – at hand.

For the body fat data, such a *conditional inference tree* can be computed using the `ctree` function

```
R> library("party")
R> bodyfat_ctree <- ctree(DEXfat ~ age + waistcirc + hipcirc +
+      elbowbreadth + kneebreadth, data = bodyfat)
```

This tree doesn't require a pruning procedure because an internal stop criterion based on formal statistical tests prevents the procedure from overfitting the data. The tree structure is shown in Figure 9.6. Although the structure of this tree and the tree depicted in Figure 9.2 are rather different, the corresponding predictions don't vary too much.

Very much the same code is needed to grow a tree on the glaucoma data:

```
R> glaucoma_ctree <- ctree(Class ~ ., data = GlaucomaM)
```

```
R> plot(glaucoma_ctree)
```

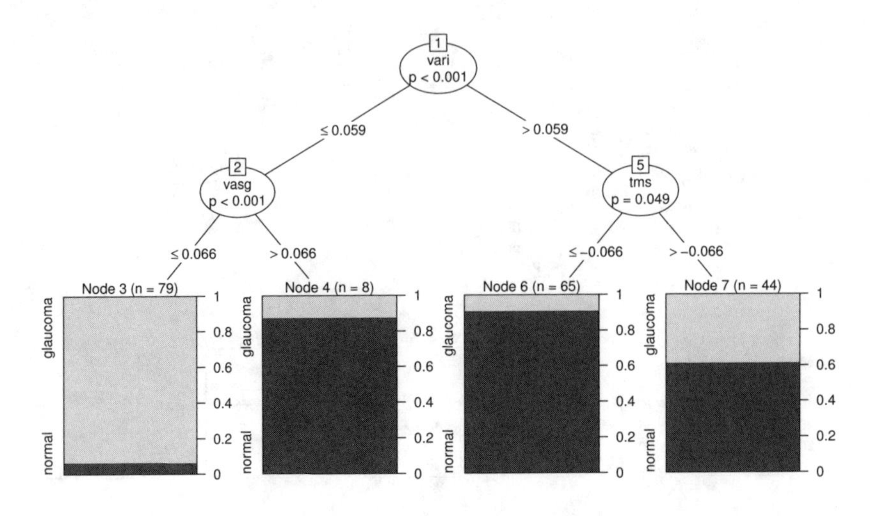


**Figure 9.7** Conditional inference tree with the distribution of glaucomateous eyes shown for each terminal leaf.

and a graphical representation is depicted in Figure 9.7 showing both the cutpoints and the $p$-values of the associated independence tests for each node. The first split is performed using a cutpoint defined with respect to the volume of the optic nerve above some reference plane, but in the inferior part of the eye only (`vari`).

## 9.4 Summary

Recursive partitioning procedures are rather simple non-parametric tools for regression modelling. The main structures of regression relationship can be visualised in a straightforward way. However, one should bear in mind that the nature of those models is very simple and can serve only as a rough approximation to reality. When multiple simple models are averaged, powerful predictors can be constructed.

## Exercises

Ex. 9.1 Construct a regression tree for the Boston Housing data reported by Harrison and Rubinfeld (1978) which are available as *data.frame* `BostonHousing` from package **mlbench** (Leisch and Dimitriadou, 2009). Compare the predictions of the tree with the predictions obtained from `randomForest`. Which method is more accurate?

Ex. 9.2 For each possible cutpoint in `varg` of the glaucoma data, compute the test statistic of the chi-square test of independence (see Chapter 3) and plot them against the values of `varg`. Is a simple cutpoint for this variable appropriate for discriminating between healthy and glaucomateous eyes?

Ex. 9.3 Compare the tree models fitted to the glaucoma data with a logistic regression model (see Chapter 7).

CHAPTER 10

# Scatterplot Smoothers and Generalised Additive Models: The Men's Olympic 1500m, Air Pollution in the USA, and Risk Factors for Kyphosis

## 10.1 Introduction

The modern Olympics began in 1896 in Greece and have been held every four years since, apart from interruptions due to the two world wars. On the track the blue ribbon event has always been the 1500m for men since competitors that want to win must have a unique combination of speed, strength and stamina combined with an acute tactical awareness. For the spectator the event lasts long enough to be interesting (unlike say the 100m dash) but not too long so as to become boring (as do most 10,000m races). The event has been witness to some of the most dramatic scenes in Olympic history; who can forget Herb Elliott winning by a street in 1960, breaking the world record and continuing his sequence of never being beaten in a 1500m or mile race in his career? And remembering the joy and relief etched on the face of Seb Coe when winning and beating his arch rival Steve Ovett still brings a tear to the eye of many of us.

The complete record of winners of the men's 1500m from 1896 to 2004 is given in Table 10.1. Can we use these winning times as the basis of a suitable statistical model that will enable us to predict the winning times for future Olympics?

**Table 10.1**: `men1500m` data. Olympic Games 1896 to 2004 winners of the men's 1500m.

| year | venue | winner | country | time |
|---|---|---|---|---|
| 1896 | Athens | E. Flack | Australia | 273.20 |
| 1900 | Paris | C. Bennett | Great Britain | 246.20 |
| 1904 | St. Louis | J. Lightbody | USA | 245.40 |
| 1908 | London | M. Sheppard | USA | 243.40 |
| 1912 | Stockholm | A. Jackson | Great Britain | 236.80 |
| 1920 | Antwerp | A. Hill | Great Britain | 241.80 |
| 1924 | Paris | P. Nurmi | Finland | 233.60 |
| 1928 | Amsterdam | H. Larva | Finland | 233.20 |
| 1932 | Los Angeles | L. Beccali | Italy | 231.20 |

**Table 10.1**: `men1500m` data (continued).

| year | venue | winner | country | time |
|---|---|---|---|---|
| 1936 | Berlin | J. Lovelock | New Zealand | 227.80 |
| 1948 | London | H. Eriksson | Sweden | 229.80 |
| 1952 | Helsinki | J. Barthel | Luxemborg | 225.10 |
| 1956 | Melbourne | R. Delaney | Ireland | 221.20 |
| 1960 | Rome | H. Elliott | Australia | 215.60 |
| 1964 | Tokyo | P. Snell | New Zealand | 218.10 |
| 1968 | Mexico City | K. Keino | Kenya | 214.90 |
| 1972 | Munich | P. Vasala | Finland | 216.30 |
| 1976 | Montreal | J. Walker | New Zealand | 219.17 |
| 1980 | Moscow | S. Coe | Great Britain | 218.40 |
| 1984 | Los Angeles | S. Coe | Great Britain | 212.53 |
| 1988 | Seoul | P. Rono | Kenya | 215.95 |
| 1992 | Barcelona | F. Cacho | Spain | 220.12 |
| 1996 | Atlanta | N. Morceli | Algeria | 215.78 |
| 2000 | Sydney | K. Ngenyi | Kenya | 212.07 |
| 2004 | Athens | H. El Guerrouj | Morocco | 214.18 |

The data in Table 10.2 relate to air pollution in 41 US cities as reported by Sokal and Rohlf (1981). The annual mean concentration of sulphur dioxide, in micrograms per cubic metre, is a measure of the air pollution of the city. The question of interest here is what aspects of climate and human ecology as measured by the other six variables in the table determine pollution. Thus, we are interested in a regression model from which we can infer the relationship between each of the exploratory variables to the response ($SO_2$ content). Details of the seven measurements are;

`SO2`: $SO_2$ content of air in micrograms per cubic metre,

`temp`: average annual temperature in Fahrenheit,

`manu`: number of manufacturing enterprises employing 20 or more workers,

`popul`: population size (1970 census); in thousands,

`wind`: average annual wind speed in miles per hour,

`precip`: average annual precipitation in inches,

`predays`: average number of days with precipitation per year.

**Table 10.2**: `USairpollution` data. Air pollution in 41 US cities.

| | SO2 | temp | manu | popul | wind | precip | predays |
|---|---|---|---|---|---|---|---|
| Albany | 46 | 47.6 | 44 | 116 | 8.8 | 33.36 | 135 |
| Albuquerque | 11 | 56.8 | 46 | 244 | 8.9 | 7.77 | 58 |

**Table 10.2**: `USairpollution` data (continued).

| | SO2 | temp | manu | popul | wind | precip | predays |
|---|---|---|---|---|---|---|---|
| Atlanta | 24 | 61.5 | 368 | 497 | 9.1 | 48.34 | 115 |
| Baltimore | 47 | 55.0 | 625 | 905 | 9.6 | 41.31 | 111 |
| Buffalo | 11 | 47.1 | 391 | 463 | 12.4 | 36.11 | 166 |
| Charleston | 31 | 55.2 | 35 | 71 | 6.5 | 40.75 | 148 |
| Chicago | 110 | 50.6 | 3344 | 3369 | 10.4 | 34.44 | 122 |
| Cincinnati | 23 | 54.0 | 462 | 453 | 7.1 | 39.04 | 132 |
| Cleveland | 65 | 49.7 | 1007 | 751 | 10.9 | 34.99 | 155 |
| Columbus | 26 | 51.5 | 266 | 540 | 8.6 | 37.01 | 134 |
| Dallas | 9 | 66.2 | 641 | 844 | 10.9 | 35.94 | 78 |
| Denver | 17 | 51.9 | 454 | 515 | 9.0 | 12.95 | 86 |
| Des Moines | 17 | 49.0 | 104 | 201 | 11.2 | 30.85 | 103 |
| Detroit | 35 | 49.9 | 1064 | 1513 | 10.1 | 30.96 | 129 |
| Hartford | 56 | 49.1 | 412 | 158 | 9.0 | 43.37 | 127 |
| Houston | 10 | 68.9 | 721 | 1233 | 10.8 | 48.19 | 103 |
| Indianapolis | 28 | 52.3 | 361 | 746 | 9.7 | 38.74 | 121 |
| Jacksonville | 14 | 68.4 | 136 | 529 | 8.8 | 54.47 | 116 |
| Kansas City | 14 | 54.5 | 381 | 507 | 10.0 | 37.00 | 99 |
| Little Rock | 13 | 61.0 | 91 | 132 | 8.2 | 48.52 | 100 |
| Louisville | 30 | 55.6 | 291 | 593 | 8.3 | 43.11 | 123 |
| Memphis | 10 | 61.6 | 337 | 624 | 9.2 | 49.10 | 105 |
| Miami | 10 | 75.5 | 207 | 335 | 9.0 | 59.80 | 128 |
| Milwaukee | 16 | 45.7 | 569 | 717 | 11.8 | 29.07 | 123 |
| Minneapolis | 29 | 43.5 | 699 | 744 | 10.6 | 25.94 | 137 |
| Nashville | 18 | 59.4 | 275 | 448 | 7.9 | 46.00 | 119 |
| New Orleans | 9 | 68.3 | 204 | 361 | 8.4 | 56.77 | 113 |
| Norfolk | 31 | 59.3 | 96 | 308 | 10.6 | 44.68 | 116 |
| Omaha | 14 | 51.5 | 181 | 347 | 10.9 | 30.18 | 98 |
| Philadelphia | 69 | 54.6 | 1692 | 1950 | 9.6 | 39.93 | 115 |
| Phoenix | 10 | 70.3 | 213 | 582 | 6.0 | 7.05 | 36 |
| Pittsburgh | 61 | 50.4 | 347 | 520 | 9.4 | 36.22 | 147 |
| Providence | 94 | 50.0 | 343 | 179 | 10.6 | 42.75 | 125 |
| Richmond | 26 | 57.8 | 197 | 299 | 7.6 | 42.59 | 115 |
| Salt Lake City | 28 | 51.0 | 137 | 176 | 8.7 | 15.17 | 89 |
| San Francisco | 12 | 56.7 | 453 | 716 | 8.7 | 20.66 | 67 |
| Seattle | 29 | 51.1 | 379 | 531 | 9.4 | 38.79 | 164 |
| St. Louis | 56 | 55.9 | 775 | 622 | 9.5 | 35.89 | 105 |
| Washington | 29 | 57.3 | 434 | 757 | 9.3 | 38.89 | 111 |
| Wichita | 8 | 56.6 | 125 | 277 | 12.7 | 30.58 | 82 |
| Wilmington | 36 | 54.0 | 80 | 80 | 9.0 | 40.25 | 114 |

*Source*: From Sokal, R. R., Rohlf, F. J., *Biometry*, W. H. Freeman, San Francisco, USA, 1981. With permission.

The final data set to be considered in this chapter is taken from Hastie and Tibshirani (1990). The data are shown in Table 10.3 and involve observations on 81 children undergoing corrective surgery of the spine. There are a number of risk factors for kyphosis, or outward curvature of the spine in excess of 40 degrees from the vertical following surgery; these are age in months (`Age`), the starting vertebral level of the surgery (`Start`) and the number of vertebrae involved (`Number`). Here we would like to model the data to determine which risk factors are of most importance for the occurrence of kyphosis.

**Table 10.3**: `kyphosis` data (package **rpart**). Children who have had corrective spinal surgery.

| Kyphosis | Age | Number | Start | Kyphosis | Age | Number | Start |
|---|---|---|---|---|---|---|---|
| absent | 71 | 3 | 5 | absent | 35 | 3 | 13 |
| absent | 158 | 3 | 14 | absent | 143 | 9 | 3 |
| present | 128 | 4 | 5 | absent | 61 | 4 | 1 |
| absent | 2 | 5 | 1 | absent | 97 | 3 | 16 |
| absent | 1 | 4 | 15 | present | 139 | 3 | 10 |
| absent | 1 | 2 | 16 | absent | 136 | 4 | 15 |
| absent | 61 | 2 | 17 | absent | 131 | 5 | 13 |
| absent | 37 | 3 | 16 | present | 121 | 3 | 3 |
| absent | 113 | 2 | 16 | absent | 177 | 2 | 14 |
| present | 59 | 6 | 12 | absent | 68 | 5 | 10 |
| present | 82 | 5 | 14 | absent | 9 | 2 | 17 |
| absent | 148 | 3 | 16 | present | 139 | 10 | 6 |
| absent | 18 | 5 | 2 | absent | 2 | 2 | 17 |
| absent | 1 | 4 | 12 | absent | 140 | 4 | 15 |
| absent | 168 | 3 | 18 | absent | 72 | 5 | 15 |
| absent | 1 | 3 | 16 | absent | 2 | 3 | 13 |
| absent | 78 | 6 | 15 | present | 120 | 5 | 8 |
| absent | 175 | 5 | 13 | absent | 51 | 7 | 9 |
| absent | 80 | 5 | 16 | absent | 102 | 3 | 13 |
| absent | 27 | 4 | 9 | present | 130 | 4 | 1 |
| absent | 22 | 2 | 16 | present | 114 | 7 | 8 |
| present | 105 | 6 | 5 | absent | 81 | 4 | 1 |
| present | 96 | 3 | 12 | absent | 118 | 3 | 16 |
| absent | 131 | 2 | 3 | absent | 118 | 4 | 16 |
| present | 15 | 7 | 2 | absent | 17 | 4 | 10 |
| absent | 9 | 5 | 13 | absent | 195 | 2 | 17 |
| absent | 8 | 3 | 6 | absent | 159 | 4 | 13 |
| absent | 100 | 3 | 14 | absent | 18 | 4 | 11 |
| absent | 4 | 3 | 16 | absent | 15 | 5 | 16 |
| absent | 151 | 2 | 16 | absent | 158 | 5 | 14 |
| absent | 31 | 3 | 16 | absent | 127 | 4 | 12 |
| absent | 125 | 2 | 11 | absent | 87 | 4 | 16 |

**Table 10.3**: `kyphosis` data (continued).

| Kyphosis | Age | Number | Start | Kyphosis | Age | Number | Start |
|---|---|---|---|---|---|---|---|
| absent | 130 | 5 | 13 | absent | 206 | 4 | 10 |
| absent | 112 | 3 | 16 | absent | 11 | 3 | 15 |
| absent | 140 | 5 | 11 | absent | 178 | 4 | 15 |
| absent | 93 | 3 | 16 | present | 157 | 3 | 13 |
| absent | 1 | 3 | 9 | absent | 26 | 7 | 13 |
| present | 52 | 5 | 6 | absent | 120 | 2 | 13 |
| absent | 20 | 6 | 9 | present | 42 | 7 | 6 |
| present | 91 | 5 | 12 | absent | 36 | 4 | 13 |
| present | 73 | 5 | 1 | | | | |

## 10.2 Scatterplot Smoothers and Generalised Additive Models

Each of the three data sets described in the Introduction appear to be perfect candidates to be analysed by one of the methods described in earlier chapters. Simple linear regression could, for example, be applied to the 1500m times and multiple linear regression to the pollution data; the kyphosis data could be analysed using logistic regression. But instead of assuming we know the linear functional form for a regression model we might consider an alternative approach in which the appropriate functional form is estimated from the data. How is this achieved? The secret is to replace the global estimates from the regression models considered in earlier chapters with local estimates, in which the statistical dependency between two variables is described, not with a single parameter such as a regression coefficient, but with a series of local estimates. For example, a regression might be estimated between the two variables for some restricted range of values for each variable and the process repeated across the range of each variable. The series of local estimates is then aggregated by drawing a line to summarise the relationship between the two variables. In this way no particular functional form is imposed on the relationship. Such an approach is particularly useful when

- the relationship between the variables is expected to be of a complex form, not easily fitted by standard linear or nonlinear models;
- there is no a priori reason for using a particular model;
- we would like the data themselves to suggest the appropriate functional form.

The starting point for a local estimation approach to fitting relationships between variables is *scatterplot smoothers*, which are described in the next subsection.

### *10.2.1 Scatterplot Smoothers*

The scatterplot is an excellent first exploratory graph to study the dependence of two variables and all readers will be familiar with plotting the outcome of a simple linear regression fit onto the graph to help in a better understanding of the pattern of dependence. But many readers will probably be less familiar with some non-parametric alternatives to linear regression fits that may be more useful than the latter in many situations. These alternatives are labelled non-parametric since unlike parametric techniques such as linear regression they do not summarise the relationship between two variables with a parameter such as a regression or correlation coefficient. Instead non-parametric 'smoothers' summarise the relationship between two variables with a line drawing. The simplest of this collection of non-parametric smoothers is a *locally weighted regression* or *lowess* fit, first suggested by Cleveland (1979). In essence this approach assumes that the independent variable $x_i$ and a response $y_i$ are related by

$$y_i = g(x_i) + \varepsilon_i, \quad i = 1, \dots, n$$

where $g$ is a locally defined $p$-degree polynomial function in the predictor variable, $x_i$, and $\varepsilon_i$ are random variables with mean zero and constant scale. Values $\hat{y}_i = g(x_i)$ are used to estimate the $y_i$ at each $x_i$ and are found by fitting the polynomials using weighted least squares with large weights for points near to $x_i$ and small otherwise. Two parameters control the shape of a lowess curve; the first is a smoothing parameter, $\alpha$, (often know as the span, the width of the local neighbourhood) with larger values leading to smoother curves – typical values are 0.25 to 1. In essence the span decides the amount of the tradeoff between reduction in bias and increase in variance. If the span is too large, the non-parametric regression estimate will be biased, but if the span is too small, the estimate will be overfitted with inflated variance. Keele (2008) gives an extended discussion of the influence of the choice of span on the non-parametric regression. The second parameter, $\lambda$ , is the degree of the polynomials that are fitted by the method; $\lambda$ can be 0, 1, or 2. In any specific application, the change of the two parameters must be based on a combination of judgement and of trial and error. Residual plots may be helpful in judging a particular combination of values.

An alternative smoother that can often be usefully applied to bivariate data is some form of *spline function*. (A spline is a term for a flexible strip of metal or rubber used by a draftsman to draw curves.) Spline functions are polynomials within intervals of the $x$-variable that are smoothly connected across different values of $x$. Figure 10.1 for example shows a linear spline function, i.e., a piecewise linear function, of the form

$$f(x) = \beta_0 + \beta_1 x + \beta_2 (x - a)_+ + \beta_3 (x - b)_+ + \beta_4 (x - c)_+$$

where $(u)_+ = u$ for $u > 0$ and zero otherwise. The interval endpoints, $a$, $b$, and $c$, are called knots. The number of knots can vary according to the amount of data available for fitting the function.

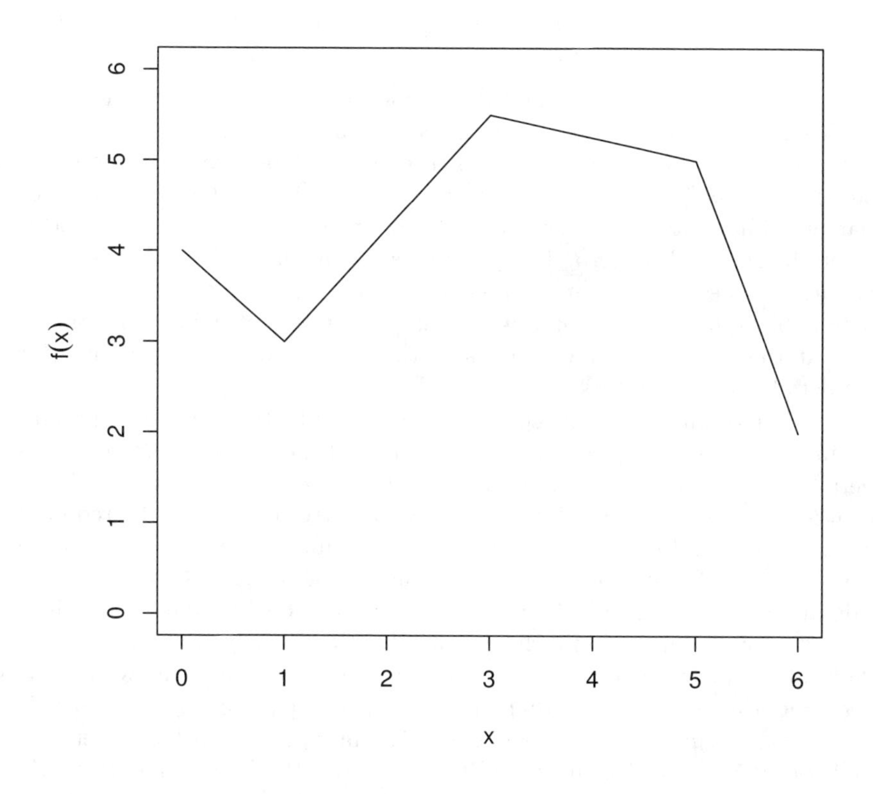


**Figure 10.1** A linear spline function with knots at $a = 1$, $b = 3$ and $c = 5$.

The linear spline is simple and can approximate some relationships, but it is not smooth and so will not fit highly curved functions well. The problem is overcome by using smoothly connected piecewise polynomials – in particular, cubics, which have been found to have nice properties with good ability to fit a variety of complex relationships. The result is a *cubic spline*. Again we wish to fit a smooth curve, $g(x)$, that summarises the dependence of $y$ on $x$. A natural first attempt might be to try to determine $g$ by least squares as the curve that minimises

$$\sum_{i=1}^{n}(y_i - g(x_i))^2. \tag{10.1}$$

But this would simply result in very wiggly curve interpolating the observations. Instead of (10.1) the criterion used to determine $g$ is

$$\sum_{i=1}^{n}(y_i - g(x_i))^2 + \lambda \int g''(x)^2 \, dx \tag{10.2}$$

where $g''(x)$ represents the second derivation of $g(x)$ with respect to $x$. Although written formally this criterion looks a little formidable, it is really nothing more than an effort to govern the trade-off between the goodness-of-fit of the data (as measured by $\sum(y_i - g(x_i))^2$ ) and the 'wiggliness' or departure of linearity of $g$ measured by $\int g''(x)^2 \, dx$; for a linear function, this part of (10.2) would be zero. The parameter $\lambda$ governs the smoothness of $g$, with larger values resulting in a smoother curve.

The cubic spline which minimises (10.2) is a series of cubic polynomials joined at the unique observed values of the explanatory variables, $x_i$, (for more details, see Keele, 2008).

The 'effective number of parameters' (analogous to the number of parameters in a parametric fit) or degrees of freedom of a cubic spline smoother is generally used to specify its smoothness rather than $\lambda$ directly. A numerical search is then used to determine the value of $\lambda$ corresponding to the required degrees of freedom. Roughly, the complexity of a cubic spline is about the same as a polynomial of degree one less than the degrees of freedom (see Keele, 2008, for details). But the cubic spline smoother 'spreads out' its parameters in a more even way and hence is much more flexible than is polynomial regression.

The spline smoother does have a number of technical advantages over the lowess smoother such as providing the best mean square error and avoiding overfitting that can cause smoothers to display unimportant variation between $x$ and $y$ that is of no real interest. But in practise the lowess smoother and the cubic spline smoother will give very similar results on many examples.

### *10.2.2 Generalised Additive Models*

The scatterplot smoothers described above are the basis of a more general, semi-parametric approach to modelling situations where there is more than a single explanatory variable, such as the air pollution data in Table 10.2 and the kyphosis data in Table 10.3. These models are usually called generalised additive models (GAMs) and allow the investigator to model the relationship between the response variable and some of the explanatory variables using the non-parametric lowess or cubic splines smoothers, with this relationship for other explanatory variables being estimated in the usual parametric fashion. So returning for a moment to the multiple linear regression model described in Chapter 6 in which there is a dependent variable, $y$, and a set of explanatory variables, $x_1, \dots, x_q$, and the model assumed is

$$y = \beta_0 + \sum_{j=1}^{q} \beta_j x_j + \varepsilon.$$

Additive models replace the linear function, $\beta_j x_j$, by a smooth non-parametric function, $g$, to give the model

$$y = \beta_0 + \sum_{j=1}^{q} g_j(x_j) + \varepsilon. \tag{10.3}$$

where $g_j$ can be one of the scatterplot smoothers described in the previous sub-section, or, if the investigator chooses, it can also be a linear function for particular explanatory variables.

A generalised additive model arises from (10.3) in the same way as a generalised linear model arises from a multiple regression model (see Chapter 7), namely that some function of the expectation of the response variable is now modelled by a sum of non-parametric and parametric functions. So, for example, the logistic additive model with binary response variable $y$ is

$$\text{logit}(\pi) = \beta_0 + \sum_{j=1}^{q} g_j(x_j)$$

where $\pi$ is the probability that the response variable takes the value one.

Fitting a generalised additive model involves either iteratively weighted least squares, an optimisation algorithm similar to the algorithm used to fit generalised linear models, or what is known as a *backfitting algorithm*. The smooth functions $g_j$ are fitted one at a time by taking the residuals

$$y - \sum_{k \neq j} g_k(x_k)$$

and fitting them against $x_j$ using one of the scatterplot smoothers described previously. The process is repeated until it converges. Linear terms in the model are fitted by least squares. The **mgcv** package fits generalised additive models using the iteratively weighted least squares algorithm, which in this case has the advantage that inference procedures, such as confidence intervals, can be derived more easily. Full details are given in Hastie and Tibshirani (1990), Wood (2006), and Keele (2008).

Various tests are available to assess the non-linear contributions of the fitted smoothers, and generalised additive models can be compared with, say linear models fitted to the same data, by means of an $F$-test on the residual sum of squares of the competing models. In this process the fitted smooth curve is assigned an estimated equivalent number of degrees of freedom. However, such a procedure has to be used with care. For full details, again, see Wood (2006) and Keele (2008).

Two alternative approaches to the variable selection and model choice problem are helpful. As always, a graphical inspection of the model properties, ideally guided by subject-matter knowledge, helps to identify the most important aspects of the fitted regression function. A more formal approach is to fit the model using algorithms that, implicitly or explicitly, have nice variable selection properties, one of which is mentioned in the following section.

### *10.2.3 Variable Selection and Model Choice*

Quantifying the influence of covariates on the response variable in generalised additive models does not merely relate to the problem of estimating regression coefficients but more generally calls for careful implementation of variable selection (determination of the relevant subset of covariates to enter the model) and model choice (specifying the particular form of the influence of a variable). The latter task requires choosing between linear and nonlinear modelling of covariate effects. While variable selection and model choice issues are already complicated in linear models (see Chapter 6) and generalised linear models (see Chapter 7) and still receive considerable attention in the statistical literature, they become even more challenging in generalised additive models. Here, variable selection and model choice needs to provide and answer on the complicated question: Should a continuous covariate be included into the model at all and, if so, as a linear effect or as a flexible, smooth effect? Methods to deal with this problem are currently actively researched. Two general approaches can be distinguished: One can fit models using a target function incorporating a penalty term which will increase for increasingly complex models (similar to 10.2) or one can iteratively fit simple, univariate models which sum to a more complex generalised additive model. The latter approach is called *boosting* and requires a careful determination of the stop criterion for the iterative model fitting algorithms. The technical details are far too complex to be sketched here, and we refer the interested reader to the review paper by Bühlmann and Hothorn (2007).

## 10.3 Analysis Using R

### *10.3.1 Olympic 1500m Times*

To begin we will construct a scatterplot of winning time against year the games were held. The R code and the resulting plot are shown in Figure 10.2. There is very clear downward trend in the times over the years, and, in addition there is a very clear outlier namely the winning time for 1896. We shall remove this time from the data set and now concentrate on the remaining times. First we will fit a simple linear regression to the data and plot the fit onto the scatterplot. The code and the resulting plot are shown in Figure 10.3. Clearly the linear regression model captures in general terms the downward trend in the times. Now we can add the fits given by the lowess smoother and by a cubic spline smoother; the resulting graph and the extra R code needed are shown in Figure 10.4.

Both non-parametric fits suggest some distinct departure from linearity, and clearly point to a quadratic model being more sensible than a linear model here. And fitting a parametric model that includes both a linear and a quadratic effect for year gives a prediction curve very similar to the non-parametric curves; see Figure 10.5.

Here use of the non-parametric smoothers has effectively diagnosed our

```
R> plot(time ~ year, data = men1500m)
```

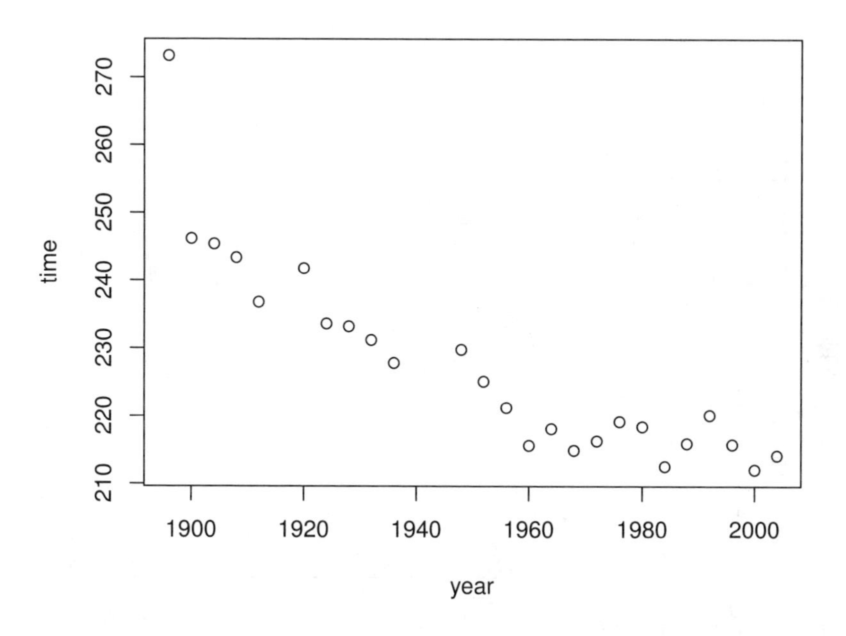


**Figure 10.2** Scatterplot of year and winning time.

linear model and pointed the way to using a more suitable parametric model; this is often how such non-parametric models can be used most effectively. For these data, of course, it is clear that the simple linear model cannot be suitable if the investigator is interested in predicting future times since even the most basic knowledge of human physiology will tell us that times cannot continue to go down. There must be some lower limit to the time man can run 1500m. But in other situations use of the non-parametric smoothers may point to a parametric model that could not have been identified *a priori*.

It is of some interest to look at the predictions of winning times in future Olympics from both the linear and quadratic models. For example, for 2008 and 2012 the predicted times and their 95% confidence intervals can be found using the following code

```
R> predict(men1500m_lm,
+          newdata = data.frame(year = c(2008, 2012)),
+          interval = "confidence")

       fit      lwr      upr
1 208.1293 204.8961 211.3624
2 206.8451 203.4325 210.2577
```

```
R> men1500m1900 <- subset(men1500m, year >= 1900)
R> men1500m_lm <- lm(time ~ year, data = men1500m1900)
R> plot(time ~ year, data = men1500m1900)
R> abline(men1500m_lm)
```

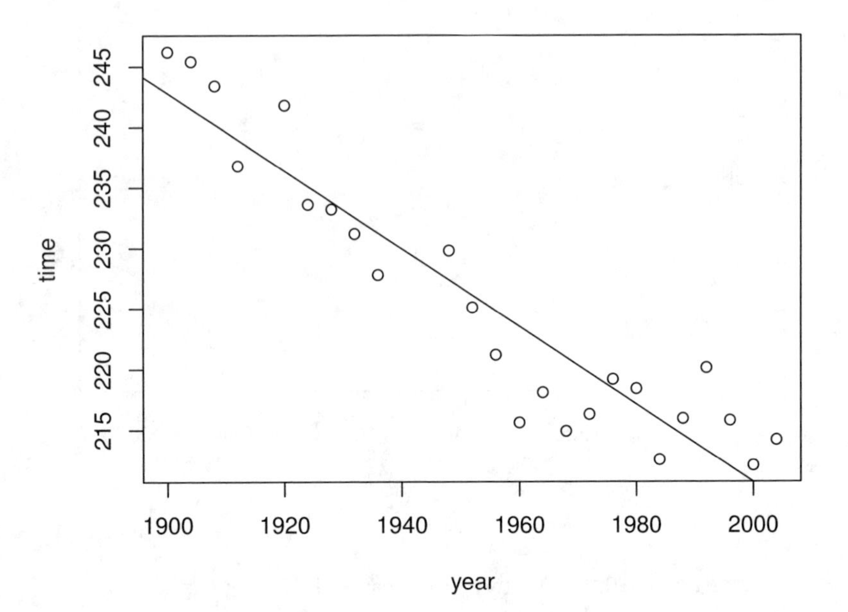


**Figure 10.3** Scatterplot of year and winning time with fitted values from a simple linear model.

```
R> predict(men1500m_lm2,
+          newdata = data.frame(year = c(2008, 2012)),
+          interval = "confidence")

       fit      lwr      upr
1 214.2709 210.3930 218.1488
2 214.3314 209.8441 218.8187
```

For predictions far into the future both the quadratic and the linear model fail; we leave readers to get some more predictions to see what happens. We can compare the first prediction with the time actually recorded by the winner of the men's 1500m in Beijing 2008, Rashid Ramzi from Brunei, who won the event in 212.94 seconds. The confidence interval obtained from the simple linear model does not include this value but the confidence interval for the prediction derived from the quadratic model does.

```
R> x <- men1500m1900$year
R> y <- men1500m1900$time
R> men1500m_lowess <- lowess(x, y)
R> plot(time ~ year, data = men1500m1900)
R> lines(men1500m_lowess, lty = 2)
R> men1500m_cubic <- gam(y ~ s(x, bs = "cr"))
R> lines(x, predict(men1500m_cubic), lty = 3)
```

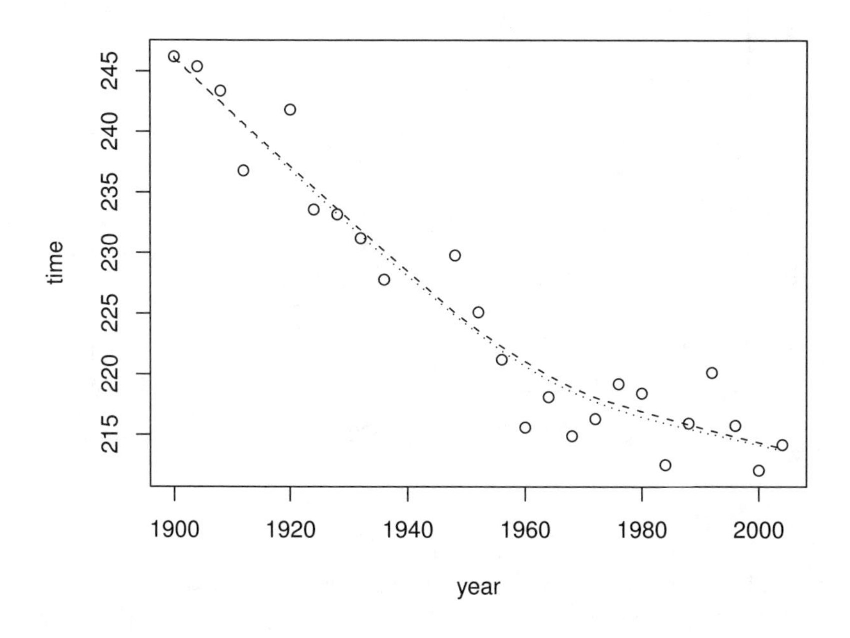


**Figure 10.4** Scatterplot of year and winning time with fitted values from a smooth non-parametric model.

### *10.3.2 Air Pollution in US Cities*

Unfortunately, we cannot fit an additive model for describing the $SO_2$ concentration based on all six covariates because this leads to more parameters than cities, i.e., more parameters than observations when using the default parameterisation of **mgcv**. Thus, before we can apply the `gam` function from package **mgcv**, we have to decide which covariates should enter the model and which subset of these covariates should be allowed to deviate from a linear regression relationship.

As briefly discussed in Section 10.2.3, we can fit an additive model using the iterative boosting algorithm as described by Bühlmann and Hothorn (2007).

```
R> men1500m_lm2 <- lm(time ~ year + I(year^2),
+                     data = men1500m1900)
R> plot(time ~ year, data = men1500m1900)
R> lines(men1500m1900$year, predict(men1500m_lm2))
```

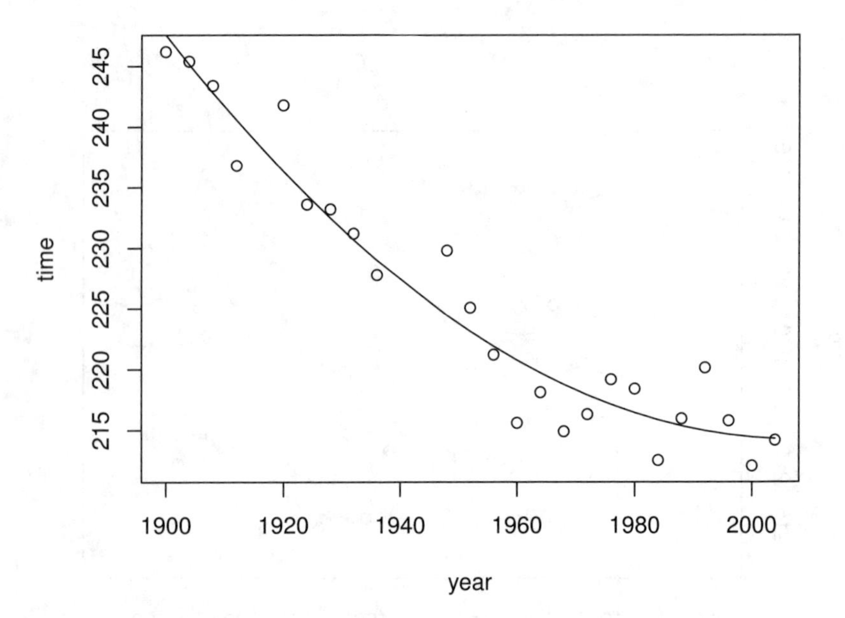


**Figure 10.5** Scatterplot of year and winning time with fitted values from a quadratic model.

The complexity of the model is determined by an AIC criterion, which can also be used to determine an appropriate number of boosting iterations to choose. The methodology is available from package **mboost** (Hothorn et al., 2009b). We start with a small number of boosting iterations (100 by default) and compute the AIC of the corresponding 100 models:

```
R> library("mboost")
R> USair_boost <- gamboost(SO2 ~ ., data = USairpollution)
R> USair_aic <- AIC(USair_boost)
R> USair_aic
```

```
[1] 6.809066
Optimal number of boosting iterations: 40
Degrees of freedom (for mstop = 40): 9.048771
```

The AIC suggests that the boosting algorithm should be stopped after 40

```
R> USair_gam <- USair_boost[mstop(USair_aic)]
R> layout(matrix(1:6, ncol = 3))
R> plot(USair_gam, ask = FALSE)
```

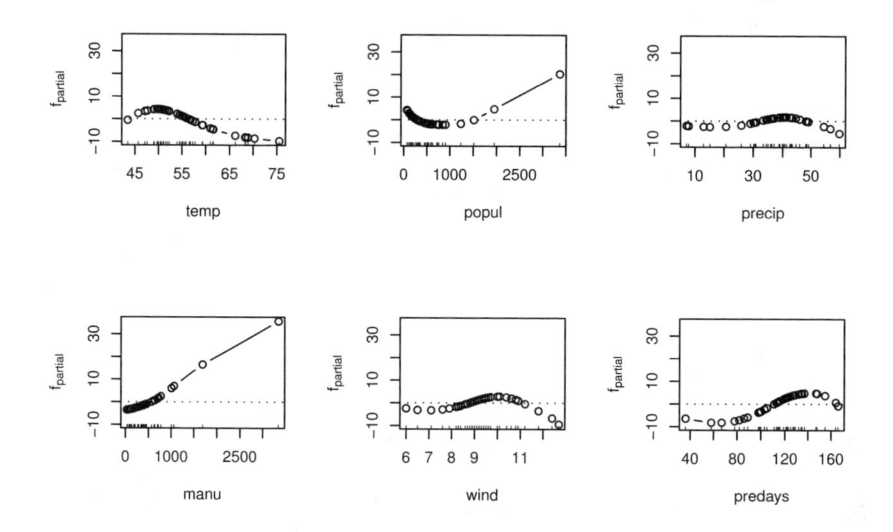


**Figure 10.6** Partial contributions of six exploratory covariates to the predicted $SO_2$ concentration.

iterations. The partial contributions of each covariate to the predicted $SO_2$ concentration are given in Figure 10.6. The plot indicates that all six covariates enter the model and the selection of a subset of covariates for modelling isn't appropriate in this case. However, the number of manufacturing enterprises seems to add linearly to the $SO_2$ concentration, which simplifies the model. Moreover, the average annual precipitation contribution seems to deviate from zero only for some extreme observations and one might refrain from using the covariate at all.

As always, an inspection of the model fit via a residual plot is worth the effort. Here, we plot the fitted values against the residuals and label the points with the name of the corresponding city. Figure 10.7 shows at least two extreme observations. Chicago has a very large observed and fitted $SO_2$ concentration, which is due to the huge number of inhabitants and manufacturing plants (see Figure 10.6 also). One smaller city, Providence, is associated with a rather large positive residual indicating that the actual $SO_2$ concentration is underestimated by the model. In fact, this small town has a rather high $SO_2$ concentration which is hardly explained by our model. Overall, the model doesn't fit the data very well, so we should avoid overinterpreting the model structure too much. In addition, since each of the six covariates contributes

```
R> SO2hat <- predict(USair_gam)
R> SO2 <- USairpollution$SO2
R> plot(SO2hat, SO2 - SO2hat, type = "n", xlim = c(0, 110))
R> text(SO2hat, SO2 - SO2hat, labels = rownames(USairpollution),
+       adj = 0)
R> abline(h = 0, lty = 2, col = "grey")
```

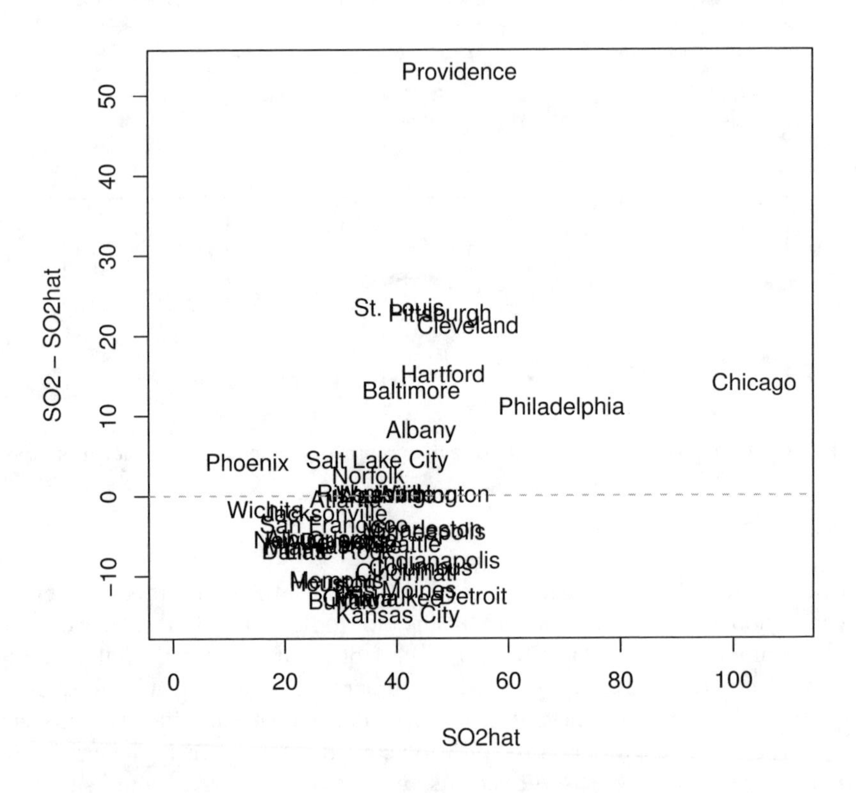


**Figure 10.7** Residual plot of $SO_2$ concentration.

to the model, we aren't able to select a smaller subset of the covariates for modelling and thus fitting a model using gam is still complicated (and will not add much knowledge anyway).

### *10.3.3 Risk Factors for Kyphosis*

Before modelling the relationship between kyphosis and the three exploratory variables age, starting vertebral level of the surgery and number of vertebrae

```
R> layout(matrix(1:3, nrow = 1))
R> spineplot(Kyphosis ~ Age, data = kyphosis,
+            ylevels = c("present", "absent"))
R> spineplot(Kyphosis ~ Number, data = kyphosis,
+            ylevels = c("present", "absent"))
R> spineplot(Kyphosis ~ Start, data = kyphosis,
+           ylevels = c("present", "absent"))
```

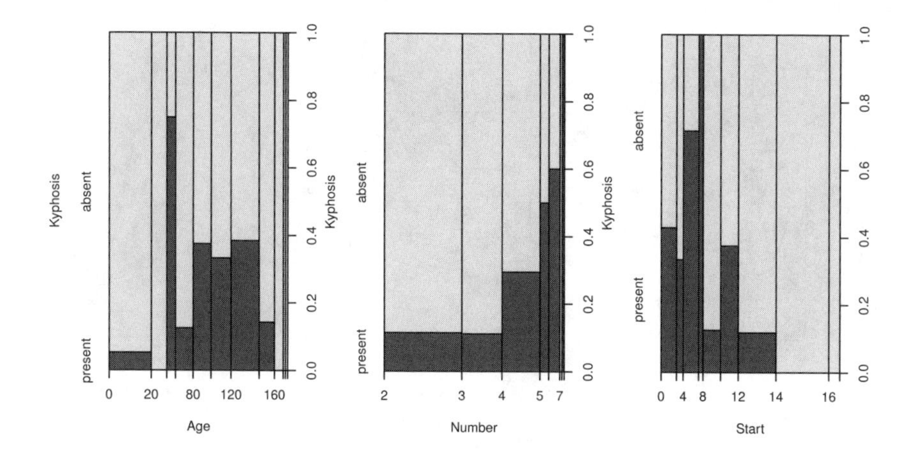


**Figure 10.8** Spinograms of the three exploratory variables and response variable kyphosis.

involved, we investigate the partial associations by so-called *spinograms*, as introduced in Chapter 2. The numeric exploratory covariates are discretised and their empirical relative frequencies are plotted against the conditional frequency of kyphosis in the corresponding group. Figure 10.8 shows that kyphosis is absent in very young or very old children, children with a small starting vertebral level and high number of vertebrae involved.

The logistic additive model needed to describe the conditional probability of kyphosis given the exploratory variables can be fitted using function `gam`. Here, the dimension of the basis ($k$) has to be modified for `Number` and `Start` since these variables are heavily tied. As for generalised linear models, the `family` argument determines the type of model to be fitted, a logistic model in our case:

```
R> kyphosis_gam <- gam(Kyphosis ~ s(Age, bs = "cr") +
+      s(Number, bs = "cr", k = 3) + s(Start, bs = "cr", k = 3),
+      family = binomial, data = kyphosis)
R> kyphosis_gam

Family: binomial
Link function: logit
```

```
R> trans <- function(x)
+      binomial()$linkinv(x)
R> layout(matrix(1:3, nrow = 1))
R> plot(kyphosis_gam, select = 1, shade = TRUE, trans = trans)
R> plot(kyphosis_gam, select = 2, shade = TRUE, trans = trans)
R> plot(kyphosis_gam, select = 3, shade = TRUE, trans = trans)
```

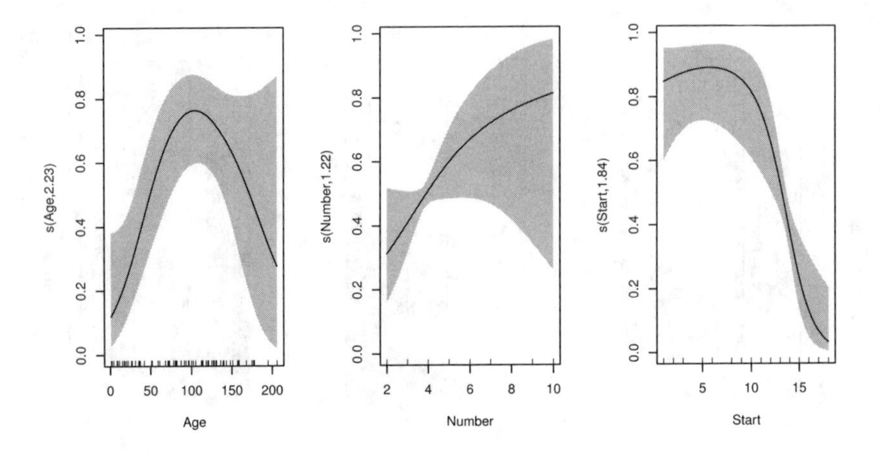


**Figure 10.9** Partial contributions of three exploratory variables with confidence bands.

```
Formula:
Kyphosis ~ s(Age, bs = "cr") + s(Number, bs = "cr", k = 3) +
    s(Start, bs = "cr", k = 3)

Estimated degrees of freedom:
2.2267 1.2190 1.8420  total = 6.287681

UBRE score: -0.2335850
```

The partial contributions of each covariate to the conditional probability of kyphosis with confidence bands are shown in Figure 10.9. In essence, the same conclusions as drawn from Figure 10.8 can be stated here. The risk of kyphosis being present increases with higher starting vertebral level and lower number of vertebrae involved.

## Summary

Additive models offer flexible modelling tools for regression problems. They stand between generalised linear models, where the regression relationship is assumed to be linear, and more complex models like random forests (see Chapter 9) where the regression relationship remains unspecified. Smooth functions describing the influence of covariates on the response can be easily interpreted. Variable selection is a technically difficult problem in this class of models; boosting methods are one possibility to deal with this problem.

## Exercises

Ex. 10.1 Consider the body fat data introduced in Chapter 9, Table 9.1. First fit a generalised additive model assuming normal errors using function `gam`. Are all potential covariates informative? Check the results against a generalised additive model that underwent AIC-based variable selection (fitted using function `gamboost`).

Ex. 10.2 Try to fit a logistic additive model to the glaucoma data discussed in Chapter 9. Which covariates should enter the model and how is their influence on the probability of suffering from glaucoma?

CHAPTER 11

# Survival Analysis: Glioma Treatment and Breast Cancer Survival

## 11.1 Introduction

Grana et al. (2002) report results of a non-randomised clinical trial investigating a novel radioimmunotherapy in malignant glioma patients. The overall survival, i.e., the time from the beginning of the therapy to the disease-caused death of the patient, is compared for two groups of patients. A control group underwent the standard therapy and another group of patients was treated with radioimmunotherapy in addition. The data, extracted from Tables 1 and 2 in Grana et al. (2002), are given in Table 11.1. The main interest is to investigate whether the patients treated with the novel radioimmunotherapy have, on average, longer survival times than patients in the control group.

**Table 11.1**: `glioma` data. Patients suffering from two types of glioma treated with the standard therapy or a novel radioimmunotherapy (`RIT`).

| `age` | `sex` | `histology` | `group` | `event` | `time` |
|---|---|---|---|---|---|
| 41 | Female | Grade3 | RIT | TRUE | 53 |
| 45 | Female | Grade3 | RIT | FALSE | 28 |
| 48 | Male | Grade3 | RIT | FALSE | 69 |
| 54 | Male | Grade3 | RIT | FALSE | 58 |
| 40 | Female | Grade3 | RIT | FALSE | 54 |
| 31 | Male | Grade3 | RIT | TRUE | 25 |
| 53 | Male | Grade3 | RIT | FALSE | 51 |
| 49 | Male | Grade3 | RIT | FALSE | 61 |
| 36 | Male | Grade3 | RIT | FALSE | 57 |
| 52 | Male | Grade3 | RIT | FALSE | 57 |
| 57 | Male | Grade3 | RIT | FALSE | 50 |
| 55 | Female | GBM | RIT | FALSE | 43 |
| 70 | Male | GBM | RIT | TRUE | 20 |
| 39 | Female | GBM | RIT | TRUE | 14 |
| 40 | Female | GBM | RIT | FALSE | 36 |
| 47 | Female | GBM | RIT | FALSE | 59 |
| 58 | Male | GBM | RIT | TRUE | 31 |

**Table 11.1:** `glioma` data (continued).

| age | sex | histology | group | event | time |
|---|---|---|---|---|---|
| 40 | Female | GBM | RIT | TRUE | 14 |
| 36 | Male | GBM | RIT | TRUE | 36 |
| 27 | Male | Grade3 | Control | TRUE | 34 |
| 32 | Male | Grade3 | Control | TRUE | 32 |
| 53 | Female | Grade3 | Control | TRUE | 9 |
| 46 | Male | Grade3 | Control | TRUE | 19 |
| 33 | Female | Grade3 | Control | FALSE | 50 |
| 19 | Female | Grade3 | Control | FALSE | 48 |
| 32 | Female | GBM | Control | TRUE | 8 |
| 70 | Male | GBM | Control | TRUE | 8 |
| 72 | Male | GBM | Control | TRUE | 11 |
| 46 | Male | GBM | Control | TRUE | 12 |
| 44 | Male | GBM | Control | TRUE | 15 |
| 83 | Female | GBM | Control | TRUE | 5 |
| 57 | Female | GBM | Control | TRUE | 8 |
| 71 | Female | GBM | Control | TRUE | 8 |
| 61 | Male | GBM | Control | TRUE | 6 |
| 65 | Male | GBM | Control | TRUE | 14 |
| 50 | Male | GBM | Control | TRUE | 13 |
| 42 | Female | GBM | Control | TRUE | 25 |

*Source*: From Grana, C., et. al., *Br. J. Cancer*, 86, 207–212, 2002. With permission.

The effects of hormonal treatment with Tamoxifen in women suffering from node-positive breast cancer were investigated in a randomised clinical trial as reported by Schumacher et al. (1994). Data from randomised patients from this trial and additional non-randomised patients (from the German Breast Cancer Study Group 2, GBSG2) are analysed by Sauerbrei and Royston (1999). Complete data of seven prognostic factors of 686 women are used in Sauerbrei and Royston (1999) for prognostic modelling. Observed hypothetical prognostic factors are age, menopausal status, tumour size, tumour grade, number of positive lymph nodes, progesterone receptor, estrogen receptor and the information of whether or not a hormonal therapy was applied. We are interested in an assessment of the impact of the covariates on the survival time of the patients. A subset of the patient data are shown in Table 11.2.

## 11.2 Survival Analysis

In many medical studies, the main outcome variable is the time to the occurrence of a particular event. In a randomised controlled trial of cancer, for example, surgery, radiation and chemotherapy might be compared with re-

**Table 11.2**: GBSG2 data (package **ipred**). Randomised clinical trial data from patients suffering from node-positive breast cancer. Only the data of the first 20 patients are shown here.

| horTh | age | menostat | tsize | tgrade | pnodes | progrec | estrec | time | cens |
|---|---|---|---|---|---|---|---|---|---|
| no | 70 | Post | 21 | II | 3 | 48 | 66 | 1814 | 1 |
| yes | 56 | Post | 12 | II | 7 | 61 | 77 | 2018 | 1 |
| yes | 58 | Post | 35 | II | 9 | 52 | 271 | 712 | 1 |
| yes | 59 | Post | 17 | II | 4 | 60 | 29 | 1807 | 1 |
| no | 73 | Post | 35 | II | 1 | 26 | 65 | 772 | 1 |
| no | 32 | Pre | 57 | III | 24 | 0 | 13 | 448 | 1 |
| yes | 59 | Post | 8 | II | 2 | 181 | 0 | 2172 | 0 |
| no | 65 | Post | 16 | II | 1 | 192 | 25 | 2161 | 0 |
| no | 80 | Post | 39 | II | 30 | 0 | 59 | 471 | 1 |
| no | 66 | Post | 18 | II | 7 | 0 | 3 | 2014 | 0 |
| yes | 68 | Post | 40 | II | 9 | 16 | 20 | 577 | 1 |
| yes | 71 | Post | 21 | II | 9 | 0 | 0 | 184 | 1 |
| yes | 59 | Post | 58 | II | 1 | 154 | 101 | 1840 | 0 |
| no | 50 | Post | 27 | III | 1 | 16 | 12 | 1842 | 0 |
| yes | 70 | Post | 22 | II | 3 | 113 | 139 | 1821 | 0 |
| no | 54 | Post | 30 | II | 1 | 135 | 6 | 1371 | 1 |
| no | 39 | Pre | 35 | I | 4 | 79 | 28 | 707 | 1 |
| yes | 66 | Post | 23 | II | 1 | 112 | 225 | 1743 | 0 |
| yes | 69 | Post | 25 | I | 1 | 131 | 196 | 1781 | 0 |
| no | 55 | Post | 65 | I | 4 | 312 | 76 | 865 | 1 |
| ⋮ | ⋮ | ⋮ | ⋮ | ⋮ | ⋮ | ⋮ | ⋮ | ⋮ | ⋮ |

*Source*: From Sauerbrei, W. and Royston, P., *J. Roy. Stat. Soc. A*, 162, 71–94, 1999. With permission.

spect to time from randomisation and the start of therapy until death. In this case, the event of interest is the death of a patient, but in other situations, it might be remission from a disease, relief from symptoms or the recurrence of a particular condition. Other censored response variables are the time to credit failure in financial applications or the time a roboter needs to successfully perform a certain task in engineering. Such observations are generally referred to by the generic term *survival data* even when the endpoint or event being considered is not death but something else. Such data generally require special techniques for analysis for two main reasons:

1. Survival data are generally not symmetrically distributed – they will often appear positively skewed, with a few people surviving a very long time compared with the majority; so assuming a normal distribution will not be reasonable.
2. At the completion of the study, some patients may not have reached the endpoint of interest (death, relapse, etc.). Consequently, the exact survival times are not known. All that is known is that the survival times are greater than the amount of time the individual has been in the study. The survival times of these individuals are said to be *censored* (precisely, they are right-censored).

Of central importance in the analysis of survival time data are two functions used to describe their distribution, namely the *survival* (or *survivor*) *function* and the *hazard function.*

### *11.2.1 The Survivor Function*

The survivor function, $S(t)$, is defined as the probability that the survival time, $T$, is greater than or equal to some time $t$, i.e.,

$$S(t) = \mathsf{P}(T \geq t).$$

A plot of an estimate $\hat{S}(t)$ of $S(t)$ against the time $t$ is often a useful way of describing the survival experience of a group of individuals. When there are no censored observations in the sample of survival times, a non-parametric survivor function can be estimated simply as

$$\hat{S}(t) = \frac{\text{number of individuals with survival times} \geq t}{n}$$

where $n$ is the total number of observations. Because this is simply a proportion, confidence intervals can be obtained for each time $t$ by using the variance estimate

$$\hat{S}(t)(1 - \hat{S}(t))/n.$$

The simple method used to estimate the survivor function when there are no censored observations cannot now be used for survival times when censored observations are present. In the presence of censoring, the survivor function is typically estimated using the *Kaplan-Meier* estimator (Kaplan and Meier,

1958). This involves first ordering the survival times from the smallest to the largest such that $t_{(1)} \leq t_{(2)} \leq \cdots \leq t_{(n)}$, where $t_{(j)}$ is the $j$th largest unique survival time. The Kaplan-Meier estimate of the survival function is obtained as

$$\hat{S}(t) = \prod_{j:t_{(j)} \leq t} \left(1 - \frac{d_j}{r_j}\right)$$

where $r_j$ is the number of individuals at risk just before $t_{(j)}$ (including those censored at $t_{(j)}$), and $d_j$ is the number of individuals who experience the event of interest (death, etc.) at time $t_{(j)}$. So, for example, the survivor function at the second death time, $t_{(2)}$, is equal to the estimated probability of not dying at time $t_{(2)}$, conditional on the individual being still at risk at time $t_{(2)}$. The estimated variance of the Kaplan-Meier estimate of the survivor function is found from

$$\mathsf{Var}(\hat{S}(t)) = \left(\hat{S}(t)\right)^2 \sum_{j:t_{(j)} \leq t} \frac{d_j}{r_j(r_j - d_j)}.$$

A formal test of the equality of the survival curves for the two groups can be made using the *log-rank test*. First, the expected number of deaths is computed for each unique death time, or *failure time* in the data set, assuming that the chances of dying, given that subjects are at risk, are the same for both groups. The total number of expected deaths is then computed for each group by adding the expected number of deaths for each failure time. The test then compares the observed number of deaths in each group with the expected number of deaths using a chi-squared test. Full details and formulae are given in Therneau and Grambsch (2000) or Everitt and Rabe-Hesketh (2001), for example.

### *11.2.2 The Hazard Function*

In the analysis of survival data it is often of interest to assess which periods have high or low chances of death (or whatever the event of interest may be), among those still active at the time. A suitable approach to characterise such risks is the hazard function, $h(t)$, defined as the probability that an individual experiences the event in a small time interval, $s$, given that the individual has survived up to the beginning of the interval, when the size of the time interval approaches zero; mathematically this is written as

$$h(t) = \lim_{s \to 0} \frac{\mathsf{P}(t \leq T \leq t + s | T \geq t)}{s}$$

where $T$ is the individual's survival time. The conditioning feature of this definition is very important. For example, the probability of dying at age 100 is very small because most people die before that age; in contrast, the probability of a person dying at age 100 who has reached that age is much greater.

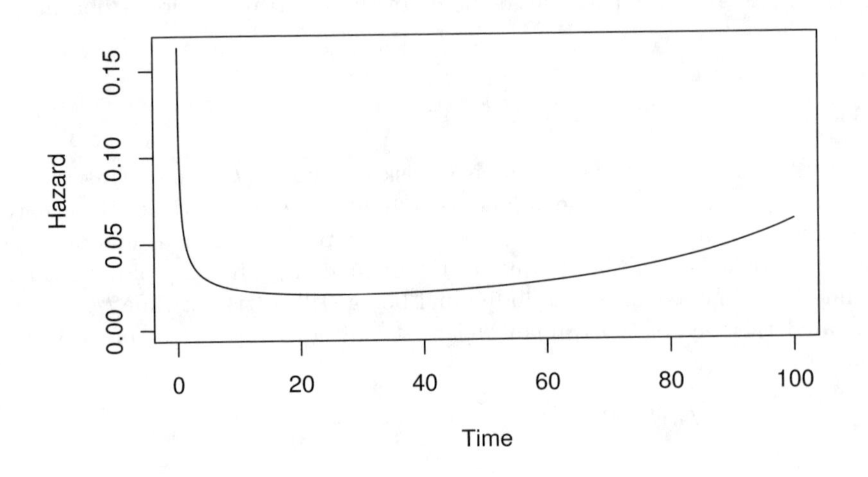


**Figure 11.1** 'Bath tub' shape of a hazard function.

The hazard function and survivor function are related by the formula

$$S(t) = \exp(-H(t))$$

where $H(t)$ is known as the *integrated hazard* or *cumulative hazard*, and is defined as follows:

$$H(t) = \int_0^t h(u)du;$$

details of how this relationship arises are given in Everitt and Pickles (2000).

In practise the hazard function may increase, decrease, remain constant or have a more complex shape. The hazard function for death in human beings, for example, has the 'bath tub' shape shown in Figure 11.1. It is relatively high immediately after birth, declines rapidly in the early years and then remains approximately constant before beginning to rise again during late middle age.

The hazard function can be estimated as the proportion of individuals experiencing the event of interest in an interval per unit time, given that they have survived to the beginning of the interval, that is

$$\hat{h}(t) = \frac{d_j}{n_j(t_{(j+1)} - t_{(j)})}.$$

The sampling variation in the estimate of the hazard function within each interval is usually considerable and so it is rarely plotted directly. Instead the integrated hazard is used. Everitt and Rabe-Hesketh (2001) show that this can be estimated as follows:

$$\hat{H}(t) = \sum_j \frac{d_j}{n_j}.$$

### 11.2.3 Cox's Regression

When the response variable of interest is a possibly censored survival time, we need special regression techniques for modelling the relationship of the response to explanatory variables of interest. A number of procedures are available but the most widely used by some margin is that known as *Cox's proportional hazards model*, or *Cox's regression* for short. Introduced by Sir David Cox in 1972 (see Cox, 1972), the method has become one of the most commonly used in medical statistics and the original paper one of the most heavily cited.

The main vehicle for modelling in this case is the hazard function rather than the survivor function, since it does not involve the cumulative history of events. But modelling the hazard function directly as a linear function of explanatory variables is not appropriate since $h(t)$ is restricted to being positive. A more suitable model might be

$$\log(h(t)) = \beta_0 + \beta_1 x_1 + \cdots + \beta_q x_q. \tag{11.1}$$

But this would only be suitable for a hazard function that is constant over time; this is very restrictive since hazards that increase or decrease with time, or have some more complex form are far more likely to occur in practise. In general it may be difficult to find the appropriate explicit function of time to include in (11.1). The problem is overcome in the proportional hazards model proposed by Cox (1972) by allowing the form of dependence of $h(t)$ on $t$ to remain unspecified, so that

$$\log(h(t)) = \log(h_0(t)) + \beta_1 x_1 + \cdots + \beta_q x_q$$

where $h_0(t)$ is known as the *baseline hazard function*, being the hazard function for individuals with all explanatory variables equal to zero. The model can be rewritten as

$$h(t) = h_0(t) \exp(\beta_1 x_1 + \cdots + \beta_q x_q).$$

Written in this way we see that the model forces the hazard ratio between two individuals to be constant over time since

$$\frac{h(t|\mathbf{x}_1)}{h(t|\mathbf{x}_2)} = \frac{\exp(\beta^\top \mathbf{x}_1)}{\exp(\beta^\top \mathbf{x}_2)}$$

where $\mathbf{x}_1$ and $\mathbf{x}_2$ are vectors of covariate values for two individuals. In other words, if an individual has a risk of death at some initial time point that is twice as high as another individual, then at all later times, the risk of death remains twice as high. Hence the term proportional hazards.

In the Cox model, the baseline hazard describes the common shape of the survival time distribution for all individuals, while the *relative risk function*, $\exp(\beta^\top \mathbf{x})$, gives the level of each individual's hazard. The interpretation of the parameter $\beta_j$ is that $\exp(\beta_j)$ gives the relative risk change associated with an increase of one unit in covariate $x_j$, all other explanatory variables remaining constant.

The parameters in a Cox model can be estimated by maximising what is known as a *partial likelihood.* Details are given in Kalbfleisch and Prentice (1980). The partial likelihood is derived by assuming continuous survival times. In reality, however, survival times are measured in discrete units and there are often ties. There are three common methods for dealing with ties which are described briefly in Everitt and Rabe-Hesketh (2001).

## 11.3 Analysis Using R

### *11.3.1 Glioma Radioimmunotherapy*

The survival times for patients from the control group and the group treated with the novel therapy can be compared graphically by plotting the Kaplan-Meier estimates of the survival times. Here, we plot the Kaplan-Meier estimates stratified for patients suffering from grade III glioma and glioblastoma (GBM, grade IV) separately; the results are given in Figure 11.2. The Kaplan-Meier estimates are computed by the `survfit` function from package **survival** (Therneau and Lumley, 2009) which takes a model *formula* of the form

```
Surv(time, event) ~ group
```

where `time` are the survival times, `event` is a logical variable being `TRUE` when the event of interest, death for example, has been observed and `FALSE` when in case of censoring. The right hand side variable `group` is a grouping factor.

Figure 11.2 leads to the impression that patients treated with the novel radioimmunotherapy survive longer, regardless of the tumour type. In order to assess if this informal finding is reliable, we may perform a log-rank test via

```
R> survdiff(Surv(time, event) ~ group, data = g3)

Call:
survdiff(formula = Surv(time, event) ~ group, data = g3)

               N Observed Expected (O-E)^2/E (O-E)^2/V
group=Control  6        4     1.49      4.23      6.06
group=RIT     11        2     4.51      1.40      6.06

 Chisq= 6.1  on 1 degrees of freedom, p= 0.0138
```

which indicates that the survival times are indeed different in both groups. However, the number of patients is rather limited and so it might be dangerous to rely on asymptotic tests. As shown in Chapter 4, conditioning on the data and computing the distribution of the test statistics without additional

```
R> data("glioma", package = "coin")
R> library("survival")
R> layout(matrix(1:2, ncol = 2))
R> g3 <- subset(glioma, histology == "Grade3")
R> plot(survfit(Surv(time, event) ~ group, data = g3),
+        main = "Grade III Glioma", lty = c(2, 1),
+        ylab = "Probability", xlab = "Survival Time in Month",
+        legend.text = c("Control", "Treated"),
+        legend.bty = "n")
R> g4 <- subset(glioma, histology == "GBM")
R> plot(survfit(Surv(time, event) ~ group, data = g4),
+        main = "Grade IV Glioma", ylab = "Probability",
+        lty = c(2, 1), xlab = "Survival Time in Month",
+        xlim = c(0, max(glioma$time) * 1.05))
```

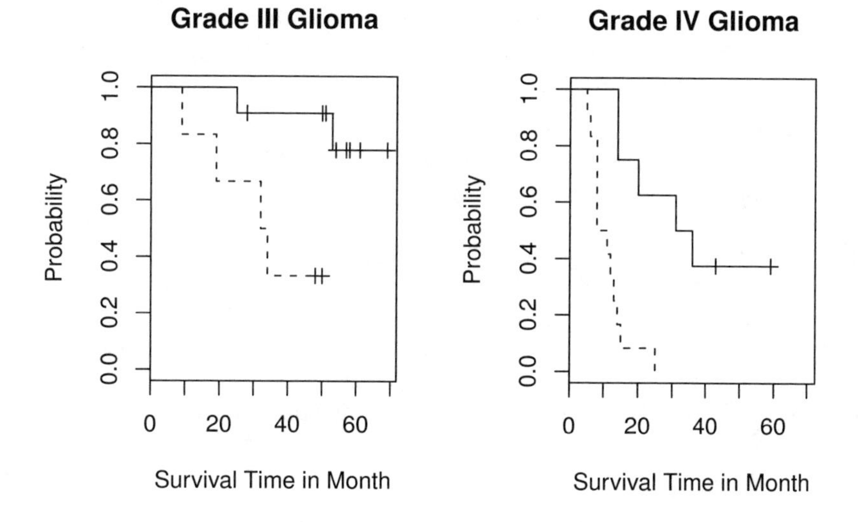


**Figure 11.2** Survival times comparing treated and control patients.

assumptions are one alternative. The function `surv_test` from package **coin** (Hothorn et al., 2006a, 2008b) can be used to compute an exact conditional test answering the question whether the survival times differ for grade III patients. For all possible permutations of the groups on the censored response variable, the test statistic is computed and the fraction of whose being greater than the observed statistic defines the exact $p$-value:

```
R> library("coin")
R> surv_test(Surv(time, event) ~ group, data = g3,
+            distribution = "exact")
```

```
        Exact Logrank Test

data:  Surv(time, event) by group (Control, RIT)
Z = 2.1711, p-value = 0.02877
alternative hypothesis: two.sided
```

which, in this case, confirms the above results. The same exercise can be performed for patients with grade IV glioma

```
R> surv_test(Surv(time, event) ~ group, data = g4,
+            distribution = "exact")

        Exact Logrank Test

data:  Surv(time, event) by group (Control, RIT)
Z = 3.2215, p-value = 0.0001588
alternative hypothesis: two.sided
```

which shows a difference as well. However, it might be more appropriate to answer the question whether the novel therapy is superior for both groups of tumours simultaneously. This can be implemented by *stratifying*, or *blocking*, with respect to tumour grading:

```
R> surv_test(Surv(time, event) ~ group | histology,
+      data = glioma, distribution = approximate(B = 10000))

        Approximative Logrank Test

data:  Surv(time, event) by
         group (Control, RIT)
         stratified by histology
Z = 3.6704, p-value = 1e-04
alternative hypothesis: two.sided
```

Here, we need to approximate the exact conditional distribution since the exact distribution is hard to compute. The result supports the initial impression implied by Figure 11.2.

### *11.3.2 Breast Cancer Survival*

Before fitting a Cox model to the GBSG2 data, we again derive a Kaplan-Meier estimate of the survival function of the data, here stratified with respect to whether a patient received a hormonal therapy or not (see Figure 11.3).

Fitting a Cox model follows roughly the same rules as shown for linear models in Chapter 6 with the exception that the response variable is again coded as a *Surv* object. For the GBSG2 data, the model is fitted via

```
R> GBSG2_coxph <- coxph(Surv(time, cens) ~ ., data = GBSG2)
```

and the results as given by the summary method are given in Figure 11.4. Since we are especially interested in the relative risk for patients who underwent a hormonal therapy, we can compute an estimate of the relative risk and a corresponding confidence interval via

```
R> data("GBSG2", package = "ipred")
R> plot(survfit(Surv(time, cens) ~ horTh, data = GBSG2),
+      lty = 1:2, mark.time = FALSE,  ylab = "Probability",
+      xlab = "Survival Time in Days")
R> legend(250, 0.2, legend = c("yes", "no"), lty = c(2, 1),
+        title = "Hormonal Therapy", bty = "n")
```

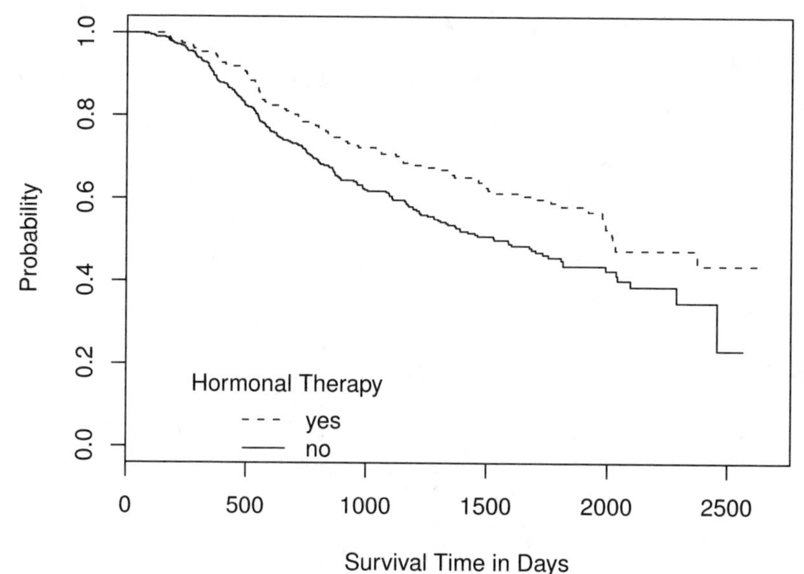


**Figure 11.3** Kaplan-Meier estimates for breast cancer patients who either received a hormonal therapy or not.

```
R> ci <- confint(GBSG2_coxph)
R> exp(cbind(coef(GBSG2_coxph), ci))["horThyes",]
              2.5 %    97.5 %
0.7073155 0.5492178 0.9109233
```

This result implies that patients treated with a hormonal therapy had a lower risk and thus survived longer compared to women who were not treated this way.

Model checking and model selection for proportional hazards models are complicated by the fact that easy-to-use residuals, such as those discussed in Chapter 6 for linear regression models, are not available, but several possibilities do exist. A check of the proportional hazards assumption can be done by looking at the parameter estimates $\beta_1, \dots, \beta_q$ over time. We can safely assume

```
R> summary(GBSG2_coxph)

Call:
coxph(formula = Surv(time, cens) ~ ., data = GBSG2)

  n= 686

                  coef exp(coef)  se(coef)      z Pr(>|z|)
horThyes    -0.3462784 0.7073155 0.1290747 -2.683 0.007301
age         -0.0094592 0.9905854 0.0093006 -1.017 0.309126
menostatPost 0.2584448 1.2949147 0.1834765  1.409 0.158954
tsize        0.0077961 1.0078266 0.0039390  1.979 0.047794
tgrade.L     0.5512988 1.7355056 0.1898441  2.904 0.003685
tgrade.Q    -0.2010905 0.8178384 0.1219654 -1.649 0.099199
pnodes       0.0487886 1.0499984 0.0074471  6.551  5.7e-11
progrec     -0.0022172 0.9977852 0.0005735 -3.866 0.000111
estrec       0.0001973 1.0001973 0.0004504  0.438 0.661307

             exp(coef) exp(-coef) lower .95 upper .95
horThyes        0.7073     1.4138    0.5492     0.911
age             0.9906     1.0095    0.9727     1.009
menostatPost    1.2949     0.7723    0.9038     1.855
tsize           1.0078     0.9922    1.0001     1.016
tgrade.L        1.7355     0.5762    1.1963     2.518
tgrade.Q        0.8178     1.2227    0.6439     1.039
pnodes          1.0500     0.9524    1.0348     1.065
progrec         0.9978     1.0022    0.9967     0.999
estrec          1.0002     0.9998    0.9993     1.001

Rsquare= 0.142   (max possible= 0.995 )
Likelihood ratio test= 104.8  on 9 df,   p=0
Wald test            = 114.8  on 9 df,   p=0
Score (logrank) test = 120.7  on 9 df,   p=0
```

**Figure 11.4** R output of the `summary` method for `GBSG2_coxph`.

proportional hazards when the estimates don't vary much over time. The null hypothesis of constant regression coefficients can be tested, both globally as well as for each covariate, by using the `cox.zph` function

```
R> GBSG2_zph <- cox.zph(GBSG2_coxph)
R> GBSG2_zph

                   rho    chisq       p
horThyes     -2.54e-02 1.96e-01 0.65778
age           9.40e-02 2.96e+00 0.08552
menostatPost -1.19e-05 3.75e-08 0.99985
tsize        -2.50e-02 1.88e-01 0.66436
tgrade.L     -1.30e-01 4.85e+00 0.02772
```

```
R> plot(GBSG2_zph, var = "age")
```

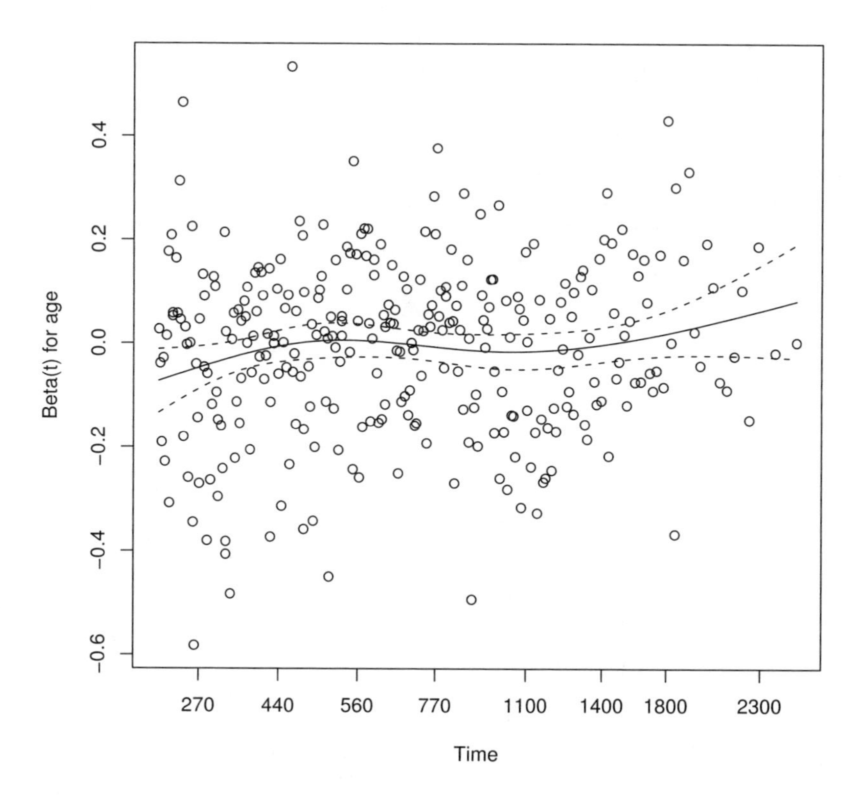


**Figure 11.5** Estimated regression coefficient for `age` depending on time for the `GBSG2` data.

```
tgrade.Q         3.22e-03 3.14e-03 0.95530
pnodes           5.84e-02 5.98e-01 0.43941
progrec          5.65e-02 1.20e+00 0.27351
estrec           5.46e-02 1.03e+00 0.30967
GLOBAL                 NA 2.27e+01 0.00695
```

There seems to be some evidence of time-varying effects, especially for age and tumour grading. A graphical representation of the estimated regression coefficient over time is shown in Figure 11.5. We refer to Therneau and Grambsch (2000) for a detailed theoretical description of these topics.

Martingale residuals as computed by the `residuals` method applied to `coxph` objects can be used to check the model fit. When evaluated at the true regression coefficient the expectation of the martingale residuals is zero. Thus, one way to check for systematic deviations is an inspection of scatter-

```
R> layout(matrix(1:3, ncol = 3))
R> res <- residuals(GBSG2_coxph)
R> plot(res ~ age, data = GBSG2, ylim = c(-2.5, 1.5),
+       pch = ".", ylab = "Martingale Residuals")
R> abline(h = 0, lty = 3)
R> plot(res ~ pnodes, data = GBSG2, ylim = c(-2.5, 1.5),
+       pch = ".", ylab = "")
R> abline(h = 0, lty = 3)
R> plot(res ~ log(progrec), data = GBSG2, ylim = c(-2.5, 1.5),
+       pch = ".", ylab = "")
R> abline(h = 0, lty = 3)
```

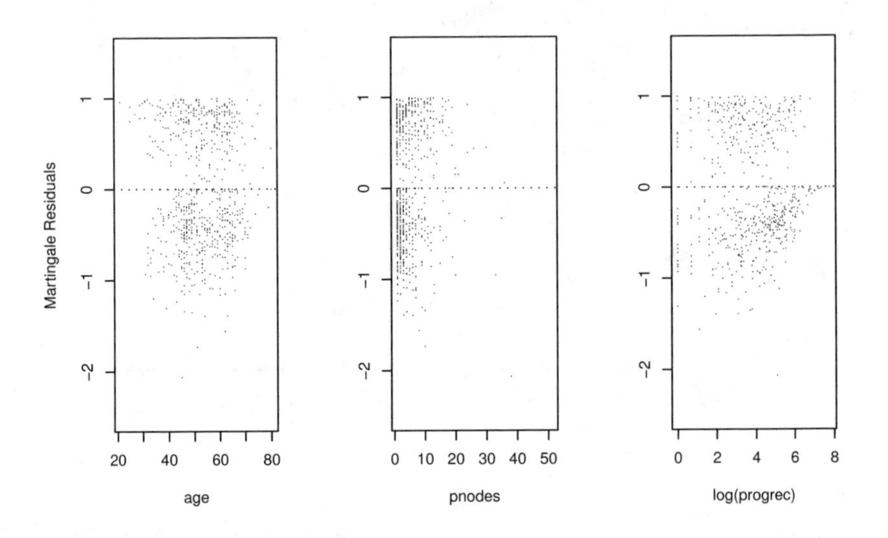


**Figure 11.6** Martingale residuals for the GBSG2 data.

plots plotting covariates against the martingale residuals. For the GBSG2 data, Figure 11.6 does not indicate severe and systematic deviations from zero.

The tree-structured regression models applied to continuous and binary responses in Chapter 9 are applicable to censored responses in survival analysis as well. Such a simple prognostic model with only a few terminal nodes might be helpful for relating the risk to certain subgroups of patients. Both rpart and the ctree function from package **party** can be applied to the GBSG2 data, where the conditional trees of the latter select cutpoints based on log-rank statistics

```
R> GBSG2_ctree <- ctree(Surv(time, cens) ~ ., data = GBSG2)
```

and the plot method applied to this tree produces the graphical representation in Figure 11.7. The number of positive lymph nodes (pnodes) is the most

```
R> plot(GBSG2_ctree)
```

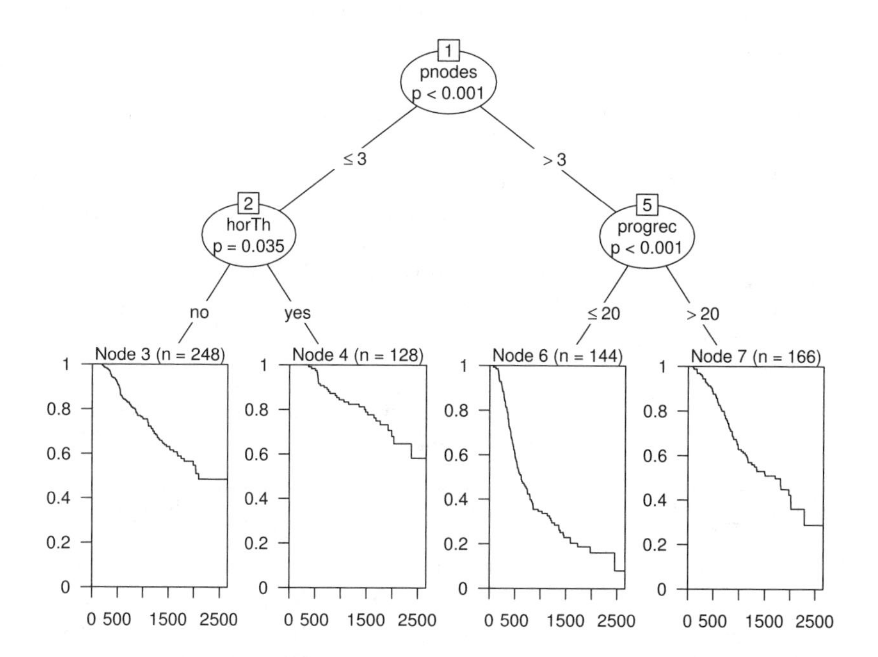


**Figure 11.7** Conditional inference tree for the GBSG2 data with the survival function, estimated by Kaplan-Meier, shown for every subgroup of patients identified by the tree.

important variable in the tree, corresponding to the $p$-value associated with this variable in Cox's regression; see Figure 11.4. Women with not more than three positive lymph nodes who have undergone a hormonal therapy seem to have the best prognosis whereas a large number of positive lymph nodes and a small value of the progesterone receptor indicates a bad prognosis.

## 11.4 Summary

The analysis of life-time data is complicated by the fact that the time to some event is not observable for all observations due to censoring. Survival times are analysed by some estimates of the survival function, for example by a non-parametric Kaplan-Meier estimate or by semi-parametric proportional hazards regression models.

## Exercises

Ex. 11.1 Sauerbrei and Royston (1999) analyse the GBSG2 data using multivariable fractional polynomials, a flexibilisation for many linear regression models including Cox's model. In R, this methodology is available by the **mfp** package (Ambler and Benner, 2009). Try to reproduce the analysis presented by Sauerbrei and Royston (1999), i.e., fit a multivariable fractional polynomial to the `GBSG2` data!

Ex. 11.2 The data in Table 11.3 (Everitt and Rabe-Hesketh, 2001) are the survival times (in months) after mastectomy of women with breast cancer. The cancers are classified as having metastasised or not based on a histochemical marker. Censoring is indicated by the `event` variable being `TRUE` in case of death. Plot the survivor functions of each group, estimated using the Kaplan-Meier estimate, on the same graph and comment on the differences. Use a log-rank test to compare the survival experience of each group more formally.

**Table 11.3**: `mastectomy` data. Survival times in months after mastectomy of women with breast cancer.

| time | event | metastasised | time | event | metastasised |
|---|---|---|---|---|---|
| 23 | TRUE | no | 40 | TRUE | yes |
| 47 | TRUE | no | 41 | TRUE | yes |
| 69 | TRUE | no | 48 | TRUE | yes |
| 70 | FALSE | no | 50 | TRUE | yes |
| 100 | FALSE | no | 59 | TRUE | yes |
| 101 | FALSE | no | 61 | TRUE | yes |
| 148 | TRUE | no | 68 | TRUE | yes |
| 181 | TRUE | no | 71 | TRUE | yes |
| 198 | FALSE | no | 76 | FALSE | yes |
| 208 | FALSE | no | 105 | FALSE | yes |
| 212 | FALSE | no | 107 | FALSE | yes |
| 224 | FALSE | no | 109 | FALSE | yes |
| 5 | TRUE | yes | 113 | TRUE | yes |
| 8 | TRUE | yes | 116 | FALSE | yes |
| 10 | TRUE | yes | 118 | TRUE | yes |
| 13 | TRUE | yes | 143 | TRUE | yes |
| 18 | TRUE | yes | 145 | FALSE | yes |
| 24 | TRUE | yes | 162 | FALSE | yes |
| 26 | TRUE | yes | 188 | FALSE | yes |
| 26 | TRUE | yes | 212 | FALSE | yes |
| 31 | TRUE | yes | 217 | FALSE | yes |
| 35 | TRUE | yes | 225 | FALSE | yes |

CHAPTER 12

# Analysing Longitudinal Data I: Computerised Delivery of Cognitive Behavioural Therapy – Beat the Blues

## 12.1 Introduction

Depression is a major public health problem across the world. Antidepressants are the front line treatment, but many patients either do not respond to them, or do not like taking them. The main alternative is psychotherapy, and the modern 'talking treatments' such as *cognitive behavioural therapy* (CBT) have been shown to be as effective as drugs, and probably more so when it comes to relapse. But there is a problem, namely availability–there are simply not enough skilled therapists to meet the demand, and little prospect at all of this situation changing.

A number of alternative modes of delivery of CBT have been explored, including interactive systems making use of the new computer technologies. The principles of CBT lend themselves reasonably well to computerisation, and, perhaps surprisingly, patients adapt well to this procedure, and do not seem to miss the physical presence of the therapist as much as one might expect. The data to be used in this chapter arise from a clinical trial of an interactive, multimedia program known as 'Beat the Blues' designed to deliver cognitive behavioural therapy to depressed patients via a computer terminal. Full details are given in Proudfoot et al. (2003), but in essence Beat the Blues is an interactive program using multimedia techniques, in particular video vignettes. The computer-based intervention consists of nine sessions, followed by eight therapy sessions, each lasting about 50 minutes. Nurses are used to explain how the program works, but are instructed to spend no more than 5 minutes with each patient at the start of each session, and are there simply to assist with the technology. In a randomised controlled trial of the program, patients with depression recruited in primary care were randomised to either the Beat the Blues program or to 'Treatment as Usual' (TAU). Patients randomised to Beat the Blues also received pharmacology and/or general practise (GP) support and practical/social help, offered as part of treatment as usual, with the exception of any face-to-face counselling or psychological intervention. Patients allocated to TAU received whatever treatment their GP prescribed. The latter included, besides any medication, discussion of problems with GP, provision of practical/social help, referral to a counsellor, referral to a practise nurse, referral to mental health professionals (psychologist, psychiatrist, community psychiatric nurse, counsellor), or further physical examination.

A number of outcome measures were used in the trial, but here we concentrate on the *Beck Depression Inventory II* (BDI, Beck et al., 1996). Measurements on this variable were made on the following five occasions:

- Prior to treatment,
- Two months after treatment began and
- At one, three and six months follow-up, i.e., at three, five and eight months after treatment.

**Table 12.1**: `BtheB` data. Data of a randomised trial evaluating the effects of Beat the Blues.

| drug | length | treatment | bdi.pre | bdi.2m | bdi.3m | bdi.5m | bdi.8m |
|---|---|---|---|---|---|---|---|
| No | >6m | TAU | 29 | 2 | 2 | NA | NA |
| Yes | >6m | BtheB | 32 | 16 | 24 | 17 | 20 |
| Yes | <6m | TAU | 25 | 20 | NA | NA | NA |
| No | >6m | BtheB | 21 | 17 | 16 | 10 | 9 |
| Yes | >6m | BtheB | 26 | 23 | NA | NA | NA |
| Yes | <6m | BtheB | 7 | 0 | 0 | 0 | 0 |
| Yes | <6m | TAU | 17 | 7 | 7 | 3 | 7 |
| No | >6m | TAU | 20 | 20 | 21 | 19 | 13 |
| Yes | <6m | BtheB | 18 | 13 | 14 | 20 | 11 |
| Yes | >6m | BtheB | 20 | 5 | 5 | 8 | 12 |
| No | >6m | TAU | 30 | 32 | 24 | 12 | 2 |
| Yes | <6m | BtheB | 49 | 35 | NA | NA | NA |
| No | >6m | TAU | 26 | 27 | 23 | NA | NA |
| Yes | >6m | TAU | 30 | 26 | 36 | 27 | 22 |
| Yes | >6m | BtheB | 23 | 13 | 13 | 12 | 23 |
| No | <6m | TAU | 16 | 13 | 3 | 2 | 0 |
| No | >6m | BtheB | 30 | 30 | 29 | NA | NA |
| No | <6m | BtheB | 13 | 8 | 8 | 7 | 6 |
| No | >6m | TAU | 37 | 30 | 33 | 31 | 22 |
| Yes | <6m | BtheB | 35 | 12 | 10 | 8 | 10 |
| No | >6m | BtheB | 21 | 6 | NA | NA | NA |
| No | <6m | TAU | 26 | 17 | 17 | 20 | 12 |
| No | >6m | TAU | 29 | 22 | 10 | NA | NA |
| No | >6m | TAU | 20 | 21 | NA | NA | NA |
| No | >6m | TAU | 33 | 23 | NA | NA | NA |
| No | >6m | BtheB | 19 | 12 | 13 | NA | NA |
| Yes | <6m | TAU | 12 | 15 | NA | NA | NA |
| Yes | >6m | TAU | 47 | 36 | 49 | 34 | NA |
| Yes | >6m | BtheB | 36 | 6 | 0 | 0 | 2 |
| No | <6m | BtheB | 10 | 8 | 6 | 3 | 3 |

**Table 12.1**: BtheB data (continued).

| drug | length | treatment | bdi.pre | bdi.2m | bdi.3m | bdi.5m | bdi.8m |
|---|---|---|---|---|---|---|---|
| No | <6m | TAU | 27 | 7 | 15 | 16 | 0 |
| No | <6m | BtheB | 18 | 10 | 10 | 6 | 8 |
| Yes | <6m | BtheB | 11 | 8 | 3 | 2 | 15 |
| Yes | <6m | BtheB | 6 | 7 | NA | NA | NA |
| Yes | >6m | BtheB | 44 | 24 | 20 | 29 | 14 |
| No | <6m | TAU | 38 | 38 | NA | NA | NA |
| No | <6m | TAU | 21 | 14 | 20 | 1 | 8 |
| Yes | >6m | TAU | 34 | 17 | 8 | 9 | 13 |
| Yes | <6m | BtheB | 9 | 7 | 1 | NA | NA |
| Yes | >6m | TAU | 38 | 27 | 19 | 20 | 30 |
| Yes | <6m | BtheB | 46 | 40 | NA | NA | NA |
| No | <6m | TAU | 20 | 19 | 18 | 19 | 18 |
| Yes | >6m | TAU | 17 | 29 | 2 | 0 | 0 |
| No | >6m | BtheB | 18 | 20 | NA | NA | NA |
| Yes | >6m | BtheB | 42 | 1 | 8 | 10 | 6 |
| No | <6m | BtheB | 30 | 30 | NA | NA | NA |
| Yes | <6m | BtheB | 33 | 27 | 16 | 30 | 15 |
| No | <6m | BtheB | 12 | 1 | 0 | 0 | NA |
| Yes | <6m | BtheB | 2 | 5 | NA | NA | NA |
| No | >6m | TAU | 36 | 42 | 49 | 47 | 40 |
| No | <6m | TAU | 35 | 30 | NA | NA | NA |
| No | <6m | BtheB | 23 | 20 | NA | NA | NA |
| No | >6m | TAU | 31 | 48 | 38 | 38 | 37 |
| Yes | <6m | BtheB | 8 | 5 | 7 | NA | NA |
| Yes | <6m | TAU | 23 | 21 | 26 | NA | NA |
| Yes | <6m | BtheB | 7 | 7 | 5 | 4 | 0 |
| No | <6m | TAU | 14 | 13 | 14 | NA | NA |
| No | <6m | TAU | 40 | 36 | 33 | NA | NA |
| Yes | <6m | BtheB | 23 | 30 | NA | NA | NA |
| No | >6m | BtheB | 14 | 3 | NA | NA | NA |
| No | >6m | TAU | 22 | 20 | 16 | 24 | 16 |
| No | >6m | TAU | 23 | 23 | 15 | 25 | 17 |
| No | <6m | TAU | 15 | 7 | 13 | 13 | NA |
| No | >6m | TAU | 8 | 12 | 11 | 26 | NA |
| No | >6m | BtheB | 12 | 18 | NA | NA | NA |
| No | >6m | TAU | 7 | 6 | 2 | 1 | NA |
| Yes | <6m | TAU | 17 | 9 | 3 | 1 | 0 |
| Yes | <6m | BtheB | 33 | 18 | 16 | NA | NA |
| No | <6m | TAU | 27 | 20 | NA | NA | NA |
| No | <6m | BtheB | 27 | 30 | NA | NA | NA |
| No | <6m | BtheB | 9 | 6 | 10 | 1 | 0 |
| No | >6m | BtheB | 40 | 30 | 12 | NA | NA |

**Table 12.1**: BtheB data (continued).

| drug | length | treatment | bdi.pre | bdi.2m | bdi.3m | bdi.5m | bdi.8m |
|---|---|---|---|---|---|---|---|
| No | >6m | TAU | 11 | 8 | 7 | NA | NA |
| No | <6m | TAU | 9 | 8 | NA | NA | NA |
| No | >6m | TAU | 14 | 22 | 21 | 24 | 19 |
| Yes | >6m | BtheB | 28 | 9 | 20 | 18 | 13 |
| No | >6m | BtheB | 15 | 9 | 13 | 14 | 10 |
| Yes | >6m | BtheB | 22 | 10 | 5 | 5 | 12 |
| No | <6m | TAU | 23 | 9 | NA | NA | NA |
| No | >6m | TAU | 21 | 22 | 24 | 23 | 22 |
| No | >6m | TAU | 27 | 31 | 28 | 22 | 14 |
| Yes | >6m | BtheB | 14 | 15 | NA | NA | NA |
| No | >6m | TAU | 10 | 13 | 12 | 8 | 20 |
| Yes | <6m | TAU | 21 | 9 | 6 | 7 | 1 |
| Yes | >6m | BtheB | 46 | 36 | 53 | NA | NA |
| No | >6m | BtheB | 36 | 14 | 7 | 15 | 15 |
| Yes | >6m | BtheB | 23 | 17 | NA | NA | NA |
| Yes | >6m | TAU | 35 | 0 | 6 | 0 | 1 |
| Yes | <6m | BtheB | 33 | 13 | 13 | 10 | 8 |
| No | <6m | BtheB | 19 | 4 | 27 | 1 | 2 |
| No | <6m | TAU | 16 | NA | NA | NA | NA |
| Yes | <6m | BtheB | 30 | 26 | 28 | NA | NA |
| Yes | <6m | BtheB | 17 | 8 | 7 | 12 | NA |
| No | >6m | BtheB | 19 | 4 | 3 | 3 | 3 |
| No | >6m | BtheB | 16 | 11 | 4 | 2 | 3 |
| Yes | >6m | BtheB | 16 | 16 | 10 | 10 | 8 |
| Yes | <6m | TAU | 28 | NA | NA | NA | NA |
| No | >6m | BtheB | 11 | 22 | 9 | 11 | 11 |
| No | <6m | TAU | 13 | 5 | 5 | 0 | 6 |
| Yes | <6m | TAU | 43 | NA | NA | NA | NA |

The resulting data from a subset of 100 patients are shown in Table 12.1. (The data are used with the kind permission of Dr. Judy Proudfoot.) In addition to assessing the effects of treatment, there is interest here in assessing the effect of taking antidepressant drugs (drug, yes or no) and length of the current episode of depression (length, less or more than six months).

## 12.2 Analysing Longitudinal Data

The distinguishing feature of a longitudinal study is that the response variable of interest and a set of explanatory variables are measured several times on each individual in the study. The main objective in such a study is to characterise change in the repeated values of the response variable and to determine the explanatory variables most associated with any change. Because several observations of the response variable are made on the same individual, it is likely that the measurements will be correlated rather than independent, even after conditioning on the explanatory variables. Consequently repeated measures data require special methods of analysis and models for such data need to include parameters linking the explanatory variables to the repeated measurements, parameters analogous to those in the usual multiple regression model (see Chapter 6), and, in addition parameters that account for the correlational structure of the repeated measurements. It is the former parameters that are generally of most interest with the latter often being regarded as *nuisance parameters*. But providing an adequate description for the correlational structure of the repeated measures is necessary to avoid misleading inferences about the parameters that *are* of real interest to the researcher.

Over the last decade methodology for the analysis of repeated measures data has been the subject of much research and development, and there are now a variety of powerful techniques available. A comprehensive account of these methods is given in Diggle et al. (2003) and Davis (2002). In this chapter we will concentrate on a single class of methods, *linear mixed effects models* suitable when, conditional on the explanatory variables, the response has a normal distribution. In Chapter 13 two other classes of models which can deal with non-normal responses will be described.

## 12.3 Linear Mixed Effects Models for Repeated Measures Data

Linear mixed effects models for repeated measures data formalise the sensible idea that an individual's pattern of responses is likely to depend on many characteristics of that individual, including some that are unobserved. These unobserved variables are then included in the model as random variables, i.e., random effects. The essential feature of such models is that correlation amongst the repeated measurements on the same unit arises from shared, unobserved variables. Conditional on the values of the random effects, the repeated measurements are assumed to be independent, the so-called *local independence* assumption.

Two commonly used linear mixed effect models, the *random intercept* and the *random intercept and slope* models, will now be described in more detail.

Let $y_{ij}$ represent the observation made at time $t_j$ on individual $i$. A possible model for the observation $y_{ij}$ might be

$$y_{ij} = \beta_0 + \beta_1 t_j + u_i + \varepsilon_{ij}. \tag{12.1}$$

Here the total residual that would be present in the usual linear regression model has been partitioned into a subject-specific random component $u_i$ which is constant over time plus a residual $\varepsilon_{ij}$ which varies randomly over time. The $u_i$ are assumed to be normally distributed with zero mean and variance $\sigma_u^2$. Similarly the residuals $\varepsilon_{ij}$ are assumed normally distributed with zero mean and variance $\sigma^2$. The $u_i$ and $\varepsilon_{ij}$ are assumed to be independent of each other and of the time $t_j$. The model in (12.1) is known as a *random intercept model*, the $u_i$ being the random intercepts. The repeated measurements for an individual vary about that individual's own regression line which can differ in intercept but not in slope from the regression lines of other individuals. The random effects model possible heterogeneity in the intercepts of the individuals whereas time has a fixed effect, $\beta_1$.

The random intercept model implies that the total variance of each repeated measurement is $\mathsf{Var}(y_{ij}) = \mathsf{Var}(u_i + \varepsilon_{ij}) = \sigma_u^2 + \sigma^2$. Due to this decomposition of the total residual variance into a between-subject component, $\sigma_u^2$, and a within-subject component, $\sigma^2$, the model is sometimes referred to as a *variance component model.*

The covariance between the total residuals at two time points $j$ and $k$ in the same individual is $\mathsf{Cov}(u_i + \varepsilon_{ij}, u_i + \varepsilon_{ik}) = \sigma_u^2$. Note that these covariances are induced by the shared random intercept; for individuals with $u_i > 0$, the total residuals will tend to be greater than the mean, for individuals with $u_i < 0$ they will tend to be less than the mean. It follows from the two relations above that the residual correlations are given by

$$\mathsf{Cor}(u_i + \varepsilon_{ij}, u_i + \varepsilon_{ik}) = \frac{\sigma_u^2}{\sigma_u^2 + \sigma^2}.$$

This is an *intra-class correlation* interpreted as the proportion of the total residual variance that is due to residual variability between subjects. A random intercept model constrains the variance of each repeated measure to be the same and the covariance between any pair of measurements to be equal. This is usually called the *compound symmetry* structure. These constraints are often not realistic for repeated measures data. For example, for longitudinal data it is more common for measures taken closer to each other in time to be more highly correlated than those taken further apart. In addition the variances of the later repeated measures are often greater than those taken earlier. Consequently for many such data sets the random intercept model will not do justice to the observed pattern of covariances between the repeated measures. A model that allows a more realistic structure for the covariances is one that allows heterogeneity in both slopes and intercepts, the *random slope and intercept model.*

In this model there are two types of random effects, the first modelling heterogeneity in intercepts, $u_i$, and the second modelling heterogeneity in slopes, $v_i$. Explicitly the model is

$$y_{ij} = \beta_0 + \beta_1 t_j + u_i + v_i t_j + \varepsilon_{ij} \tag{12.2}$$

where the parameters are not, of course, the same as in (12.1). The two random effects are assumed to have a bivariate normal distribution with zero means for both variables and variances $\sigma_u^2$ and $\sigma_v^2$ with covariance $\sigma_{uv}$. With this model the total residual is $u_i + u_i t_j + \varepsilon_{ij}$ with variance

$$\mathsf{Var}(u_i + v_i t_j + \varepsilon_{ij}) = \sigma_u^2 + 2\sigma_{uv} t_j + \sigma_v^2 t_j^2 + \sigma^2$$

which is no longer constant for different values of $t_j$. Similarly the covariance between two total residuals of the same individual

$$\mathsf{Cov}(u_i + v_i t_j + \varepsilon_{ij}, u_i + v_i t_k + \varepsilon_{ik}) = \sigma_u^2 + \sigma_{uv}(t_j - t_k) + \sigma_v^2 t_j t_k$$

is not constrained to be the same for all pairs $t_j$ and $t_k$.

(It should also be noted that re-estimating the model after adding or subtracting a constant from $t_j$, e.g., its mean, will lead to different variance and covariance estimates, but will not affect fixed effects.)

Linear mixed-effects models can be estimated by maximum likelihood. However, this method tends to underestimate the variance components. A modified version of maximum likelihood, known as *restricted maximum likelihood* is therefore often recommended; this provides consistent estimates of the variance components. Details are given in Diggle et al. (2003) and Longford (1993). Competing linear mixed-effects models can be compared using a likelihood ratio test. If however the models have been estimated by restricted maximum likelihood this test can be used only if both models have the same set of fixed effects, see Longford (1993). (It should be noted that there are some technical problems with the likelihood ratio test which are discussed in detail in Rabe-Hesketh and Skrondal, 2008).

## 12.4 Analysis Using R

Almost all statistical analyses should begin with some graphical representation of the data and here we shall construct the boxplots of each of the five repeated measures separately for each treatment group. The data are available as the data frame `BtheB` and the necessary R code is given along with Figure 12.1. The boxplots show that there is decline in BDI values in both groups with perhaps the values in the group of patients treated in the Beat the Blues arm being lower at each post-randomisation visit.

We shall fit both random intercept and random intercept and slope models to the data including the baseline BDI values (`pre.bdi`), `treatment` group, `drug` and `length` as fixed effect covariates. Linear mixed effects models are fitted in R by using the `lmer` function contained in the **lme4** package (Bates and Sarkar, 2008, Pinheiro and Bates, 2000, Bates, 2005), but an essential first step is to rearrange the data from the 'wide form' in which they appear in the `BtheB` data frame into the 'long form' in which each separate repeated measurement and associated covariate values appear as a separate row in a *data.frame*. This rearrangement can be made using the following code:

```
R> data("BtheB", package = "HSAUR2")
R> BtheB$subject <- factor(rownames(BtheB))
R> nobs <- nrow(BtheB)
R> BtheB_long <- reshape(BtheB, idvar = "subject",
+      varying = c("bdi.2m", "bdi.3m", "bdi.5m", "bdi.8m"),
+      direction = "long")
R> BtheB_long$time <- rep(c(2, 3, 5, 8), rep(nobs, 4))
```

```
R> data("BtheB", package = "HSAUR2")
R> layout(matrix(1:2, nrow = 1))
R> ylim <- range(BtheB[,grep("bdi", names(BtheB))],
+                na.rm = TRUE)
R> tau <- subset(BtheB, treatment == "TAU")[,
+      grep("bdi", names(BtheB))]
R> boxplot(tau, main = "Treated as Usual", ylab = "BDI",
+          xlab = "Time (in months)", names = c(0, 2, 3, 5, 8),
+          ylim = ylim)
R> btheb <- subset(BtheB, treatment == "BtheB")[,
+      grep("bdi", names(BtheB))]
R> boxplot(btheb, main = "Beat the Blues", ylab = "BDI",
+          xlab = "Time (in months)", names = c(0, 2, 3, 5, 8),
+          ylim = ylim)
```

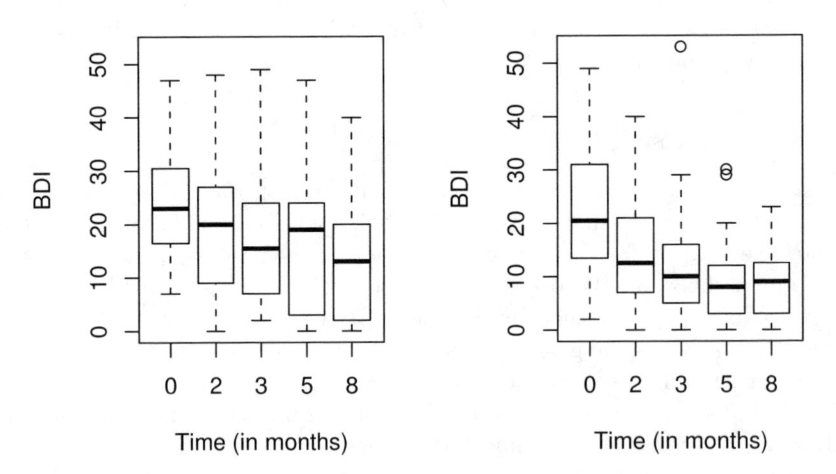
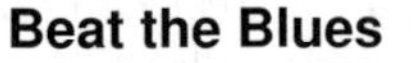


**Figure 12.1** Boxplots for the repeated measures by treatment group for the BtheB data.

such that the data are now in the form (here shown for the first three subjects)

```
R> subset(BtheB_long, subject %in% c("1", "2", "3"))
```

```
     drug length treatment bdi.pre subject time bdi
1.2m   No    >6m       TAU      29       1    2   2
2.2m  Yes    >6m     BtheB      32       2    2  16
3.2m  Yes    <6m       TAU      25       3    2  20
1.3m   No    >6m       TAU      29       1    3   2
2.3m  Yes    >6m     BtheB      32       2    3  24
3.3m  Yes    <6m       TAU      25       3    3  NA
```

```
1.5m    No   >6m    TAU   29   1   5  NA
2.5m   Yes   >6m  BtheB   32   2   5  17
3.5m   Yes   <6m    TAU   25   3   5  NA
1.8m    No   >6m    TAU   29   1   8  NA
2.8m   Yes   >6m  BtheB   32   2   8  20
3.8m   Yes   <6m    TAU   25   3   8  NA
```

The resulting *data.frame* `BtheB_long` contains a number of missing values and in applying the `lmer` function these will be dropped. But notice it is only the missing values that are removed, *not* participants that have at least one missing value. All the available data is used in the model fitting process. The `lmer` function is used in a similar way to the `lm` function met in Chapter 6 with the addition of a random term to identify the source of the repeated measurements, here `subject`. We can fit the two models (12.1) and (12.2) and test which is most appropriate using

```
R> library("lme4")
R> BtheB_lmer1 <- lmer(bdi ~ bdi.pre + time + treatment + drug +
+       length + (1 | subject), data = BtheB_long,
+       REML = FALSE, na.action = na.omit)
R> BtheB_lmer2 <- lmer(bdi ~ bdi.pre + time + treatment + drug +
+       length + (time | subject), data = BtheB_long,
+       REML = FALSE, na.action = na.omit)
R> anova(BtheB_lmer1, BtheB_lmer2)
```

```
Data: BtheB_long
Models:
BtheB_lmer1: bdi ~ bdi.pre + time + treatment + drug + length +
BtheB_lmer1:     (1 | subject)
BtheB_lmer2: bdi ~ bdi.pre + time + treatment + drug + length +
BtheB_lmer2:     (time | subject)
            Df     AIC     BIC  logLik  Chisq Chi Df
BtheB_lmer1  8 1887.49 1916.57 -935.75
BtheB_lmer2 10 1891.04 1927.39 -935.52 0.4542      2
            Pr(>Chisq)
BtheB_lmer1
BtheB_lmer2     0.7969
```

The log-likelihood test indicates that the simpler random intercept model is adequate for these data. More information about the fitted random intercept model can be extracted from object `BtheB_lmer1` using `summary` by the R code in Figure 12.2. We see that the regression coefficients for `time` and the *Beck Depression Inventory II* values measured at baseline (`bdi.pre`) are highly significant, but there is no evidence that the coefficients for the other three covariates differ from zero. In particular, there is no clear evidence of a treatment effect.

The `summary` method for *lmer* objects doesn't print $p$-values for Gaussian mixed models because the degrees of freedom of the $t$ reference distribution are not obvious. However, one can rely on the asymptotic normal distribution for

```
R> summary(BtheB_lmer1)

Linear mixed model fit by maximum likelihood
Formula: bdi ~ bdi.pre + time + treatment + drug + length +
         (1 | subject)
   Data: BtheB_long
  AIC  BIC logLik deviance REMLdev
 1887 1917 -935.7     1871    1867
Random effects:
 Groups   Name        Variance Std.Dev.
 subject  (Intercept) 48.777   6.9841
 Residual             25.140   5.0140
Number of obs: 280, groups: subject, 97

Fixed effects:
               Estimate Std. Error t value
(Intercept)     5.59244    2.24232   2.494
bdi.pre         0.63967    0.07789   8.213
time           -0.70477    0.14639  -4.814
treatmentBtheB -2.32912    1.67026  -1.394
drugYes        -2.82497    1.72674  -1.636
length>6m       0.19712    1.63823   0.120

Correlation of Fixed Effects:
            (Intr) bdi.pr time   trtmBB drugYs
bdi.pre     -0.682
time        -0.238  0.020
tretmntBthB -0.390  0.121  0.018
drugYes     -0.073 -0.237 -0.022 -0.323
length>6m   -0.243 -0.242 -0.036  0.002  0.157
```

**Figure 12.2** R output of the linear mixed-effects model fit for the BtheB data.

computing univariate $p$-values for the fixed effects using the `cftest` function from package **multcomp**. The asymptotic $p$-values are given in Figure 12.3.

We can check the assumptions of the final model fitted to the BtheB data, i.e., the normality of the random effect terms and the residuals, by first using the `ranef` method to *predict* the former and the `residuals` method to calculate the differences between the observed data values and the fitted values, and then using normal probability plots on each. How the random effects are predicted is explained briefly in Section 12.5. The necessary R code to obtain the effects, residuals and plots is shown with Figure 12.4. There appear to be no large departures from linearity in either plot.

```
R> cftest(BtheB_lmer1)

	 Simultaneous Tests for General Linear Hypotheses

Fit: lmer(formula = bdi ~ bdi.pre + time + treatment + drug +
    length + (1 | subject), data = BtheB_long, REML = FALSE,
    na.action = na.omit)

Linear Hypotheses:
                      Estimate Std. Error z value Pr(>|z|)
(Intercept) == 0       5.59244    2.24232   2.494   0.0126
bdi.pre == 0           0.63967    0.07789   8.213 2.22e-16
time == 0             -0.70477    0.14639  -4.814 1.48e-06
treatmentBtheB == 0 -2.32912    1.67026  -1.394   0.1632
drugYes == 0          -2.82497    1.72674  -1.636   0.1018
length>6m == 0         0.19712    1.63823   0.120   0.9042
(Univariate p values reported)
```

**Figure 12.3** R output of the asymptotic $p$-values for linear mixed-effects model fit for the BtheB data.

## 12.5 Prediction of Random Effects

The random effects are not estimated as part of the model. However, having estimated the model, we can *predict* the values of the random effects. According to Bayes' Theorem, the *posterior probability* of the random effects is given by

$$\mathsf{P}(u|y,x) = f(y|u,x)g(u)$$

where $f(y|u,x)$ is the conditional density of the responses given the random effects and covariates (a product of normal densities) and $g(u)$ is the *prior* density of the random effects (multivariate normal). The means of this posterior distribution can be used as estimates of the random effects and are known as *empirical Bayes estimates*. The empirical Bayes estimator is also known as a shrinkage estimator because the predicted random effects are smaller in absolute value than their fixed effect counterparts. *Best linear unbiased predictions* (BLUP) are linear combinations of the responses that are unbiased estimators of the random effects and minimise the mean square error.

## 12.6 The Problem of Dropouts

We now need to consider briefly how the dropouts may affect the analyses reported above. To understand the problems that patients dropping out can cause for the analysis of data from a longitudinal trial we need to consider a classification of dropout mechanisms first introduced by Rubin (1976). The type of mechanism involved has implications for which approaches to analysis

```
R> layout(matrix(1:2, ncol = 2))
R> qint <- ranef(BtheB_lmer1)$subject[["(Intercept)"]]
R> qres <- residuals(BtheB_lmer1)
R> qqnorm(qint, ylab = "Estimated random intercepts",
+        xlim = c(-3, 3), ylim = c(-20, 20),
+        main = "Random intercepts")
R> qqline(qint)
R> qqnorm(qres, xlim = c(-3, 3), ylim = c(-20, 20),
+        ylab = "Estimated residuals",
+        main = "Residuals")
R> qqline(qres)
```

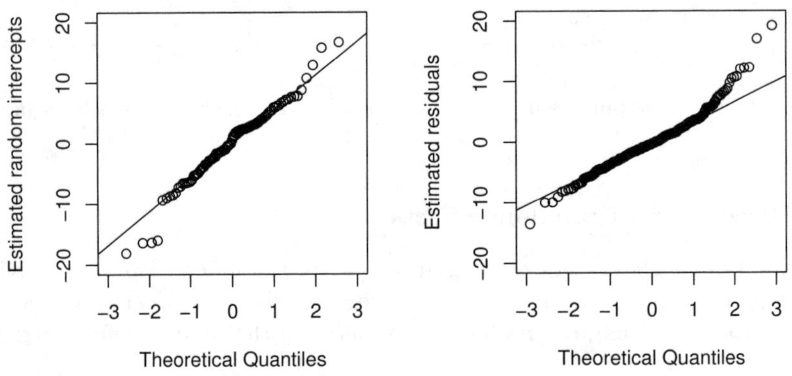


**Figure 12.4** Quantile-quantile plots of predicted random intercepts and residuals for the random intercept model `BtheB_lmer1` fitted to the `BtheB` data.

are suitable and which are not. Rubin's suggested classification involves three types of dropout mechanism:

*Dropout completely at random* (DCAR): here the probability that a patient drops out does not depend on either the observed or missing values of the response. Consequently the observed (non-missing) values effectively constitute a simple random sample of the values for all subjects. Possible examples include missing laboratory measurements because of a dropped test-tube (if it was not dropped because of the knowledge of any measurement), the accidental death of a participant in a study, or a participant moving to another area. Intermittent missing values in a longitudinal data set, whereby a patient misses a clinic visit for transitory reasons ('went shopping instead' or the like) can reasonably be assumed to be DCAR.

Completely random dropout causes least problem for data analysis, but it is a strong assumption.

*Dropout at random* (DAR): The dropout at random mechanism occurs when the probability of dropping out depends on the outcome measures that have been observed in the past, but given this information is conditionally independent of all the future (unrecorded) values of the outcome variable following dropout. Here 'missingness' depends only on the observed data with the distribution of future values for a subject who drops out at a particular time being the same as the distribution of the future values of a subject who remains in at that time, if they have the same covariates and the same past history of outcome up to and including the specific time point. Murray and Findlay (1988) provide an example of this type of missing value from a study of hypertensive drugs in which the outcome measure was diastolic blood pressure. The protocol of the study specified that the participant was to be removed from the study when his/her blood pressure got too large. Here blood pressure at the time of dropout was observed before the participant dropped out, so although the dropout mechanism is not DCAR since it depends on the values of blood pressure, it *is* DAR, because dropout depends only on the observed part of the data. A further example of a DAR mechanism is provided by Heitjan (1997), and involves a study in which the response measure is body mass index (BMI). Suppose that the measure is missing because subjects who had high body mass index values at earlier visits avoided being measured at later visits out of embarrassment, regardless of whether they had gained or lost weight in the intervening period. The missing values here are DAR but *not* DCAR; consequently methods applied to the data that assumed the latter might give misleading results (see later discussion).

*Non-ignorable* (sometimes referred to as *informative*): The final type of dropout mechanism is one where the probability of dropping out depends on the unrecorded missing values – observations are likely to be missing when the outcome values that would have been observed had the patient not dropped out, are systematically higher or lower than usual (corresponding perhaps to their condition becoming worse or improving). A non-medical example is when individuals with lower income levels or very high incomes are less likely to provide their personal income in an interview. In a medical setting possible examples are a participant dropping out of a longitudinal study when his/her blood pressure became too high and this value was not observed, or when their pain become intolerable and we did not record the associated pain value. For the BDI example introduced above, if subjects were more likely to avoid being measured if they had put on extra weight since the last visit, then the data are non-ignorably missing. Dealing with data containing missing values that result from this type of dropout mechanism is difficult. The correct analyses for such data must estimate the dependence of the missingness probability on the missing values. Models and software that attempt this are available (see, for example, Diggle and

Kenward, 1994) but their use is not routine and, in addition, it must be remembered that the associated parameter estimates can be unreliable.

Under what type of dropout mechanism are the mixed effects models considered in this chapter valid? The good news is that such models can be shown to give valid results under the relatively weak assumption that the dropout mechanism is DAR (see Carpenter et al., 2002). When the missing values are thought to be informative, any analysis is potentially problematical. But Diggle and Kenward (1994) have developed a modelling framework for longitudinal data with informative dropouts, in which random or completely random dropout mechanisms are also included as explicit models. The essential feature of the procedure is a logistic regression model for the probability of dropping out, in which the explanatory variables can include previous values of the response variable, and, in addition, the *unobserved* value at dropout as a *latent* variable (i.e., an unobserved variable). In other words, the dropout probability is allowed to depend on both the *observed* measurement history and the unobserved value at dropout. This allows both a formal assessment of the type of dropout mechanism in the data, and the estimation of effects of interest, for example, treatment effects under different assumptions about the dropout mechanism. A full technical account of the model is given in Diggle and Kenward (1994) and a detailed example that uses the approach is described in Carpenter et al. (2002).

One of the problems for an investigator struggling to identify the dropout mechanism in a data set is that there are no routine methods to help, although a number of largely ad hoc graphical procedures can be used as described in Diggle (1998), Everitt (2002b) and Carpenter et al. (2002). One very simple procedure for assessing the dropout mechanism suggested in Carpenter et al. (2002) involves plotting the observations for each treatment group, at each time point, differentiating between two categories of patients; those who do and those who do not attend their next scheduled visit. Any clear difference between the distributions of values for these two categories indicates that dropout is not completely at random. For the Beat the Blues data, such a plot is shown in Figure 12.5. When comparing the distribution of BDI values for patients that do (circles) and do not (bullets) attend the next scheduled visit, there is no apparent difference and so it is reasonable to assume dropout completely at random.

## 12.7 Summary

Linear mixed effects models are extremely useful for modelling longitudinal data. The models allow the correlations between the repeated measurements to be accounted for so that correct inferences can be drawn about the effects of covariates of interest on the repeated response values. In this chapter we have concentrated on responses that are continuous and conditional on the explanatory variables and random effects have a normal distribution. But ran-

```
R> bdi <- BtheB[, grep("bdi", names(BtheB))]
R> plot(1:4, rep(-0.5, 4), type = "n", axes = FALSE,
+       ylim = c(0, 50), xlab = "Months", ylab = "BDI")
R> axis(1, at = 1:4, labels = c(0, 2, 3, 5))
R> axis(2)
R> for (i in 1:4) {
+      dropout <- is.na(bdi[,i + 1])
+      points(rep(i, nrow(bdi)) + ifelse(dropout, 0.05, -0.05),
+             jitter(bdi[,i]), pch = ifelse(dropout, 20, 1))
+ }
```

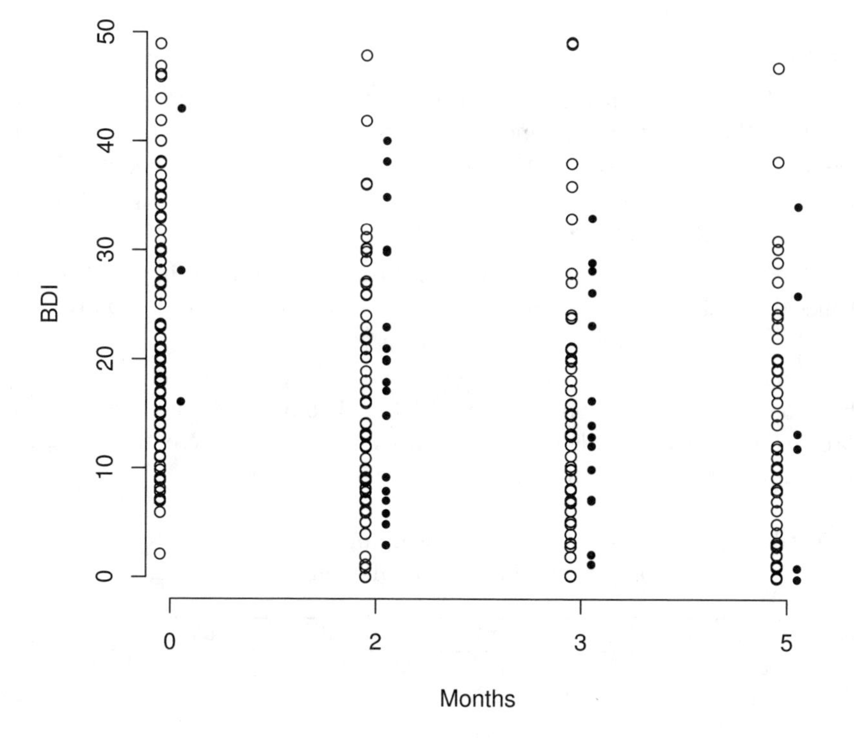


**Figure 12.5** Distribution of BDI values for patients that do (circles) and do not (bullets) attend the next scheduled visit.

dom effects models can also be applied to non-normal responses, for example binary variables – see, for example, Everitt (2002b).

The lack of independence of repeated measures data is what makes the modelling of such data a challenge. But even when only a single measurement of a response is involved, correlation can, in some circumstances, occur between the response values of different individuals and cause similar problems. As an example consider a randomised clinical trial in which subjects are recruited at multiple study centres. The multicentre design can help to provide adequate sample sizes and enhance the generalisability of the results. However factors that vary by centre, including patient characteristics and medical practise patterns, may exert a sufficiently powerful effect to make inferences that ignore the 'clustering' seriously misleading. Consequently it may be necessary to incorporate random effects for centres into the analysis.

## Exercises

Ex. 12.1 Use the `lm` function to fit a model to the Beat the Blues data that assumes that the repeated measurements are independent. Compare the results to those from fitting the random intercept model `BtheB_lmer1`.

Ex. 12.2 Investigate whether there is any evidence of an interaction between treatment and time for the Beat the Blues data.

Ex. 12.3 Construct a plot of the mean profiles of both groups in the Beat the Blues data, showing also standard deviation bars at each time point.

Ex. 12.4 The `phosphate` data given in Table 12.2 show the plasma inorganic phosphate levels for 33 subjects, 20 of whom are controls and 13 of whom have been classified as obese (Davis, 2002). Produce separate plots of the profiles of the individuals in each group, and guided by these plots fit what you think might be sensible linear mixed effects models.

**Table 12.2**: `phosphate` data. Plasma inorganic phosphate levels for various time points after glucose challenge.

| group | t0 | t0.5 | t1 | t1.5 | t2 | t3 | t4 | t5 |
|---|---|---|---|---|---|---|---|---|
| control | 4.3 | 3.3 | 3.0 | 2.6 | 2.2 | 2.5 | 3.4 | 4.4 |
| control | 3.7 | 2.6 | 2.6 | 1.9 | 2.9 | 3.2 | 3.1 | 3.9 |
| control | 4.0 | 4.1 | 3.1 | 2.3 | 2.9 | 3.1 | 3.9 | 4.0 |
| control | 3.6 | 3.0 | 2.2 | 2.8 | 2.9 | 3.9 | 3.8 | 4.0 |
| control | 4.1 | 3.8 | 2.1 | 3.0 | 3.6 | 3.4 | 3.6 | 3.7 |
| control | 3.8 | 2.2 | 2.0 | 2.6 | 3.8 | 3.6 | 3.0 | 3.5 |
| control | 3.8 | 3.0 | 2.4 | 2.5 | 3.1 | 3.4 | 3.5 | 3.7 |
| control | 4.4 | 3.9 | 2.8 | 2.1 | 3.6 | 3.8 | 4.0 | 3.9 |
| control | 5.0 | 4.0 | 3.4 | 3.4 | 3.3 | 3.6 | 4.0 | 4.3 |
| control | 3.7 | 3.1 | 2.9 | 2.2 | 1.5 | 2.3 | 2.7 | 2.8 |
| control | 3.7 | 2.6 | 2.6 | 2.3 | 2.9 | 2.2 | 3.1 | 3.9 |
| control | 4.4 | 3.7 | 3.1 | 3.2 | 3.7 | 4.3 | 3.9 | 4.8 |

**Table 12.2:** `phosphate` data (continued).

| group | t0 | t0.5 | t1 | t1.5 | t2 | t3 | t4 | t5 |
|---|---|---|---|---|---|---|---|---|
| control | 4.7 | 3.1 | 3.2 | 3.3 | 3.2 | 4.2 | 3.7 | 4.3 |
| control | 4.3 | 3.3 | 3.0 | 2.6 | 2.2 | 2.5 | 2.4 | 3.4 |
| control | 5.0 | 4.9 | 4.1 | 3.7 | 3.7 | 4.1 | 4.7 | 4.9 |
| control | 4.6 | 4.4 | 3.9 | 3.9 | 3.7 | 4.2 | 4.8 | 5.0 |
| control | 4.3 | 3.9 | 3.1 | 3.1 | 3.1 | 3.1 | 3.6 | 4.0 |
| control | 3.1 | 3.1 | 3.3 | 2.6 | 2.6 | 1.9 | 2.3 | 2.7 |
| control | 4.8 | 5.0 | 2.9 | 2.8 | 2.2 | 3.1 | 3.5 | 3.6 |
| control | 3.7 | 3.1 | 3.3 | 2.8 | 2.9 | 3.6 | 4.3 | 4.4 |
| obese | 5.4 | 4.7 | 3.9 | 4.1 | 2.8 | 3.7 | 3.5 | 3.7 |
| obese | 3.0 | 2.5 | 2.3 | 2.2 | 2.1 | 2.6 | 3.2 | 3.5 |
| obese | 4.9 | 5.0 | 4.1 | 3.7 | 3.7 | 4.1 | 4.7 | 4.9 |
| obese | 4.8 | 4.3 | 4.7 | 4.6 | 4.7 | 3.7 | 3.6 | 3.9 |
| obese | 4.4 | 4.2 | 4.2 | 3.4 | 3.5 | 3.4 | 3.8 | 4.0 |
| obese | 4.9 | 4.3 | 4.0 | 4.0 | 3.3 | 4.1 | 4.2 | 4.3 |
| obese | 5.1 | 4.1 | 4.6 | 4.1 | 3.4 | 4.2 | 4.4 | 4.9 |
| obese | 4.8 | 4.6 | 4.6 | 4.4 | 4.1 | 4.0 | 3.8 | 3.8 |
| obese | 4.2 | 3.5 | 3.8 | 3.6 | 3.3 | 3.1 | 3.5 | 3.9 |
| obese | 6.6 | 6.1 | 5.2 | 4.1 | 4.3 | 3.8 | 4.2 | 4.8 |
| obese | 3.6 | 3.4 | 3.1 | 2.8 | 2.1 | 2.4 | 2.5 | 3.5 |
| obese | 4.5 | 4.0 | 3.7 | 3.3 | 2.4 | 2.3 | 3.1 | 3.3 |
| obese | 4.6 | 4.4 | 3.8 | 3.8 | 3.8 | 3.6 | 3.8 | 3.8 |

*Source*: From Davis, C. S., *Statistical Methods for the Analysis of Repeated Measurements*, Springer, New York, 2002. With kind permission of Springer Science and Business Media.

CHAPTER 13

# Analysing Longitudinal Data II – Generalised Estimation Equations and Linear Mixed Effect Models: Treating Respiratory Illness and Epileptic Seizures

## 13.1 Introduction

The data in Table 13.1 were collected in a clinical trial comparing two treatments for a respiratory illness (Davis, 1991).

**Table 13.1**: `respiratory` data. Randomised clinical trial data from patients suffering from respiratory illness. Only the data of the first seven patients are shown here.

| centre | treatment | gender | age | status | month | subject |
|---:|---:|---:|---:|---:|---:|---:|
| 1 | placebo | female | 46 | poor | 0 | 1 |
| 1 | placebo | female | 46 | poor | 1 | 1 |
| 1 | placebo | female | 46 | poor | 2 | 1 |
| 1 | placebo | female | 46 | poor | 3 | 1 |
| 1 | placebo | female | 46 | poor | 4 | 1 |
| 1 | placebo | female | 28 | poor | 0 | 2 |
| 1 | placebo | female | 28 | poor | 1 | 2 |
| 1 | placebo | female | 28 | poor | 2 | 2 |
| 1 | placebo | female | 28 | poor | 3 | 2 |
| 1 | placebo | female | 28 | poor | 4 | 2 |
| 1 | treatment | female | 23 | good | 0 | 3 |
| 1 | treatment | female | 23 | good | 1 | 3 |
| 1 | treatment | female | 23 | good | 2 | 3 |
| 1 | treatment | female | 23 | good | 3 | 3 |
| 1 | treatment | female | 23 | good | 4 | 3 |
| 1 | placebo | female | 44 | good | 0 | 4 |
| 1 | placebo | female | 44 | good | 1 | 4 |
| 1 | placebo | female | 44 | good | 2 | 4 |
| 1 | placebo | female | 44 | good | 3 | 4 |
| 1 | placebo | female | 44 | poor | 4 | 4 |
| 1 | placebo | male | 13 | good | 0 | 5 |

**Table 13.1**: `respiratory` data (continued).

| centre | treatment | gender | age | status | month | subject |
|---|---|---|---|---|---|---|
| 1 | placebo | male | 13 | good | 1 | 5 |
| 1 | placebo | male | 13 | good | 2 | 5 |
| 1 | placebo | male | 13 | good | 3 | 5 |
| 1 | placebo | male | 13 | good | 4 | 5 |
| 1 | treatment | female | 34 | poor | 0 | 6 |
| 1 | treatment | female | 34 | poor | 1 | 6 |
| 1 | treatment | female | 34 | poor | 2 | 6 |
| 1 | treatment | female | 34 | poor | 3 | 6 |
| 1 | treatment | female | 34 | poor | 4 | 6 |
| 1 | placebo | female | 43 | poor | 0 | 7 |
| 1 | placebo | female | 43 | good | 1 | 7 |
| 1 | placebo | female | 43 | poor | 2 | 7 |
| 1 | placebo | female | 43 | good | 3 | 7 |
| 1 | placebo | female | 43 | good | 4 | 7 |
| ⋮ | ⋮ | ⋮ | ⋮ | ⋮ | ⋮ | ⋮ |

In each of two centres, eligible patients were randomly assigned to active treatment or placebo. During the treatment, the respiratory status (categorised `poor` or `good`) was determined at each of four, monthly visits. The trial recruited 111 participants (54 in the active group, 57 in the placebo group) and there were no missing data for either the responses or the covariates. The question of interest is to assess whether the treatment is effective and to estimate its effect.

**Table 13.2**: `epilepsy` data. Randomised clinical trial data from patients suffering from epilepsy. Only the data of the first seven patients are shown here.

| treatment | base | age | seizure.rate | period | subject |
|---|---|---|---|---|---|
| placebo | 11 | 31 | 5 | 1 | 1 |
| placebo | 11 | 31 | 3 | 2 | 1 |
| placebo | 11 | 31 | 3 | 3 | 1 |
| placebo | 11 | 31 | 3 | 4 | 1 |
| placebo | 11 | 30 | 3 | 1 | 2 |
| placebo | 11 | 30 | 5 | 2 | 2 |
| placebo | 11 | 30 | 3 | 3 | 2 |
| placebo | 11 | 30 | 3 | 4 | 2 |
| placebo | 6 | 25 | 2 | 1 | 3 |
| placebo | 6 | 25 | 4 | 2 | 3 |
| placebo | 6 | 25 | 0 | 3 | 3 |

**Table 13.2:** `epilepsy` data (continued).

| treatment | base | age | seizure.rate | period | subject |
|---|---|---|---|---|---|
| placebo | 6 | 25 | 5 | 4 | 3 |
| placebo | 8 | 36 | 4 | 1 | 4 |
| placebo | 8 | 36 | 4 | 2 | 4 |
| placebo | 8 | 36 | 1 | 3 | 4 |
| placebo | 8 | 36 | 4 | 4 | 4 |
| placebo | 66 | 22 | 7 | 1 | 5 |
| placebo | 66 | 22 | 18 | 2 | 5 |
| placebo | 66 | 22 | 9 | 3 | 5 |
| placebo | 66 | 22 | 21 | 4 | 5 |
| placebo | 27 | 29 | 5 | 1 | 6 |
| placebo | 27 | 29 | 2 | 2 | 6 |
| placebo | 27 | 29 | 8 | 3 | 6 |
| placebo | 27 | 29 | 7 | 4 | 6 |
| placebo | 12 | 31 | 6 | 1 | 7 |
| placebo | 12 | 31 | 4 | 2 | 7 |
| placebo | 12 | 31 | 0 | 3 | 7 |
| placebo | 12 | 31 | 2 | 4 | 7 |
| ⋮ | ⋮ | ⋮ | ⋮ | ⋮ | ⋮ |

In a clinical trial reported by Thall and Vail (1990), 59 patients with epilepsy were randomised to groups receiving either the antiepileptic drug Progabide or a placebo in addition to standard chemotherapy. The numbers of seizures suffered in each of four, two-week periods were recorded for each patient along with a baseline seizure count for the 8 weeks prior to being randomised to treatment and age. The main question of interest is whether taking Progabide reduced the number of epileptic seizures compared with placebo. A subset of the data is given in Table 13.2.

Note that the two data sets are shown in their 'long form' i.e., one measurement per row in the corresponding *data.frame*s.

## 13.2 Methods for Non-normal Distributions

The data sets `respiratory` and `epilepsy` arise from longitudinal clinical trials, the same type of study that was the subject of consideration in Chapter 12. But in each case the repeatedly measured response variable is clearly not normally distributed making the models considered in the previous chapter unsuitable. In Table 13.1 we have a binary response observed on four occasions, and in Table 13.2 a count response also observed on four occasions. If we choose to ignore the repeated measurements aspects of the two data sets we could use the methods of Chapter 7 applied to the data arranged in the 'long' form introduced in Chapter 12. For the `respiratory` data in Table 13.1 we could then apply logistic regression and for `epilepsy` in Table 13.2, Poisson regression. It can be shown that this approach will give *consistent* estimates of the regression coefficients, i.e., with large samples these point estimates should be close to the true population values. But the assumption of the independence of the repeated measurements will lead to estimated standard errors that are too small for the between-subjects covariates (at least when the correlation between the repeated measurements are positive) as a result of assuming that there are more independent data points than are justified.

We might begin by asking if there is something relatively simple that can be done to 'fix-up' these standard errors so that we can still apply the R `glm` function to get reasonably satisfactory results on longitudinal data with a non-normal response? Two approaches which can often help to get more suitable estimates of the required standard errors are *bootstrapping* and use of the *robust/sandwich, Huber-White variance estimator.*

The idea underlying the bootstrap (see Chapter 8 and Chapter 9), a technique described in detail in Efron and Tibshirani (1993), is to resample from the observed data with replacement to achieve a sample of the same size each time, and to use the variation in the estimated parameters across the set of bootstrap samples in order to get a value for the sampling variability of the estimate (see Chapter 8 also). With correlated data, the bootstrap sample needs to be drawn with replacement from the set of independent subjects, so that intra-subject correlation is preserved in the bootstrap samples. We shall not consider this approach any further here.

The sandwich or robust estimate of variance (see Everitt and Pickles, 2000, for complete details including an explicit definition), involves, unlike the bootstrap which is computationally intensive, a closed-form calculation, based on an asymptotic (large-sample) approximation; it is known to provide good results in many situations. We shall illustrate its use in later examples.

But perhaps more satisfactory would be an approach that fully utilises information on the data's structure, including dependencies over time. In the linear mixed models for Gaussian responses described in Chapter 12, estimation of the regression parameters linking explanatory variables to the response variable and their standard errors needed to take account of the correlational structure of the data, but their interpretation could be undertaken independent of this structure. When modelling non-normal responses this independence of estimation and interpretation no longer holds. Different assumptions about how the correlations are generated can lead to regression coefficients with different interpretations. The essential difference is between *marginal models* and *conditional models.*

### *13.2.1 Marginal Models*

Longitudinal data can be considered as a series of cross-sections, and marginal models for such data use the generalised linear model (see Chapter 7) to fit each cross-section. In this approach the relationship of the marginal mean and the explanatory variables is modelled separately from the within-subject correlation. The marginal regression coefficients have the same interpretation as coefficients from a cross-sectional analysis, and marginal models are natural analogues for correlated data of generalised linear models for independent data. Fitting marginal models to non-normal longitudinal data involves the use of a procedure known as *generalised estimating equations* (GEE), introduced by Liang and Zeger (1986). This approach may be viewed as a multivariate extension of the generalised linear model and the quasi-likelihood method (see Chapter 7). But the problem with applying a direct analogue of the generalised linear model to longitudinal data with non-normal responses is that there is usually no suitable likelihood function with the required combination of the appropriate link function, error distribution and correlation structure. To overcome this problem Liang and Zeger (1986) introduced a general method for incorporating within-subject correlation in GLMs, which is essentially an extension of the quasi-likelihood approach mentioned briefly in Chapter 7. As in conventional generalised linear models, the variances of the responses given the covariates are assumed to be of the form $\mathsf{Var}(\text{response}) = \phi\mathbf{V}(\mu)$ where the variance function $V(\mu)$ is determined by the choice of distribution family (see Chapter 7). Since overdispersion is common in longitudinal data, the dispersion parameter $\phi$ is typically estimated even if the distribution requires $\phi = 1$. The feature of these generalised estimation equations that differs from the usual generalised linear model is that different responses on the same individual are allowed to be correlated given the covariates. These correlations are assumed to have a relatively simple structure defined by a small number of parameters. The following correlation structures are commonly used ($Y_{ij}$ represents the value of the $j$th repeated measurement of the response variable on subject $i$).

**An identity matrix** leading to the independence working model in which the generalised estimating equation reduces to the univariate estimating equation given in Chapter 7, obtained by assuming that the repeated measurements are independent.

**An exchangeable correlation matrix** with a single parameter similar to that described in Chapter 12. Here the correlation between each pair of repeated measurements is assumed to be the same, i.e., $\text{corr}(Y_{ij}, Y_{ik}) = \rho$.

**An AR-1 autoregressive correlation matrix,** also with a single parameter, but in which $\text{corr}(Y_{ij}, Y_{ik}) = \rho^{|k-j|}, j \neq k$. This can allow the correlations of measurements taken farther apart to be less than those taken closer to one another.

**An unstructured correlation matrix** with $K(K-1)/2$ parameters where $K$ is the number of repeated measurements and$\text{corr}(Y_{ij}, Y_{jk}) = \rho_{jk}$

For given values of the regression parameters $\beta_1, \ldots \beta_q$, the $\rho$-parameters of the working correlation matrix can be estimated along with the dispersion parameter $\phi$ (see Zeger and Liang, 1986, for details). These estimates can then be used in the so-called generalised estimating equations to obtain estimates of the regression parameters. The GEE algorithm proceeds by iterating between (1) estimation of the regression parameters using the correlation and dispersion parameters from the previous iteration and (2) estimation of the correlation and dispersion parameters using the regression parameters from the previous iteration.

The estimated regression coefficients are 'robust' in the sense that they are consistent from misspecified correlation structures assuming that the mean structure is correctly specified. Note however that the GEE estimates of marginal effects are not robust against misspecified regression structures, such as omitted covariates.

The use of GEE estimation on a longitudinal data set in which some subjects drop out assumes that they drop out completely at random (see Chapter 12).

### *13.2.2 Conditional Models*

The random effects approach described in the previous chapter can be extended to non-normal responses although the resulting models can be difficult to estimate because the likelihood involves integrals over the random effects distribution that generally do not have closed forms. A consequence is that it is often possible to fit only relatively simple models. In these models estimated regression coefficients have to be interpreted, conditional on the random effects. The regression parameters in the model are said to be subject-specific and such effects will differ from the marginal or population averaged effects estimated using GEE, except when using an identity link function and a normal error distribution.

Consider a set of longitudinal data in which $Y_{ij}$ is the value of a binary response for individual $i$ at say time $t_j$. The logistic regression model (see Chapter 7) for the response is now written as

$$\text{logit}\,(\mathsf{P}(y_{ij} = 1|u_i)) = \beta_0 + \beta_1 t_j + u_i \qquad (13.1)$$

where $u_i$ is a random effect assumed to be normally distributed with zero mean and variance $\sigma_u^2$. This is a simple example of a *generalised linear mixed model* because it is a generalised linear model with both a fixed effect, $\beta_1$, and a random effect, $u_i$.

Here the regression parameter $\beta_1$ again represents the change in the log odds per unit change in time, but this is now *conditional* on the random effect. We can illustrate this difference graphically by simulating the model (13.1); the result is shown in Figure 13.1. Here the thin grey curves represent subject-specific relationships between the probability that the response equals one and a covariate $t$ for model (13.1). The horizontal shifts are due to different values of the random intercept. The thick black curve represents the population averaged relationship, formed by averaging the thin curves for each value of $t$. It is, in effect, the thick curve that would be estimated in a marginal model (see

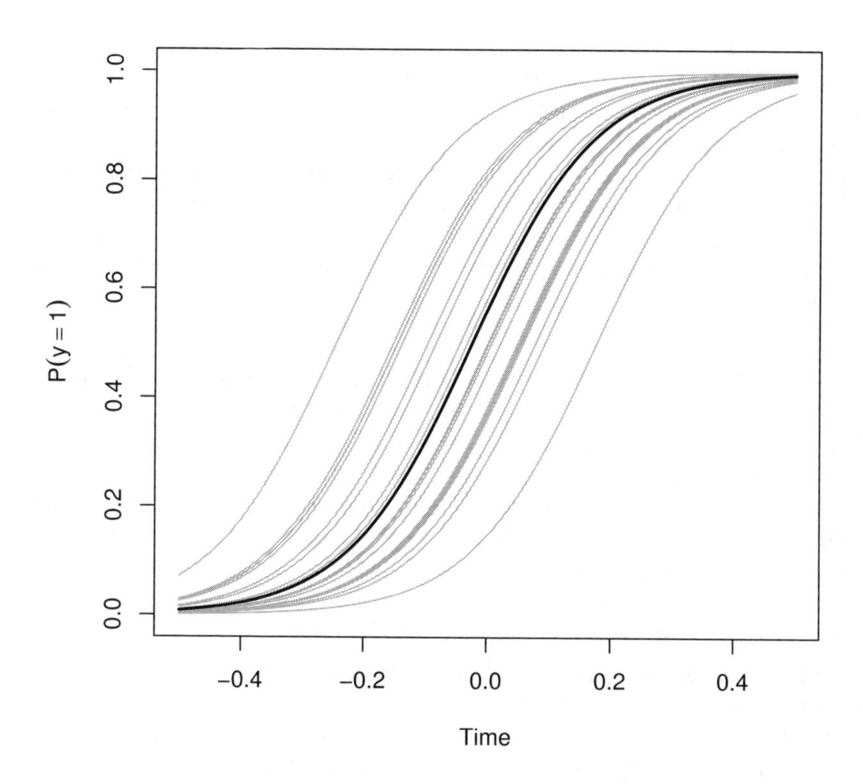


**Figure 13.1** Simulation of a positive response in a random intercept logistic regression model for 20 subjects. The thick line is the average over all 20 subjects.

previous sub-section). The population averaged regression parameters tend to be attenuated (closest to zero) relative to the subject-specific regression parameters. A marginal regression model does not address questions concerning heterogeneity between individuals.

Estimating the parameters in a logistic random effects model is undertaken by maximum likelihood. Details are given in Skrondal and Rabe-Hesketh (2004). If the model is correctly specified, maximum likelihood estimates are consistent when subjects in the study drop out at random (see Chapter 12).

## 13.3 Analysis Using R: GEE

### *13.3.1 Beat the Blues Revisited*

Although we have introduced GEE as a method for analysing longitudinal data where the response variable is non-normal, it can also be applied to data where the response can be assumed to follow a conditional normal distribution (conditioning being on the explanatory variables). Consequently we first apply the method to the data used in the previous chapter so we can compare the results we get with those obtained from using the mixed-effects models used there.

To use the `gee` function, package **gee** (Carey et al., 2008) has to be installed and attached:

```
R> library("gee")
```

The `gee` function is used in a similar way to the `lme` function met in Chapter 12 with the addition of the features of the `glm` function that specify the appropriate error distribution for the response and the implied link function, and an argument to specify the structure of the working correlation matrix. Here we will fit an independence structure and then an exchangeable structure. The R code for fitting generalised estimation equations to the `BtheB_long` data (as constructed in Chapter 12) with identity working correlation matrix is as follows (note that the `gee` function assumes the rows of the *data.frame* `BtheB_long` to be ordered with respect to subjects):

```
R> osub <- order(as.integer(BtheB_long$subject))
R> BtheB_long <- BtheB_long[osub,]
R> btb_gee <- gee(bdi ~ bdi.pre + trt + length + drug,
+      data = BtheB_long, id = subject, family = gaussian,
+      corstr = "independence")
```

and with exchangeable correlation matrix:

```
R> btb_gee1 <- gee(bdi ~ bdi.pre + trt + length + drug,
+      data = BtheB_long, id = subject, family = gaussian,
+      corstr = "exchangeable")
```

The `summary` method can be used to inspect the fitted models; the results are shown in Figures 13.2 and 13.3.

Note how the naïve and the sandwich or robust estimates of the standard errors are considerably different for the independence structure (Figure 13.2), but quite similar for the exchangeable structure (Figure 13.3). This simply reflects that using an exchangeable working correlation matrix is more realistic for these data and that the standard errors resulting from this assumption are already quite reasonable without applying the 'sandwich' procedure to them. And if we compare the results under this assumed structure with those for the random intercept model given in Chapter 12 (Figure 12.2) we see that they are almost identical, since the random intercept model also implies an exchangeable structure for the correlations of the repeated measurements.

The single estimated parameter for the working correlation matrix from the

```
R> summary(btb_gee)

...
Model:
 Link:                      Identity
 Variance to Mean Relation: Gaussian
 Correlation Structure:     Independent

...

Coefficients:
            Estimate Naive S.E. Naive z Robust S.E. Robust z
(Intercept)    3.569     1.4833    2.41      2.2695    1.572
bdi.pre        0.582     0.0564   10.32      0.0916    6.355
trtBtheB      -3.237     1.1296   -2.87      1.7746   -1.824
length>6m      1.458     1.1380    1.28      1.4826    0.983
drugYes       -3.741     1.1766   -3.18      1.7827   -2.099

Estimated Scale Parameter:  79.3
...
```

**Figure 13.2** R output of the `summary` method for the `btb_gee` model (slightly abbreviated).

GEE procedure is 0.676, very similar to the estimated intra-class correlation coefficient from the random intercept model. i.e., $7.03^2/(5.07^2 + 7.03^2) = 0.66$ – see Figure 12.2.

### *13.3.2 Respiratory Illness*

We will now apply the GEE procedure to the `respiratory` data shown in Table 13.1. Given the binary nature of the response variable we will choose a binomial error distribution and by default a logistic link function. We shall also fix the scale parameter $\phi$ described in Chapter 7 at one. (The default in the `gee` function is to estimate this parameter.) Again we will apply the procedure twice, firstly with an independence structure and then with an exchangeable structure for the working correlation matrix. We will also fit a logistic regression model to the data using `glm` so we can compare results.

The baseline status, i.e., the status for `month == 0`, will enter the models as an explanatory variable and thus we have to rearrange the *data.frame* `respiratory` in order to create a new variable `baseline`:

```
R> data("respiratory", package = "HSAUR2")
R> resp <- subset(respiratory, month > "0")
R> resp$baseline <- rep(subset(respiratory, month == "0")$status,
+                        rep(4, 111))
```

```
R> summary(btb_gee1)

...
Model:
 Link:                      Identity
 Variance to Mean Relation: Gaussian
 Correlation Structure:     Exchangeable

...

Coefficients:
            Estimate Naive S.E. Naive z Robust S.E. Robust z
(Intercept)    3.023     2.3039  1.3122      2.2320   1.3544
bdi.pre        0.648     0.0823  7.8741      0.0835   7.7583
trtBtheB      -2.169     1.7664 -1.2281      1.7361  -1.2495
length>6m     -0.111     1.7309 -0.0643      1.5509  -0.0718
drugYes       -3.000     1.8257 -1.6430      1.7316  -1.7323

Estimated Scale Parameter:  81.7
...
```

**Figure 13.3** R output of the summary method for the btb_gee1 model (slightly abbreviated).

```
R> resp$nstat <- as.numeric(resp$status == "good")
R> resp$month <- resp$month[, drop = TRUE]
```

The new variable nstat is simply a dummy coding for a poor respiratory status. Now we can use the data resp to fit a logistic regression model and GEE models with an independent and an exchangeable correlation structure as follows.

```
R> resp_glm <- glm(status ~ centre + trt + gender + baseline
+      + age, data = resp, family = "binomial")
R> resp_gee1 <- gee(nstat ~ centre + trt + gender + baseline
+      + age, data = resp, family = "binomial", id = subject,
+      corstr = "independence", scale.fix = TRUE,
+      scale.value = 1)
R> resp_gee2 <- gee(nstat ~ centre + trt + gender + baseline
+      + age, data = resp, family = "binomial", id = subject,
+      corstr = "exchangeable", scale.fix = TRUE,
+      scale.value = 1)
```

Again, summary methods can be used for an inspection of the details of the fitted models; the results are given in Figures 13.4, 13.5 and 13.6. We see that the results from applying logistic regression to the data with the glm function gives identical results to those obtained from gee with an independence correlation structure (comparing the glm standard errors with the naïve standard errors from gee). The robust standard errors for the between subject

```
R> summary(resp_glm)

Call:
glm(formula = status ~ centre + trt + gender + baseline
    + age, family = "binomial", data = resp)

Deviance Residuals:
    Min       1Q   Median       3Q      Max
-2.315   -0.855   0.434    0.895    1.925

Coefficients:
             Estimate Std. Error z value Pr(>|z|)
(Intercept)  -0.90017    0.33765   -2.67   0.0077
centre2       0.67160    0.23957    2.80   0.0051
trttrt        1.29922    0.23684    5.49  4.1e-08
gendermale    0.11924    0.29467    0.40   0.6857
baselinegood  1.88203    0.24129    7.80  6.2e-15
age          -0.01817    0.00886   -2.05   0.0404

(Dispersion parameter for binomial family taken to be 1)

    Null deviance: 608.93  on 443  degrees of freedom
Residual deviance: 483.22  on 438  degrees of freedom
AIC: 495.2

Number of Fisher Scoring iterations: 4
```

**Figure 13.4** R output of the `summary` method for the `resp_glm` model.

covariates are considerably larger than those estimated assuming independence, implying that the independence assumption is not realistic for these data. Applying the GEE procedure with an exchangeable correlation structure results in naïve and robust standard errors that are identical, and similar to the robust estimates from the independence structure. It is clear that the exchangeable structure more adequately reflects the correlational structure of the observed repeated measurements than does independence.

The estimated treatment effect taken from the exchangeable structure GEE model is 1.299 which, using the robust standard errors, has an associated 95% confidence interval

```
R> se <- summary(resp_gee2)$coefficients["trttrt",
+                                          "Robust S.E."]
R> coef(resp_gee2)["trttrt"] +
+      c(-1, 1) * se * qnorm(0.975)

[1] 0.612 1.987
```

These values reflect effects on the log-odds scale. Interpretation becomes sim-

```
R> summary(resp_gee1)

...
Model:
 Link:                      Logit
 Variance to Mean Relation: Binomial
 Correlation Structure:     Independent

...

Coefficients:
             Estimate Naive S.E. Naive z Robust S.E. Robust z
(Intercept)   -0.9002    0.33765  -2.666       0.460   -1.956
centre2        0.6716    0.23957   2.803       0.357    1.882
trttrt         1.2992    0.23684   5.486       0.351    3.704
gendermale     0.1192    0.29467   0.405       0.443    0.269
baselinegood   1.8820    0.24129   7.800       0.350    5.376
age           -0.0182    0.00886  -2.049       0.013   -1.397

Estimated Scale Parameter:  1
...
```

**Figure 13.5** R output of the `summary` method for the `resp_gee1` model (slightly abbreviated).

pler if we exponentiate the values to get the effects in terms of odds. This gives a treatment effect of 3.666 and a 95% confidence interval of

```
R> exp(coef(resp_gee2)["trttrt"] +
+      c(-1, 1) * se * qnorm(0.975))

[1] 1.84 7.29
```

The odds of achieving a 'good' respiratory status with the active treatment is between about twice and seven times the corresponding odds for the placebo.

### *13.3.3 Epilepsy*

Moving on to the count data in `epilepsy` from Table 13.2, we begin by calculating the means and variances of the number of seizures for all interactions between treatment and period:

```
R> data("epilepsy", package = "HSAUR2")
R> itp <- interaction(epilepsy$treatment, epilepsy$period)
R> tapply(epilepsy$seizure.rate, itp, mean)

  placebo.1 Progabide.1   placebo.2 Progabide.2   placebo.3
       9.36        8.58        8.29        8.42        8.79
Progabide.3   placebo.4 Progabide.4
       8.13        7.96        6.71
```

```
R> summary(resp_gee2)

...
Model:
 Link:                      Logit
 Variance to Mean Relation: Binomial
 Correlation Structure:     Exchangeable

...

Coefficients:
             Estimate Naive S.E. Naive z Robust S.E. Robust z
(Intercept)   -0.9002     0.4785  -1.881       0.460   -1.956
centre2        0.6716     0.3395   1.978       0.357    1.882
trttrt         1.2992     0.3356   3.871       0.351    3.704
gendermale     0.1192     0.4176   0.286       0.443    0.269
baselinegood   1.8820     0.3419   5.504       0.350    5.376
age           -0.0182     0.0126  -1.446       0.013   -1.397

Estimated Scale Parameter:  1
...
```

**Figure 13.6** R output of the `summary` method for the `resp_gee2` model (slightly abbreviated).

```
R> tapply(epilepsy$seizure.rate, itp, var)

  placebo.1 Progabide.1   placebo.2 Progabide.2   placebo.3
      102.8       332.7        66.7       140.7       215.3
Progabide.3   placebo.4 Progabide.4
      193.0        58.2       126.9
```

Some of the variances are considerably larger than the corresponding means, which for a Poisson variable may suggest that overdispersion may be a problem, see Chapter 7.

We will now construct some boxplots first for the numbers of seizures observed in each two-week period post randomisation. The resulting diagram is shown in Figure 13.7. Some quite extreme 'outliers' are indicated, particularly the observation in period one in the Progabide group. But given these are count data which we will model using a Poisson error distribution and a log link function, it may be more appropriate to look at the boxplots *after* taking a log transformation. (Since some observed counts are zero we will add 1 to all observations before taking logs.) To get the plots we can use the R code displayed with Figure 13.8. In Figure 13.8 the outlier problem seems less troublesome and we shall not attempt to remove any of the observations for subsequent analysis.

Before proceeding with the formal analysis of these data we have to deal with a small problem produced by the fact that the baseline counts were observed

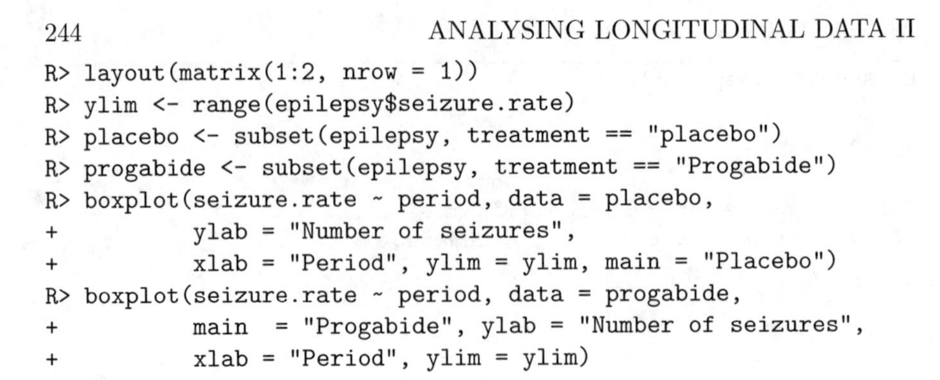

```
R> layout(matrix(1:2, nrow = 1))
R> ylim <- range(epilepsy$seizure.rate)
R> placebo <- subset(epilepsy, treatment == "placebo")
R> progabide <- subset(epilepsy, treatment == "Progabide")
R> boxplot(seizure.rate ~ period, data = placebo,
+          ylab = "Number of seizures",
+          xlab = "Period", ylim = ylim, main = "Placebo")
R> boxplot(seizure.rate ~ period, data = progabide,
+          main  = "Progabide", ylab = "Number of seizures",
+          xlab = "Period", ylim = ylim)
```

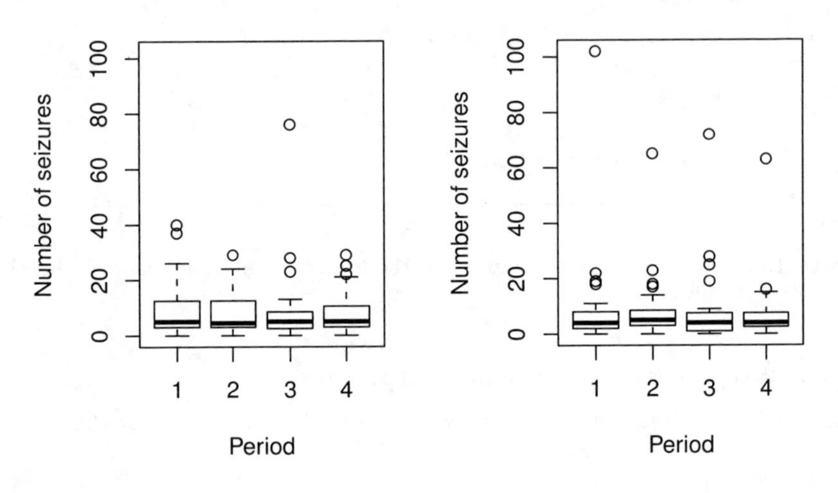


**Figure 13.7** Boxplots of numbers of seizures in each two-week period post randomisation for placebo and active treatments.

over an eight-week period whereas all subsequent counts are over two-week periods. For the baseline count we shall simply divide by eight to get an average weekly rate, but we cannot do the same for the post-randomisation counts if we are going to assume a Poisson distribution (since we will no longer have integer values for the response). But we can model the mean count for each two-week period by introducing the log of the observation period as an *offset* (a covariate with regression coefficient set to one). The model then becomes log(expected count in observation period) = linear function of explanatory variables+log(observation period), leading to the model for the rate in counts per week (assuming the observation periods are measured in weeks) as expected count in observation period/observation period = exp(linear function

```
R> layout(matrix(1:2, nrow = 1))
R> ylim <- range(log(epilepsy$seizure.rate + 1))
R> boxplot(log(seizure.rate + 1) ~ period, data = placebo,
+          main = "Placebo", ylab = "Log number of seizures",
+          xlab = "Period", ylim = ylim)
R> boxplot(log(seizure.rate + 1) ~ period, data = progabide,
+          main = "Progabide", ylab = "Log number of seizures",
+          xlab = "Period", ylim = ylim)
```

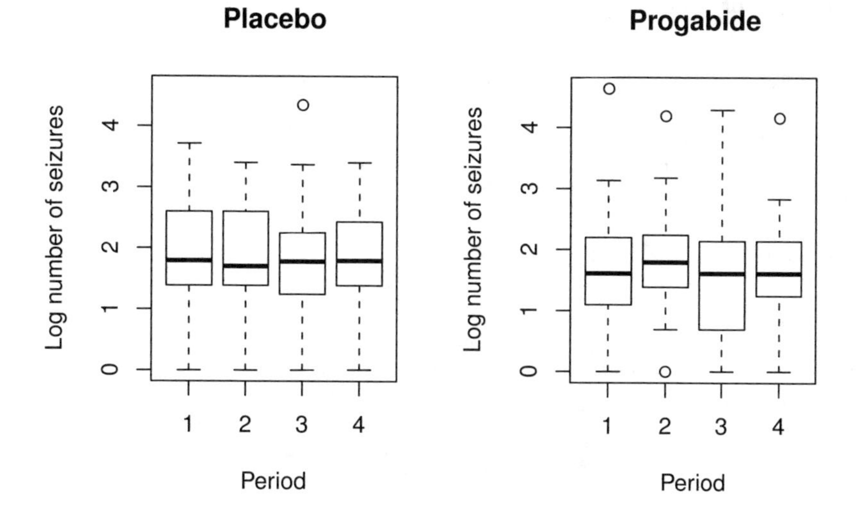


**Figure 13.8** Boxplots of log of numbers of seizures in each two-week period post randomisation for placebo and active treatments.

of explanatory variables). In our example the observation period is two weeks, so we simply need to set log(2) for each observation as the offset.

We can now fit a Poisson regression model to the data assuming independence using the `glm` function. We also use the GEE approach to fit an independence structure, followed by an exchangeable structure using the following R code:

```
R> per <- rep(log(2),nrow(epilepsy))
R> epilepsy$period <- as.numeric(epilepsy$period)
R> names(epilepsy)[names(epilepsy) == "treatment"] <- "trt"
R> fm <- seizure.rate ~ base + age + trt + offset(per)
R> epilepsy_glm <- glm(fm, data = epilepsy, family = "poisson")
R> epilepsy_gee1 <- gee(fm, data = epilepsy, family = "poisson",
+      id = subject, corstr = "independence", scale.fix = TRUE,
+      scale.value = 1)
```

```
R> epilepsy_gee2 <- gee(fm, data = epilepsy, family = "poisson",
+      id = subject, corstr = "exchangeable", scale.fix = TRUE,
+      scale.value = 1)
R> epilepsy_gee3 <- gee(fm, data = epilepsy, family = "poisson",
+      id = subject, corstr = "exchangeable", scale.fix = FALSE,
+      scale.value = 1)
```

As usual we inspect the fitted models using the summary method, the results are given in Figures 13.9, 13.10, 13.11, and 13.12.

```
R> summary(epilepsy_glm)

Call:
glm(formula = fm, family = "poisson", data = epilepsy)

Deviance Residuals:
   Min       1Q   Median       3Q      Max
-4.436   -1.403   -0.503    0.484   12.322

Coefficients:
              Estimate Std. Error z value Pr(>|z|)
(Intercept)  -0.130616   0.135619   -0.96   0.3355
base          0.022652   0.000509   44.48  < 2e-16
age           0.022740   0.004024    5.65  1.6e-08
trtProgabide -0.152701   0.047805   -3.19   0.0014

(Dispersion parameter for poisson family taken to be 1)

    Null deviance: 2521.75  on 235  degrees of freedom
Residual deviance:  958.46  on 232  degrees of freedom
AIC: 1732

Number of Fisher Scoring iterations: 5
```

**Figure 13.9** R output of the summary method for the epilepsy_glm model.

For this example, the estimates of standard errors under independence are about half of the corresponding robust estimates, and the situation improves only a little when an exchangeable structure is fitted. Using the naïve standard errors leads, in particular, to a highly significant treatment effect which disappears when the robust estimates are used. The problem with the GEE approach here, using either the independence or exchangeable correlation structure lies in constraining the scale parameter to be one. For these data there is overdispersion which has to be accommodated by allowing this parameter to be freely estimated. When this is done, it gives the last set of results shown above. The estimate of $\phi$ is 5.09 and the naïve and robust estimates of the standard errors are now very similar. It is clear that there is no evidence of a treatment effect.

```
R> summary(epilepsy_gee1)

...
Model:
 Link:                      Logarithm
 Variance to Mean Relation: Poisson
 Correlation Structure:     Independent

...

Coefficients:
             Estimate Naive S.E. Naive z Robust S.E. Robust z
(Intercept)   -0.1306   0.135619  -0.963     0.36515   -0.358
base           0.0227   0.000509  44.476     0.00124   18.332
age            0.0227   0.004024   5.651     0.01158    1.964
trtProgabide  -0.1527   0.047805  -3.194     0.17111   -0.892

Estimated Scale Parameter:  1
...
```

**Figure 13.10** R output of the summary method for the epilepsy_gee1 model (slightly abbreviated).

## 13.4 Analysis Using R: Random Effects

As an example of using generalised mixed models for the analysis of longitudinal data with a non-normal response, the following logistic model will be fitted to the respiratory illness data

$$\begin{aligned}\text{logit}(\mathsf{P}(\text{status} = \text{good})) &= \beta_0 + \beta_1\text{treatment} + \beta_2\text{time} + \beta_3\text{gender} \\ &\quad +\beta_4\text{age} + \beta_5\text{centre} + \beta_6\text{baseline} + u\end{aligned}$$

where $u$ is a subject specific random effect.

The necessary R code for fitting the model using the lmer function from package **lme4** (Bates and Sarkar, 2008, Bates, 2005) is:

```
R> library("lme4")
R> resp_lmer <- lmer(status ~ baseline + month +
+      trt + gender + age + centre + (1 | subject),
+      family = binomial(), data = resp)
R> exp(fixef(resp_lmer))

 (Intercept) baselinegood      month.L      month.Q
       0.189       22.361        0.796        0.962
     month.C       trttrt   gendermale          age
       0.691        8.881        1.227        0.975
     centre2
       2.875
```

The significance of the effects as estimated by this random effects model

```
R> summary(epilepsy_gee2)

...
Model:
 Link:                      Logarithm
 Variance to Mean Relation: Poisson
 Correlation Structure:     Exchangeable

...

Coefficients:
               Estimate Naive S.E. Naive z Robust S.E. Robust z
(Intercept)     -0.1306   0.200442  -0.652     0.36515   -0.358
base             0.0227   0.000753  30.093     0.00124   18.332
age              0.0227   0.005947   3.824     0.01158    1.964
trtProgabide    -0.1527   0.070655  -2.161     0.17111   -0.892

Estimated Scale Parameter:  1
...
```

**Figure 13.11** R output of the `summary` method for the `epilepsy_gee2` model (slightly abbreviated).

and by the GEE model described in Section 13.3.2 is generally similar. But as expected from our previous discussion the estimated coefficients are substantially larger. While the estimated effect of treatment on a randomly sampled individual, given the set of observed covariates, is estimated by the marginal model using GEE to increase the log-odds of being disease free by 1.299, the corresponding estimate from the random effects model is 2.184. These are not inconsistent results but reflect the fact that the models are estimating different parameters. The random effects estimate is conditional upon the patient's random effect, a quantity that is rarely known in practise. Were we to examine the log-odds of the average predicted probabilities with and without treatment (averaged over the random effects) this would give an estimate comparable to that estimated within the marginal model.

```
R> summary(epilepsy_gee3)

...
Model:
 Link:                      Logarithm
 Variance to Mean Relation: Poisson
 Correlation Structure:     Exchangeable

...

Coefficients:
               Estimate Naive S.E. Naive z Robust S.E. Robust z
(Intercept)     -0.1306    0.45220  -0.289     0.36515   -0.358
base             0.0227    0.00170  13.339     0.00124   18.332
age              0.0227    0.01342   1.695     0.01158    1.964
trtProgabide    -0.1527    0.15940  -0.958     0.17111   -0.892

Estimated Scale Parameter:  5.09
...
```

**Figure 13.12** R output of the summary method for the epilepsy_gee3 model (slightly abbreviated).

```
R> summary(resp_lmer)

...
Fixed effects:
               Estimate Std. Error z value Pr(>|z|)
(Intercept)     -1.6666     0.7671   -2.17     0.03
baselinegood     3.1073     0.5325    5.84  5.4e-09
month.L         -0.2279     0.2719   -0.84     0.40
month.Q         -0.0389     0.2716   -0.14     0.89
month.C         -0.3689     0.2727   -1.35     0.18
trttrt           2.1839     0.5237    4.17  3.0e-05
gendermale       0.2045     0.6688    0.31     0.76
age             -0.0257     0.0202   -1.27     0.20
centre2          1.0561     0.5381    1.96     0.05

...
```

**Figure 13.13** R output of the summary method for the resp_lmer model (abbreviated).

## 13.5 Summary

This chapter has outlined and illustrated two approaches to the analysis of non-normal longitudinal data: the marginal approach and the random effect (mixed modelling) approach. Though less unified than the methods available for normally distributed responses, these methods provide powerful and flexible tools to analyse, what until relatively recently, have been seen as almost intractable data.

## Exercises

Ex. 13.1 For the `epilepsy` data investigate what Poisson models are most suitable when subject 49 is excluded from the analysis.

Ex. 13.2 Investigate the use of other correlational structures than the independence and exchangeable structures used in the text, for both the `respiratory` and the `epilepsy` data.

Ex. 13.3 The data shown in Table 13.3 were collected in a follow-up study of women patients with schizophrenia (Davis, 2002). The binary response recorded at 0, 2, 6, 8 and 10 months after hospitalisation was thought disorder (absent or present). The single covariate is the factor indicating whether a patient had suffered early or late onset of her condition (age of onset less than 20 years or age of onset 20 years or above). The question of interest is whether the course of the illness differs between patients with early and late onset? Investigate this question using the GEE approach.

**Table 13.3**: `schizophrenia2` data. Clinical trial data from patients suffering from schizophrenia. Only the data of the first four patients are shown here.

| `subject` | `onset` | `disorder` | `month` |
|---:|---:|---:|---:|
| 1 | < 20 yrs | present | 0 |
| 1 | < 20 yrs | present | 2 |
| 1 | < 20 yrs | absent | 6 |
| 1 | < 20 yrs | absent | 8 |
| 1 | < 20 yrs | absent | 10 |
| 2 | > 20 yrs | absent | 0 |
| 2 | > 20 yrs | absent | 2 |
| 2 | > 20 yrs | absent | 6 |
| 2 | > 20 yrs | absent | 8 |
| 2 | > 20 yrs | absent | 10 |
| 3 | < 20 yrs | present | 0 |
| 3 | < 20 yrs | present | 2 |
| 3 | < 20 yrs | absent | 6 |
| 3 | < 20 yrs | absent | 8 |
| 3 | < 20 yrs | absent | 10 |
| 4 | < 20 yrs | absent | 0 |
| 4 | < 20 yrs | absent | 2 |
| 4 | < 20 yrs | absent | 6 |
| 4 | < 20 yrs | absent | 8 |
| 4 | < 20 yrs | absent | 10 |
| ⋮ | ⋮ | ⋮ | ⋮ |

*Source*: From Davis, C. S., *Statistical Methods for the Analysis of Repeated Measurements*, Springer, New York, 2002. With kind permission of Springer Science and Business Media.

CHAPTER 14

# Simultaneous Inference and Multiple Comparisons: Genetic Components of Alcoholism, Deer Browsing Intensities, and Cloud Seeding

## 14.1 Introduction

Various studies have linked alcohol dependence phenotypes to chromosome 4. One candidate gene is *NACP* (non-amyloid component of plaques), coding for alpha synuclein. Bönsch et al. (2005) found longer alleles of *NACP*-REP1 in alcohol-dependent patients and report that the allele lengths show some association with levels of expressed alpha synuclein mRNA in alcohol-dependent subjects. The data are given in Table 14.1. Allele length is measured as a sum score built from additive dinucleotide repeat length and categorised into three groups: short ($0-4$, $n=24$), intermediate ($5-9$, $n=58$), and long ($10-12$, $n=15$). Here, we are interested in comparing the distribution of the expression level of alpha synuclein mRNA in three groups of subjects defined by the allele length. A global $F$-test in an ANOVA model answers the question if there is any difference in the distribution of the expression levels among allele length groups but additional effort is needed to identify the nature of these differences. Multiple comparison procedures, i.e., tests and confidence intervals for pairwise comparisons of allele length groups, may lead to additional insight into the dependence of expression levels and allele length.

**Table 14.1:** `alpha` data (package **coin**). Allele length and levels of expressed alpha synuclein mRNA in alcohol-dependent patients.

| `alength` | `elevel` | `alength` | `elevel` | `alength` | `elevel` |
|---|---|---|---|---|---|
| short | 1.43 | intermediate | 1.63 | intermediate | 3.07 |
| short | -2.83 | intermediate | 2.53 | intermediate | 4.43 |
| short | 1.23 | intermediate | 0.10 | intermediate | 1.33 |
| short | -1.47 | intermediate | 2.53 | intermediate | 1.03 |
| short | 2.57 | intermediate | 2.27 | intermediate | 3.13 |
| short | 3.00 | intermediate | 0.70 | intermediate | 4.17 |
| short | 5.63 | intermediate | 3.80 | intermediate | 2.70 |
| short | 2.80 | intermediate | -2.37 | intermediate | 3.93 |
| short | 3.17 | intermediate | 0.67 | intermediate | 3.90 |

**Table 14.1**: alpha data (continued).

| alength | elevel | alength | elevel | alength | elevel |
|---|---|---|---|---|---|
| short | 2.00 | intermediate | -0.37 | intermediate | 2.17 |
| short | 2.93 | intermediate | 3.20 | intermediate | 3.13 |
| short | 2.87 | intermediate | 3.05 | intermediate | -2.40 |
| short | 1.83 | intermediate | 1.97 | intermediate | 1.90 |
| short | 1.05 | intermediate | 3.33 | intermediate | 1.60 |
| short | 1.00 | intermediate | 2.90 | intermediate | 0.67 |
| short | 2.77 | intermediate | 2.77 | intermediate | 0.73 |
| short | 1.43 | intermediate | 4.05 | long | 1.60 |
| short | 5.80 | intermediate | 2.13 | long | 3.60 |
| short | 2.80 | intermediate | 3.53 | long | 1.45 |
| short | 1.17 | intermediate | 3.67 | long | 4.10 |
| short | 0.47 | intermediate | 2.13 | long | 3.37 |
| short | 2.33 | intermediate | 1.40 | long | 3.20 |
| short | 1.47 | intermediate | 3.50 | long | 3.20 |
| short | 0.10 | intermediate | 3.53 | long | 4.23 |
| intermediate | -1.90 | intermediate | 2.20 | long | 3.43 |
| intermediate | 1.55 | intermediate | 4.23 | long | 4.40 |
| intermediate | 3.27 | intermediate | 2.87 | long | 3.27 |
| intermediate | 0.30 | intermediate | 3.20 | long | 1.75 |
| intermediate | 1.90 | intermediate | 3.40 | long | 1.77 |
| intermediate | 2.53 | intermediate | 4.17 | long | 3.43 |
| intermediate | 2.83 | intermediate | 4.30 | long | 3.50 |
| intermediate | 3.10 | intermediate | 3.07 | | |
| intermediate | 2.07 | intermediate | 4.03 | | |

In most parts of Germany, the natural or artificial regeneration of forests is difficult due to a high browsing intensity. Young trees suffer from browsing damage, mostly by roe and red deer. An enormous amount of money is spent for protecting these plants by fences trying to exclude game from regeneration areas. The problem is most difficult in mountain areas, where intact and regenerating forest systems play an important role to prevent damages from floods and landslides. In order to estimate the browsing intensity for several tree species, the Bavarian State Ministry of Agriculture and Forestry conducts a survey every three years. Based on the estimated percentage of damaged trees, suggestions for the implementation or modification of deer management plans are made. The survey takes place in all 756 game management districts ('Hegegemeinschaften') in Bavaria. Here, we focus on the 2006 data of the game management district number 513 'Unterer Aischgrund' (located in Frankonia between Erlangen and Höchstadt). The data of 2700 trees include the species and a binary variable indicating whether or not the tree suffered from damage caused by deer browsing; a small fraction of the data is shown in

Table 14.2 (see Hothorn et al., 2008a, also). For each of 36 points on a predefined lattice laid out over the observation area, 15 small trees are investigated on each of 5 plots located on a 100m transect line. Thus, the observations aren't independent of each other and this spatial structure has to be taken into account for our analysis. Our main target is to estimate the probability of suffering from roe deer browsing for all tree species simultaneously.

**Table 14.2**: `trees513` data (package **multcomp**).

| | damage | species | lattice | plot |
|---|---|---|---|---|
| 1 | yes | oak | 1 | 1_1 |
| 2 | no | pine | 1 | 1_1 |
| 3 | no | oak | 1 | 1_1 |
| 4 | no | pine | 1 | 1_1 |
| 5 | no | pine | 1 | 1_1 |
| 6 | no | pine | 1 | 1_1 |
| 7 | yes | oak | 1 | 1_1 |
| 8 | no | hardwood (other) | 1 | 1_1 |
| 9 | no | oak | 1 | 1_1 |
| 10 | no | hardwood (other) | 1 | 1_1 |
| 11 | no | oak | 1 | 1_1 |
| 12 | no | pine | 1 | 1_1 |
| 13 | no | pine | 1 | 1_1 |
| 14 | yes | oak | 1 | 1_1 |
| 15 | no | oak | 1 | 1_1 |
| 16 | no | pine | 1 | 1_2 |
| 17 | yes | hardwood (other) | 1 | 1_2 |
| 18 | no | oak | 1 | 1_2 |
| 19 | no | pine | 1 | 1_2 |
| 20 | no | oak | 1 | 1_2 |
| 21 | ⋮ | ⋮ | ⋮ | ⋮ |

For the cloud seeding data presented in Table 6.2 of Chapter 6, we investigated the dependency of rainfall on the suitability criterion when clouds were seeded or not (see Figure 6.6). In addition to the regression lines presented there, confidence bands for the regression lines would add further information on the variability of the predicted rainfall depending on the suitability criterion; simultaneous confidence intervals are a simple method for constructing such bands as we will see in the following section.

## 14.2 Simultaneous Inference and Multiple Comparisons

Multiplicity is an intrinsic problem of any simultaneous inference. If each of $k$, say, null hypotheses is tested at nominal level $\alpha$ on the same data set, the overall type I error rate can be substantially larger than $\alpha$. That is, the probability of at least one erroneous rejection is larger than $\alpha$ for $k \geq 2$. Simultaneous inference procedures adjust for multiplicity and thus ensure that the overall type I error remains below the pre-specified significance level $\alpha$.

The term *multiple comparison procedure* refers to simultaneous inference, i.e., simultaneous tests or confidence intervals, where the main interest is in comparing characteristics of different groups represented by a nominal factor. In fact, we have already seen such a procedure in Chapter 5 where multiple differences of mean rat weights were compared for all combinations of the mother rat's genotype (Figure 5.5). Further examples of such multiple comparison procedures include Dunnett's many-to-one comparisons, sequential pairwise contrasts, comparisons with the average, change-point analyses, dose-response contrasts, etc. These procedures are all well established for classical regression and ANOVA models allowing for covariates and/or factorial treatment structures with i.i.d. normal errors and constant variance. For a general reading on multiple comparison procedures we refer to Hochberg and Tamhane (1987) and Hsu (1996).

Here, we follow a slightly more general approach allowing for null hypotheses on arbitrary model parameters, not only mean differences. Each individual null hypothesis is specified through a linear combination of elemental model parameters and we allow for $k$ of such null hypotheses to be tested simultaneously, regardless of the number of elemental model parameters $p$. More precisely, we assume that our model contains fixed but unknown $p$-dimensional elemental parameters $\theta$. We are primarily interested in linear functions $\vartheta := \mathbf{K}\theta$ of the parameter vector $\theta$ as specified through the constant $k \times p$ matrix $\mathbf{K}$. For example, in a linear model

$$y_i = \beta_0 + \beta_1 x_{i1} + \cdots + \beta_q x_{iq} + \varepsilon_i$$

as introduced in Chapter 6, we might be interested in inference about the parameter $\beta_1$, $\beta_q$ *and* $\beta_2 - \beta_1$. Chapter 6 offers methods for answering each of these questions separately but does not provide an answer for all three questions together. We can formulate the three inference problems as a linear combination of the elemental parameter vector $\theta = (\beta_0, \beta_1, \dots, \beta_q)$ as (here for $q = 3$)

$$\mathbf{K}\theta = \begin{pmatrix} 0 & 1 & 0 & 0 \\ 0 & 0 & 0 & 1 \\ 0 & -1 & 1 & 0 \end{pmatrix} \theta = (\beta_1, \beta_q, \beta_2 - \beta_1)^\top =: \vartheta.$$

The global null hypothesis now reads

$$H_0 : \vartheta := \mathbf{K}\theta = \mathbf{m},$$

where $\theta$ are the elemental model parameters that are estimated by some estimate $\hat{\theta}$, $\mathbf{K}$ is the matrix defining linear functions of the elemental parameters resulting in our parameters of interest $\vartheta$ and $\mathbf{m}$ is a $k$-vector of constants. The null hypothesis states that $\vartheta_j = m_j$ for all $j = 1, \dots, k$, where $m_j$ is some predefined scalar being zero in most applications. The global hypothesis $H_0$ is classically tested using an $F$-test in linear and ANOVA models (see Chapter 5 and Chapter 6). Such a test procedure gives only the answer $\vartheta_j \neq m_j$ for at least one $j$ but doesn't tell us which subset of our null hypotheses actually can be rejected. Here, we are mainly interested in which of the $k$ partial hypotheses $H_0^j : \vartheta_j = m_j$ for $j = 1, \dots, k$ are actually false. A simultaneous inference procedure gives us information about which of these $k$ hypotheses can be rejected in light of the data.

The estimated elemental parameters $\hat{\theta}$ are normally distributed in classical linear models and consequently, the estimated parameters of interest $\hat{\vartheta} = \mathbf{K}\hat{\theta}$ share this property. It can be shown that the $t$-statistics

$$\left( \frac{\hat{\vartheta}_1 - m_1}{\text{se}(\hat{\vartheta}_1)}, \dots, \frac{\hat{\vartheta}_k - m_k}{\text{se}(\hat{\vartheta}_k)} \right)$$

follow a joint multivariate $k$-dimensional $t$-distribution with correlation matrix Cor. This correlation matrix and the standard deviations of our estimated parameters of interest $\hat{\vartheta}_j$ can be estimated from the data. In most other models, the parameter estimates $\hat{\theta}$ are only asymptotically normal distributed. In this situation, the joint limiting distribution of all $t$-statistics on the parameters of interest is a $k$-variate normal distribution with zero mean and correlation matrix Cor which can be estimated as well.

The key aspect of simultaneous inference procedures is to take these joint distributions and thus the correlation among the estimated parameters of interest into account when constructing $p$-values and confidence intervals. The detailed technical aspects are computationally demanding since one has to carefully evaluate multivariate distribution functions by means of numerical integration procedures. However, these difficulties are rather unimportant to the data analyst. For a detailed treatment of the statistical methodology we refer to Hothorn et al. (2008a).

## 14.3 Analysis Using R

### *14.3.1 Genetic Components of Alcoholism*

We start with a graphical display of the data. Three parallel boxplots shown in Figure 14.1 indicate increasing expression levels of alpha synuclein mRNA for longer *NACP*-REP1 alleles.

In order to model this relationship, we start fitting a simple one-way ANOVA model of the form $y_{ij} = \mu + \gamma_i + \varepsilon_{ij}$ to the data with independent normal errors $\varepsilon_{ij} \sim \mathcal{N}(0, \sigma^2)$, $j \in \{\text{short}, \text{intermediate}, \text{long}\}$, and $i = 1, \dots, n_j$. The parameters $\mu + \gamma_{\text{short}}$, $\mu + \gamma_{\text{intermediate}}$ and $\mu + \gamma_{\text{long}}$ can be interpreted as the mean expression levels in the corresponding groups. As already discussed

```
R> n <- table(alpha$alength)
R> levels(alpha$alength) <- abbreviate(levels(alpha$alength), 4)
R> plot(elevel ~ alength, data = alpha, varwidth = TRUE,
+       ylab = "Expression Level",
+       xlab = "NACP-REP1 Allele Length")
R> axis(3, at = 1:3, labels = paste("n = ", n))
```

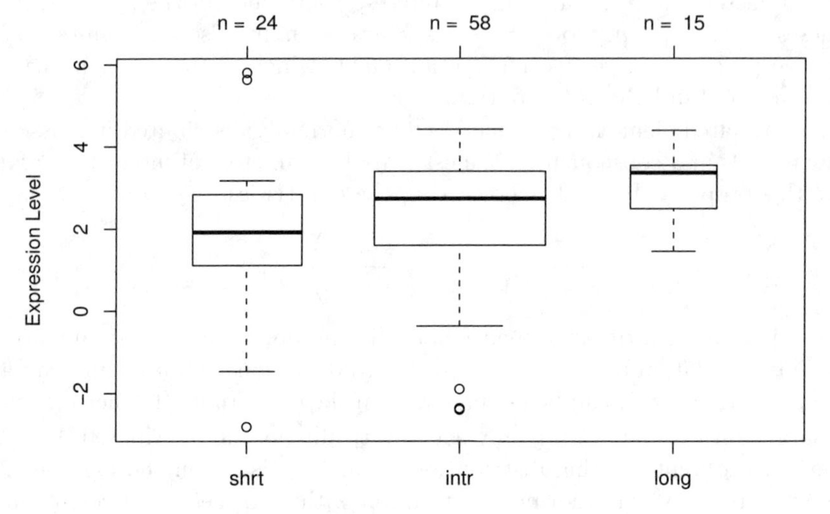


**Figure 14.1** Distribution of levels of expressed alpha synuclein mRNA in three groups defined by the *NACP*-REP1 allele lengths.

in Chapter 5, this model description is overparameterised. A standard approach is to consider a suitable re-parameterization. The so-called "treatment contrast" vector $\theta = (\mu, \gamma_{\text{intermediate}} - \gamma_{\text{short}}, \gamma_{\text{long}} - \gamma_{\text{short}})$ (the default re-parameterization used as elemental parameters in R) is one possibility and is equivalent to imposing the restriction $\gamma_{\text{short}} = 0$.

In addition, we define all comparisons among our three groups by choosing $\mathbf{K}$ such that $\mathbf{K}\theta$ contains all three group differences (Tukey's all-pairwise comparisons):

$$\mathbf{K}_{\text{Tukey}} = \begin{pmatrix} 0 & 1 & 0 \\ 0 & 0 & 1 \\ 0 & -1 & 1 \end{pmatrix}$$

with parameters of interest

$$\vartheta_{\text{Tukey}} = \mathbf{K}_{\text{Tukey}}\theta = (\gamma_{\text{intermediate}} - \gamma_{\text{short}}, \gamma_{\text{long}} - \gamma_{\text{short}}, \gamma_{\text{long}} - \gamma_{\text{intermediate}}).$$

The function `glht` (for generalised linear hypothesis) from package **multcomp** (Hothorn et al., 2009a, 2008a) takes the fitted *aov* object and a description of the matrix **K**. Here, we use the `mcp` function to set up the matrix of all pairwise differences for the model parameters associated with factor `alength`:

```
R> library("multcomp")
R> amod <- aov(elevel ~ alength, data = alpha)
R> amod_glht <- glht(amod, linfct = mcp(alength = "Tukey"))
```

The matrix **K** reads

```
R> amod_glht$linfct

            (Intercept) alengthintr alengthlong
intr - shrt           0           1           0
long - shrt           0           0           1
long - intr           0          -1           1
attr(,"type")
[1] "Tukey"
```

The `amod_glht` object now contains information about the estimated linear function $\hat{\vartheta}$ and their covariance matrix which can be inspected via the `coef` and `vcov` methods:

```
R> coef(amod_glht)

intr - shrt long - shrt long - intr
  0.4341523   1.1887500   0.7545977

R> vcov(amod_glht)

            intr - shrt long - shrt long - intr
intr - shrt  0.14717604   0.1041001 -0.04307591
long - shrt  0.10410012   0.2706603  0.16656020
long - intr -0.04307591   0.1665602  0.20963611
```

The `summary` and `confint` methods can be used to compute a summary statistic including adjusted $p$-values and simultaneous confidence intervals, respectively:

```
R> confint(amod_glht)

         Simultaneous Confidence Intervals

Multiple Comparisons of Means: Tukey Contrasts


Fit: aov(formula = elevel ~ alength, data = alpha)

Estimated Quantile = 2.3718
95% family-wise confidence level


Linear Hypotheses:
                 Estimate lwr      upr
intr - shrt == 0  0.43415 -0.47574  1.34405
```

```
long - shrt == 0  1.18875 -0.04516  2.42266
long - intr == 0  0.75460 -0.33134  1.84054
```

```
R> summary(amod_glht)

         Simultaneous Tests for General Linear Hypotheses

Multiple Comparisons of Means: Tukey Contrasts


Fit: aov(formula = elevel ~ alength, data = alpha)

Linear Hypotheses:
                 Estimate Std. Error t value Pr(>|t|)
intr - shrt == 0   0.4342     0.3836   1.132   0.4924
long - shrt == 0   1.1888     0.5203   2.285   0.0615
long - intr == 0   0.7546     0.4579   1.648   0.2270
(Adjusted p values reported -- single-step method)
```

Because of the variance heterogeneity that can be observed in Figure 14.1, one might be concerned with the validity of the above results stating that there is no difference between any combination of the three allele lengths. A sandwich estimator might be more appropriate in this situation, and the `vcov` argument can be used to specify a function to compute some alternative covariance estimator as follows:

```
R> amod_glht_sw <- glht(amod, linfct = mcp(alength = "Tukey"),
+                        vcov = sandwich)
R> summary(amod_glht_sw)

         Simultaneous Tests for General Linear Hypotheses

Multiple Comparisons of Means: Tukey Contrasts


Fit: aov(formula = elevel ~ alength, data = alpha)

Linear Hypotheses:
                 Estimate Std. Error t value Pr(>|t|)
intr - shrt == 0   0.4342     0.4239   1.024   0.5594
long - shrt == 0   1.1888     0.4432   2.682   0.0227
long - intr == 0   0.7546     0.3184   2.370   0.0501
(Adjusted p values reported -- single-step method)
```

We use the `sandwich` function from package **sandwich** (Zeileis, 2004, 2006) which provides us with a heteroscedasticity-consistent estimator of the covariance matrix. This result is more in line with previously published findings for this study obtained from non-parametric test procedures such as the Kruskal-Wallis test. A comparison of the simultaneous confidence intervals calculated based on the ordinary and sandwich estimator is given in Figure 14.2.

It should be noted that this data set is heavily unbalanced; see Figure 14.1,

```
R> par(mai = par("mai") * c(1, 2.1, 1, 0.5))
R> layout(matrix(1:2, ncol = 2))
R> ci1 <- confint(glht(amod, linfct = mcp(alength = "Tukey")))
R> ci2 <- confint(glht(amod, linfct = mcp(alength = "Tukey"),
+                     vcov = sandwich))
R> ox <- expression(paste("Tukey (ordinary ", bold(S)[n], ")"))
R> sx <- expression(paste("Tukey (sandwich ", bold(S)[n], ")"))
R> plot(ci1, xlim = c(-0.6, 2.6), main = ox,
+      xlab = "Difference", ylim = c(0.5, 3.5))
R> plot(ci2, xlim = c(-0.6, 2.6), main = sx,
+      xlab = "Difference", ylim = c(0.5, 3.5))
```

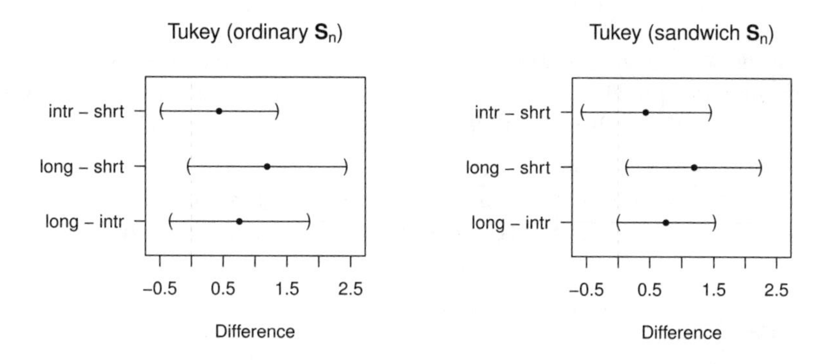


**Figure 14.2** Simultaneous confidence intervals for the alpha data based on the ordinary covariance matrix (left) and a sandwich estimator (right).

and therefore the results obtained from function TukeyHSD might be less accurate.

### *14.3.2 Deer Browsing*

Since we have to take the spatial structure of the deer browsing data into account, we cannot simply use a logistic regression model as introduced in Chapter 7. One possibility is to apply a mixed logistic regression model (using package **lme4**, Bates and Sarkar, 2008) with random intercept accounting for the spatial variation of the trees. These models have already been discussed in Chapter 13. For each plot nested within a set of five plots oriented on a 100m transect (the location of the transect is determined by a predefined equally spaced lattice of the area under test), a random intercept is included in the model. Essentially, trees that are close to each other are handled like repeated measurements in a longitudinal analysis. We are interested in probability estimates and confidence intervals for each tree species. Each of the six fixed parameters of the model corresponds to one species (in absence of a global intercept term); therefore, $\mathbf{K} = \text{diag}(6)$ is the linear function we are interested in:

```
R> mmod <- lmer(damage ~ species - 1 + (1 | lattice / plot),
+                    data = trees513, family = binomial())
R> K <- diag(length(fixef(mmod)))
R> K
```

```
     [,1] [,2] [,3] [,4] [,5]
[1,]    1    0    0    0    0
[2,]    0    1    0    0    0
[3,]    0    0    1    0    0
[4,]    0    0    0    1    0
[5,]    0    0    0    0    1
```

In order to help interpretation, the names of the tree species and the corresponding sample sizes (computed via `table`) are added to **K** as row names; this information will carry through all subsequent steps of our analysis:

```
R> colnames(K) <- rownames(K) <-
+     paste(gsub("species", "", names(fixef(mmod))),
+           " (", table(trees513$species), ")", sep = "")
R> K
```

```
               spruce (119) pine (823) beech (266) oak (1258)
spruce (119)              1          0           0          0
pine (823)                0          1           0          0
beech (266)               0          0           1          0
oak (1258)                0          0           0          1
hardwood (191)            0          0           0          0
               hardwood (191)
spruce (119)                0
pine (823)                  0
beech (266)                 0
oak (1258)                  0
hardwood (191)              1
```

Based on **K**, we first compute simultaneous confidence intervals for $\mathbf{K}\theta$ and transform these into probabilities. Note that $\left(1 + \exp(-\hat{\vartheta})\right)^{-1}$ (cf. Equation 7.2) is the vector of estimated probabilities; simultaneous confidence intervals can be transformed to the probability scale in the same way:

```
R> ci <- confint(glht(mmod, linfct = K))
R> ci$confint <- 1 - binomial()$linkinv(ci$confint)
R> ci$confint[,2:3] <- ci$confint[,3:2]
```

The result is shown in Figure 14.3. Browsing is less frequent in hardwood but especially small oak trees are severely at risk. Consequently, the local authorities increased the number of roe deers to be harvested in the following years. The large confidence interval for ash, maple, elm and lime trees is caused by the small sample size.

```
R> plot(ci, xlab = "Probability of Damage Caused by Browsing",
+       xlim = c(0, 0.5), main = "", ylim = c(0.5, 5.5))
```

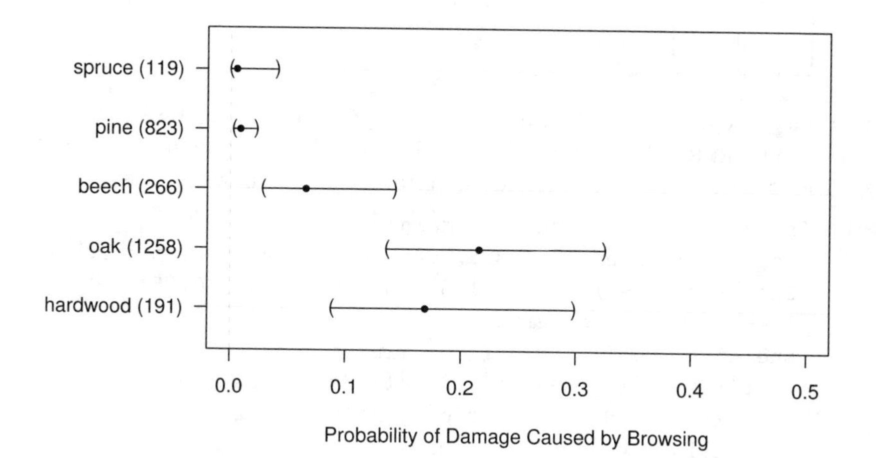


**Figure 14.3** Probability of damage caused by roe deer browsing for six tree species. Sample sizes are given in brackets.

### *14.3.3 Cloud Seeding*

In Chapter 6 we studied the dependency of rainfall on S-Ne values by means of linear models. Because the number of observations is small, an additional assessment of the variability of the fitted regression lines is interesting. Here, we are interested in a confidence band around some estimated regression line, i.e., a confidence region which covers the true but unknown regression line with probability greater or equal $1 - \alpha$. It is straightforward to compute *pointwise* confidence intervals but we have to make sure that the type I error is controlled for all $x$ values simultaneously. Consider the simple linear regression model

$$\text{rainfall}_i = \beta_0 + \beta_1 \text{sne}_i + \varepsilon_i$$

where we are interested in a confidence band for the predicted rainfall, i.e., the values $\hat{\beta}_0 + \hat{\beta}_1 \text{sne}_i$ for some observations $\text{sne}_i$. (Note that the estimates $\hat{\beta}_0$ and $\hat{\beta}_1$ are random variables.)

We can formulate the problem as a linear combination of the regression coefficients by multiplying a matrix $\mathbf{K}$ to a grid of S-Ne values (ranging from

1.5 to 4.5, say) from the left to the elemental parameters $\theta = (\beta_0, \beta_1)$:

$$\mathbf{K}\theta = \begin{pmatrix} 1 & 1.50 \\ 1 & 1.75 \\ \vdots & \vdots \\ 1 & 4.25 \\ 1 & 4.50 \end{pmatrix} \theta = (\beta_0 + \beta_1 1.50, \beta_0 + \beta_1 1.75, \dots, \beta_0 + \beta_1 4.50) = \vartheta.$$

Simultaneous confidence intervals for all the parameters of interest $\vartheta$ form a confidence band for the estimated regression line. We implement this idea for the `clouds` data writing a small reusable function as follows:

```
R> confband <- function(subset, main) {
+      mod <- lm(rainfall ~ sne, data = clouds, subset = subset)
+      sne_grid <- seq(from = 1.5, to = 4.5, by = 0.25)
+      K <- cbind(1, sne_grid)
+      sne_ci <- confint(glht(mod, linfct = K))
+      plot(rainfall ~ sne, data = clouds, subset = subset,
+           xlab = "S-Ne criterion", main = main,
+           xlim = range(clouds$sne),
+           ylim = range(clouds$rainfall))
+      abline(mod)
+      lines(sne_grid, sne_ci$confint[,2], lty = 2)
+      lines(sne_grid, sne_ci$confint[,3], lty = 2)
+ }
```

The function `confband` basically fits a linear model using `lm` to a subset of the data, sets up the matrix $\mathbf{K}$ as shown above and nicely plots both the regression line and the confidence band. Now, this function can be reused to produce plots similar to Figure 6.6 separately for days with and without cloud seeding in Figure 14.4. For the days without seeding, there is more uncertainty about the true regression line compared to the days with cloud seeding. Clearly, this is caused by the larger variability of the observations in the left part of the figure.

## 14.4 Summary

Multiple comparisons in linear models have been in use for a long time. The **multcomp** package extends much of the theory to a broad class of parametric and semi-parametric statistical models, which allows for a unified treatment of multiple comparisons and other simultaneous inference procedures in generalised linear models, mixed models, models for censored data, robust models, etc. Honest decisions based on simultaneous inference procedures maintaining a pre-specified familywise error rate (at least asymptotically) can be derived from almost all classical and modern statistical models. The technical details and more examples can be found in Hothorn et al. (2008a) and the package vignettes of package **multcomp** (Hothorn et al., 2009a).

```
R> layout(matrix(1:2, ncol = 2))
R> confband(clouds$seeding == "no", main = "No seeding")
R> confband(clouds$seeding == "yes", main = "Seeding")
```

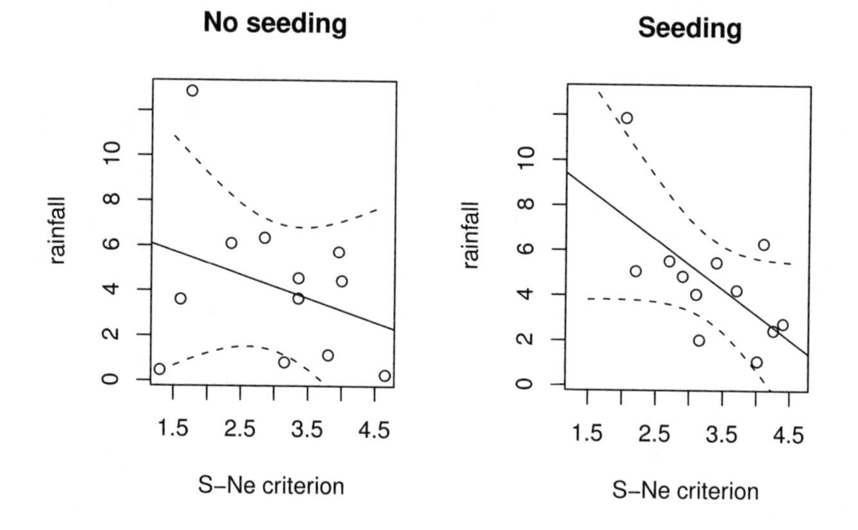


**Figure 14.4** Regression relationship between S-Ne criterion and rainfall with and without seeding. The confidence bands cover the area within the dashed curves.

## Exercises

Ex. 14.1 Compare the results of `glht` and `TukeyHSD` on the `alpha` data.

Ex. 14.2 Consider the linear model fitted to the clouds data as summarised in Figure 6.5. Set up a matrix **K** corresponding to the global null hypothesis that all interaction terms present in the model are zero. Test both the global hypothesis and all hypotheses corresponding to each of the interaction terms. Which interaction remains significant after adjustment for multiple testing?

Ex. 14.3 For the logistic regression model presented in Figure 7.7 perform a multiplicity adjusted test on all regression coefficients (except for the intercept) being zero. Do the conclusions drawn in Chapter 7 remain valid?

CHAPTER 15

# Meta-Analysis: Nicotine Gum and Smoking Cessation and the Efficacy of BCG Vaccine in the Treatment of Tuberculosis

## 15.1 Introduction

Cigarette smoking is the leading cause of preventable death in the United States and kills more Americans than AIDS, alcohol, illegal drug use, car accidents, fires, murders and suicides combined. It has been estimated that 430,000 Americans die from smoking every year. Fighting tobacco use is, consequently, one of the major public health goals of our time and there are now many programs available designed to help smokers quit. One of the major aids used in these programs is nicotine chewing gum, which acts as a substitute oral activity and provides a source of nicotine that reduces the withdrawal symptoms experienced when smoking is stopped. But separate randomised clinical trials of nicotine gum have been largely inconclusive, leading Silagy (2003) to consider combining the results from 26 such studies found from an extensive literature search. The results of these trials in terms of numbers of people in the treatment arm and the control arm who stopped smoking for at least 6 months after treatment are given in Table 15.1.

Bacille Calmette Guerin (BCG) is the most widely used vaccination in the world. Developed in the 1930s and made of a live, weakened strain of Mycobacterium bovis, the BCG is the only vaccination available against tuberculosis (TBC) today. Colditz et al. (1994) report data from 13 clinical trials of BCG vaccine each investigating its efficacy in the prevention of tuberculosis. The number of subjects suffering from TB with or without BCG vaccination are given in Table 15.2. In addition, the table contains the values of two other variables for each study, namely, the geographic latitude of the place where the study was undertaken and the year of publication. These two variables will be used to investigate and perhaps explain any heterogeneity among the studies.

**Table 15.1:** `smoking` data. Meta-analysis on nicotine gum showing the number of quitters who have been treated (`qt`), the total number of treated (`tt`) as well as the number of quitters in the control group (`qc`) with total number of smokers in the control group (`tc`).

| | qt | tt | qc | tc |
|---|---|---|---|---|
| Blondal89 | 37 | 92 | 24 | 90 |
| Campbell91 | 21 | 107 | 21 | 105 |
| Fagerstrom82 | 30 | 50 | 23 | 50 |
| Fee82 | 23 | 180 | 15 | 172 |
| Garcia89 | 21 | 68 | 5 | 38 |
| Garvey00 | 75 | 405 | 17 | 203 |
| Gross95 | 37 | 131 | 6 | 46 |
| Hall85 | 18 | 41 | 10 | 36 |
| Hall87 | 30 | 71 | 14 | 68 |
| Hall96 | 24 | 98 | 28 | 103 |
| Hjalmarson84 | 31 | 106 | 16 | 100 |
| Huber88 | 31 | 54 | 11 | 60 |
| Jarvis82 | 22 | 58 | 9 | 58 |
| Jensen91 | 90 | 211 | 28 | 82 |
| Killen84 | 16 | 44 | 6 | 20 |
| Killen90 | 129 | 600 | 112 | 617 |
| Malcolm80 | 6 | 73 | 3 | 121 |
| McGovern92 | 51 | 146 | 40 | 127 |
| Nakamura90 | 13 | 30 | 5 | 30 |
| Niaura94 | 5 | 84 | 4 | 89 |
| Pirie92 | 75 | 206 | 50 | 211 |
| Puska79 | 29 | 116 | 21 | 113 |
| Schneider85 | 9 | 30 | 6 | 30 |
| Tonnesen88 | 23 | 60 | 12 | 53 |
| Villa99 | 11 | 21 | 10 | 26 |
| Zelman92 | 23 | 58 | 18 | 58 |

**Table 15.2**: `BCG` data. Meta-analysis on BCG vaccine with the following data: the number of TBC cases after a vaccination with BCG (`BCGTB`), the total number of people who received BCG (`BCG`) as well as the number of TBC cases without vaccination (`NoVaccTB`) and the total number of people in the study without vaccination (`NoVacc`).

| Study | BCGTB | BCGVacc | NoVaccTB | NoVacc | Latitude | Year |
|---|---|---|---|---|---|---|
| 1 | 4 | 123 | 11 | 139 | 44 | 1948 |
| 2 | 6 | 306 | 29 | 303 | 55 | 1949 |
| 3 | 3 | 231 | 11 | 220 | 42 | 1960 |
| 4 | 62 | 13598 | 248 | 12867 | 52 | 1977 |
| 5 | 33 | 5069 | 47 | 5808 | 13 | 1973 |
| 6 | 180 | 1541 | 372 | 1451 | 44 | 1953 |
| 7 | 8 | 2545 | 10 | 629 | 19 | 1973 |
| 8 | 505 | 88391 | 499 | 88391 | 13 | 1980 |
| 9 | 29 | 7499 | 45 | 7277 | 27 | 1968 |
| 10 | 17 | 1716 | 65 | 1665 | 42 | 1961 |
| 11 | 186 | 50634 | 141 | 27338 | 18 | 1974 |
| 12 | 5 | 2498 | 3 | 2341 | 33 | 1969 |
| 13 | 27 | 16913 | 29 | 17854 | 33 | 1976 |

## 15.2 Systematic Reviews and Meta-Analysis

Many individual clinical trials are not large enough to answer the questions we want to answer as reliably as we would want to answer them. Often trials are too small for adequate conclusions to be drawn about potentially small advantages of particular therapies. Advocacy of large trials is a natural response to this situation, but it is not always possible to launch very large trials before therapies become widely accepted or rejected prematurely. One possible answer to this problem lies in the classical narrative review of a set of clinical trials with an accompanying informal synthesis of evidence from the different studies. It is now generally recognised, however, that such review articles can, unfortunately, be very misleading as a result of both the possible biased selection of evidence and the emphasis placed upon it by the reviewer to support his or her personal opinion.

An alternative approach that has become increasingly popular in the last decade or so is the *systematic review* which has, essentially, two components:

**Qualitative:** the description of the available trials, in terms of their relevance and methodological strengths and weaknesses.

**Quantitative:** a means of mathematically combining results from different studies, even when these studies have used different measures to assess the dependent variable.

The quantitative component of a systematic review is usually known as a *meta-analysis*, defined in the *Cambridge Dictionary of Statistics in the Medical Sciences* (Everitt, 2002a), as follows:

> A collection of techniques whereby the results of two or more independent studies are statistically combined to yield an overall answer to a question of interest. The rationale behind this approach is to provide a test with more power than is provided by the separate studies themselves. The procedure has become increasingly popular in the last decade or so, but is not without its critics, particularly because of the difficulties of knowing which studies should be included and to which population final results actually apply.

It is now generally accepted that meta-analysis gives the systematic review an objectivity that is inevitably lacking in literature reviews and can also help the process to achieve greater precision and generalisability of findings than any single study. Chalmers and Lau (1993) make the point that both the classical review article and a meta-analysis can be biased, but that at least the writer of a meta-analytic paper is required by the rudimentary standards of the discipline to give the data on which any conclusions are based, and to defend the development of these conclusions by giving evidence that all available data are included, or to give the reasons for not including the data. Chalmers and Lau (1993) conclude

> It seems obvious that a discipline that requires all available data be revealed and included in an analysis has an advantage over one that has traditionally not presented analyses of all the data in which conclusions are based.

The demand for systematic reviews of health care interventions has developed rapidly during the last decade, initiated by the widespread adoption of the principles of *evidence-based medicine* amongst both health care practitioners and policy makers. Such reviews are now increasingly used as a basis for both individual treatment decisions and the funding of health care and health care research worldwide. Systematic reviews have a number of aims:

- To review systematically the available evidence from a particular research area,
- To provide quantitative summaries of the results from each study,
- To combine the results across studies if appropriate; such combination of results typically leads to greater statistical power in estimating treatment effects,
- To assess the amount of variability between studies,
- To estimate the degree of benefit associated with a particular study treatment,
- To identify study characteristics associated with particularly effective treatments.

Perhaps the most important aspect of a meta-analysis is study selection.

Selection is a matter of inclusion and exclusion and the judgements required are, at times, problematic. But we shall say nothing about this fundamental component of a meta-analysis here since it has been comprehensively dealt with by a number of authors, including Chalmers and Lau (1993) and Petitti (2000). Instead we shall concentrate on the statistics of meta-analysis.

## 15.3 Statistics of Meta-Analysis

Two models that are frequently used in the meta-analysis of medical studies are the *fixed effects* and *random effects* models. Whilst the former assumes that each observed individual study result is estimating a common unknown overall pooled effect, the latter assumes that each individual observed result is estimating its own unknown underlying effect, which in turn is estimating a common population mean. Thus the random effects model specifically allows for the existence of *between-study heterogeneity* as well as *within-study variability*. DeMets (1987) and Bailey (1987) discuss the strengths and weaknesses of the two competing models. Bailey suggests that when the research question involves extrapolation to the future – *will* the treatment have an effect, on the average – then the random effects model for the studies is the appropriate one. The research question implicitly assumes that there is a population of studies from which those analysed in the meta-analysis were sampled, and anticipate future studies being conducted or previously unknown studies being uncovered.

When the research question concerns whether treatment *has* produced an effect, on the average, *in the set of studies being analysed*, then the fixed effects model for the studies may be the appropriate one; here there is no interest in generalising the results to other studies.

Many statisticians believe that random effects models are more appropriate than fixed effects models for meta-analysis because between-study variation is an important source of uncertainty that should not be ignored.

### *15.3.1 Fixed Effects Model – Mantel-Haenszel*

This model uses as its estimate of the common pooled effect, $\bar{Y}$, a weighted average of the individual study effects, the weights being inversely proportional to the within-study variances. Specifically

$$\bar{Y} = \frac{\sum_{i=1}^{k} W_i Y_i}{\sum_{i=1}^{k} W_i} \tag{15.1}$$

where $k$ is the number of the studies in the meta-analysis, $Y_i$ is the effect size estimated in the $i$th study (this might be a odds-ratio, log-odds ratio, relative risk or difference in means, for example), and $W_i = 1/V_i$ where $V_i$ is the within study estimate of variance for the $i$th study. The estimated variance of $\bar{Y}$ is given by

$$\mathsf{Var}(\bar{Y}) = 1/\left(\sum_{i=1}^{k} W_i\right). \tag{15.2}$$

From (15.1) and (15.2) a confidence interval for the pooled effect can be constructed in the usual way. For the Mantel-Haenszel analysis, consider a two-by-two table below.

| | | response | |
|---|---|---|---|
| | | success | failure |
| group | control | $a$ | $b$ |
| | treatment | $c$ | $d$ |

Then, the odds ratio for the $i$th study reads $Y_i = ad/bc$ and the weight is $W_i = bc/(a+b+c+d)$.

### *15.3.2 Random Effects Model – DerSimonian-Laird*

The random effects model has the form;

$$\begin{aligned} Y_i &= \mu_i + \sigma_i \varepsilon_i; \quad \varepsilon_i \sim \mathcal{N}(0,1) \\ \mu_i &\sim \mathcal{N}(\mu, \tau^2); \quad i = 1, \ldots, k. \end{aligned} \tag{15.3}$$

Unlike the fixed effects model, the individual studies are not assumed to be estimating a true single effect size; rather the true effects in each study, the $\mu_i$'s, are assumed to have been sampled from a distribution of effects, assumed to be normal with mean $\mu$ and variance $\tau^2$. The estimate of $\mu$ is that given in (15.1) but in this case the weights are given by $W_i = 1/\left(V_i + \hat{\tau}^2\right)$ where $\hat{\tau}^2$ is an estimate of the between study variance. DerSimonian and Laird (1986) derive a suitable estimator for $\hat{\tau}^2$, which is as follows;

$$\hat{\tau}^2 = \begin{cases} 0 & \text{if } Q \leq k-1 \\ (Q-k+1)/U & \text{if } Q > k-1 \end{cases}$$

where $Q = \sum_{i=1}^{k} W_i (Y_i - \bar{Y})^2$ and $U = (k-1)\left(\bar{W} - s_W^2/kW\right)$ with $\bar{W}$ and $s_W^2$ being the mean and variance of the weights, $W_i$.

A test for homogeneity of studies is provided by the statistic $Q$. The hypothesis of a common effect size is rejected if $Q$ exceeds the quantile of a $\chi^2$-distribution with $k-1$ degrees of freedom at the chosen significance level.

Allowing for this extra between-study variation has the effect of reducing the relative weighting given to the more precise studies. Hence the random effects model produces a more conservative confidence interval for the pooled effect size.

A Bayesian dimension can be added to the random effects model by allowing the parameters of the model to have prior distributions. Some examples are given in Sutton and Abrams (2001).

## 15.4 Analysis Using R

The methodology described above is implemented in package **rmeta** (Lumley, 2009) and we will utilise the functionality from this package to analyse the smoking and BCG studies introduced earlier.

The aim in collecting the results from the randomised trials of using nicotine gum to help smokers quit was to estimate the overall *odds ratio*, the odds of quitting smoking for those given the gum, divided by the odds of quitting for those not receiving the gum. Following formula (15.1), we can compute the pooled odds ratio as follows:

```
R> data("smoking", package = "HSAUR2")
R> odds <- function(x) (x[1] * (x[4] - x[3])) /
+                      ((x[2] - x[1]) * x[3])
R> weight <- function(x) ((x[2] - x[1]) * x[3]) / sum(x)
R> W <- apply(smoking, 1, weight)
R> Y <- apply(smoking, 1, odds)
R> sum(W * Y) / sum(W)
```

```
[1] 1.664159
```

Of course, the computations are more conveniently done using the functionality provided in package **rmeta**. The odds ratios and corresponding confidence intervals are computed by means of the `meta.MH` function for fixed effects meta-analysis as shown here

```
R> library("rmeta")
R> smokingOR <- meta.MH(smoking[["tt"]], smoking[["tc"]],
+                       smoking[["qt"]], smoking[["qc"]],
+                       names = rownames(smoking))
```

and the results can be inspected via a `summary` method – see Figure 15.1.

Before proceeding to the calculation of a combined effect size it will be helpful to graph the data by plotting confidence intervals for the odds ratios from each study (this is often known as a *forest plot* – see Sutton et al., 2000). The `plot` function applied to `smokingOR` produces such a plot; see Figure 15.2. It appears that the tendency in the trials considered was to favour nicotine gum but we need now to quantify this evidence in the form of an overall estimate of the odds ratio.

We shall use both the fixed effects and random effects approaches here so that we can compare results. For the fixed effects model (see Figure 15.1) the estimated overall log-odds ratio is 0.513 with a standard error of 0.066. This leads to an estimate of the overall odds ratio of 1.67, with a 95% confidence interval as given above. For the random effects model, which is fitted by applying function `meta.DSL` to the `smoking` data as follows

```
R> summary(smokingOR)

Fixed effects ( Mantel-Haenszel ) meta-analysis
Call: meta.MH(ntrt = smoking[["tt"]], nctrl = smoking[["tc"]],
  ptrt = smoking[["qt"]], pctrl = smoking[["qc"]],
  names = rownames(smoking))
------------------------------------
              OR (lower  95% upper)
Blondal89     1.85    0.99     3.46
Campbell91    0.98    0.50     1.92
Fagerstrom82 1.76    0.80     3.89
Fee82         1.53    0.77     3.05
Garcia89      2.95    1.01     8.62
Garvey00      2.49    1.43     4.34
Gross95       2.62    1.03     6.71
Hall85        2.03    0.78     5.29
Hall87        2.82    1.33     5.99
Hall96        0.87    0.46     1.64
Hjalmarson84 2.17    1.10     4.28
Huber88       6.00    2.57    14.01
Jarvis82      3.33    1.37     8.08
Jensen91      1.43    0.84     2.44
Killen84      1.33    0.43     4.15
Killen90      1.23    0.93     1.64
Malcolm80     3.52    0.85    14.54
McGovern92    1.17    0.70     1.94
Nakamura90    3.82    1.15    12.71
Niaura94      1.34    0.35     5.19
Pirie92       1.84    1.20     2.82
Puska79       1.46    0.78     2.75
Schneider85   1.71    0.52     5.62
Tonnesen88    2.12    0.93     4.86
Villa99       1.76    0.55     5.64
Zelman92      1.46    0.68     3.14
------------------------------------
Mantel-Haenszel OR =1.67 95% CI ( 1.47,1.9 )
Test for heterogeneity: X^2( 25 ) = 34.9 ( p-value 0.09 )
```

**Figure 15.1** R output of the summary method for smokingOR.

```
R> smokingDSL <- meta.DSL(smoking[["tt"]], smoking[["tc"]],
+                         smoking[["qt"]], smoking[["qc"]],
+                         names = rownames(smoking))
R> print(smokingDSL)

Random effects ( DerSimonian-Laird ) meta-analysis
Call: meta.DSL(ntrt = smoking[["tt"]], nctrl = smoking[["tc"]],
  ptrt = smoking[["qt"]], pctrl = smoking[["qc"]],
```

```
R> plot(smokingOR, ylab = "")
```

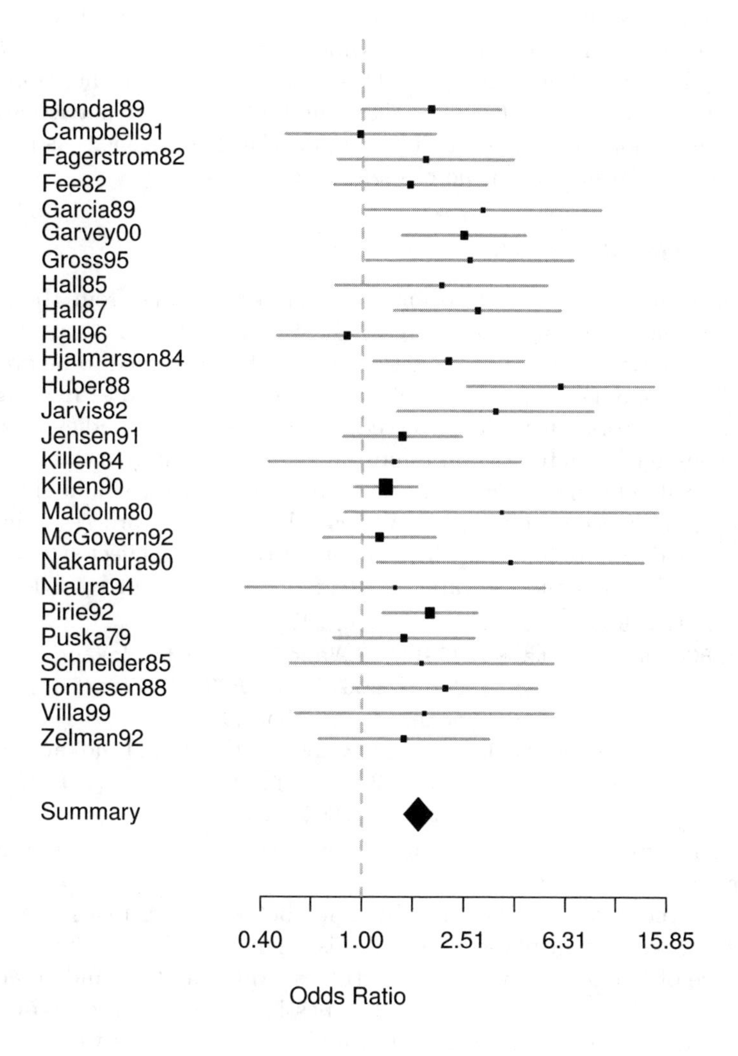


**Figure 15.2** Forest plot of observed effect sizes and 95% confidence intervals for the nicotine gum studies.

```
  names = rownames(smoking))
Summary OR= 1.75    95% CI ( 1.48, 2.07 )
Estimated random effects variance: 0.05
```

the corresponding estimate is 1.751. Both models suggest that there is clear evidence that nicotine gum increases the odds of quitting. The random effects confidence interval is considerably wider than that from the fixed effects model; here the test of homogeneity of the studies is not significant implying that we might use the fixed effects results. But the test is not particularly powerful and it is more sensible to assume a priori that heterogeneity is present and so we use the results from the random effects model.

## 15.5 Meta-Regression

The examination of heterogeneity of the effect sizes from the studies in a meta-analysis begins with the formal test for its presence, although in most meta-analyses such heterogeneity can almost be assumed to be present. There will be many possible sources of such heterogeneity and estimating how these various factors affect the observed effect sizes in the studies chosen is often of considerable interest and importance, indeed usually more important than the relatively simplistic use of meta-analysis to determine a single summary estimate of overall effect size. We can illustrate the process using the BCG vaccine data. We first find the estimate of the overall effect size from applying the fixed effects and the random effects models described previously:

```
R> data("BCG", package = "HSAUR2")
R> BCG_OR <- meta.MH(BCG[["BCGVacc"]], BCG[["NoVacc"]],
+                    BCG[["BCGTB"]], BCG[["NoVaccTB"]],
+                    names = BCG$Study)
R> BCG_DSL <- meta.DSL(BCG[["BCGVacc"]], BCG[["NoVacc"]],
+                    BCG[["BCGTB"]], BCG[["NoVaccTB"]],
+                    names = BCG$Study)
```

The results are inspected using the `summary` method as shown in Figures 15.3 and 15.4.

For these data the test statistics for heterogeneity takes the value 163.16 which, with 12 degrees of freedom, is highly significant; there is strong evidence of heterogeneity in the 13 studies. Applying the random effects model to the data gives (see Figure 15.4) an estimated odds ratio 0.474, with a 95% confidence interval of $(0.325, 0.69)$ and an estimated between-study variance of 0.366.

To assess how the two covariates, latitude and year, relate to the observed effect sizes we shall use multiple linear regression but will weight each observation by $W_i = (\hat{\sigma}^2 + V_i^2)^{-1}, i = 1, \dots, 13$, where $\hat{\sigma}^2$ is the estimated between-study variance and $V_i^2$ is the estimated variance from the $i$th study. The required R code to fit the linear model via weighted least squares is:

```
R> studyweights <- 1 / (BCG_DSL$tau2 + BCG_DSL$selogs^2)
R> y <- BCG_DSL$logs
```

```
R> summary(BCG_OR)

Fixed effects ( Mantel-Haenszel ) meta-analysis
Call: meta.MH(ntrt = BCG[["BCGVacc"]], nctrl = BCG[["NoVacc"]],
  ptrt = BCG[["BCGTB"]], pctrl = BCG[["NoVaccTB"]],
  names = BCG$Study)
------------------------------------
     OR (lower  95% upper)
1   0.39    0.12       1.26
2   0.19    0.08       0.46
3   0.25    0.07       0.91
4   0.23    0.18       0.31
5   0.80    0.51       1.26
6   0.38    0.32       0.47
7   0.20    0.08       0.50
8   1.01    0.89       1.15
9   0.62    0.39       1.00
10  0.25    0.14       0.42
11  0.71    0.57       0.89
12  1.56    0.37       6.55
13  0.98    0.58       1.66
------------------------------------
Mantel-Haenszel OR =0.62 95% CI ( 0.57,0.68 )
Test for heterogeneity: X^2( 12 ) = 163.94 ( p-value 0 )
```

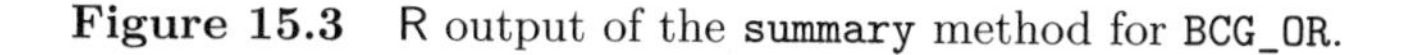

**Figure 15.3** R output of the `summary` method for `BCG_OR`.

```
R> BCG_mod <- lm(y ~ Latitude + Year, data = BCG,
+                weights = studyweights)
```

and the results of the `summary` method are shown in Figure 15.5. There is some evidence that latitude is associated with observed effect size, the log-odds ratio becoming increasingly negative as latitude increases, as we can see from a scatterplot of the two variables with the added weighted regression fit seen in Figure 15.6.

## 15.6 Publication Bias

The selection of studies to be integrated by a meta-analysis will clearly have a bearing on the conclusions reached. Selection is a matter of inclusion and exclusion and the judgements required are often difficult; Chalmers and Lau (1993) discuss the general issues involved, but here we shall concentrate on the particular potential problem of publication bias, which is a major problem, perhaps *the* major problem in meta-analysis.

Ensuring that a meta-analysis is truly representative can be problematic. It has long been known that journal articles are not a representative sample of work addressed to any particular area of research (see Sterlin, 1959, Green-

```
R> summary(BCG_DSL)

Random effects ( DerSimonian-Laird ) meta-analysis
Call: meta.DSL(ntrt = BCG[["BCGVacc"]], nctrl = BCG[["NoVacc"]],
  ptrt = BCG[["BCGTB"]], pctrl = BCG[["NoVaccTB"]],
  names = BCG$Study)
------------------------------------
      OR (lower  95% upper)
1  0.39    0.12        1.26
2  0.19    0.08        0.46
3  0.25    0.07        0.91
4  0.23    0.18        0.31
5  0.80    0.51        1.26
6  0.38    0.32        0.47
7  0.20    0.08        0.50
8  1.01    0.89        1.15
9  0.62    0.39        1.00
10 0.25    0.14        0.42
11 0.71    0.57        0.89
12 1.56    0.37        6.55
13 0.98    0.58        1.66
------------------------------------
SummaryOR= 0.47  95% CI ( 0.32,0.69 )
Test for heterogeneity: X^2( 12 ) = 163.16 ( p-value 0 )
Estimated random effects variance: 0.37
```

**Figure 15.4** R output of the summary method for BCG_DSL.

wald, 1975, Smith, 1980, for example). Research with statistically significant results is potentially more likely to be submitted and published than work with null or non-significant results (Easterbrook et al., 1991). The problem is made worse by the fact that many medical studies look at multiple outcomes, and there is a tendency for only those suggesting a significant effect to be mentioned when the study is written up. Outcomes which show no clear treatment effect are often ignored, and so will not be included in any later review of studies looking at those particular outcomes. Publication bias is likely to lead to an over-representation of positive results.

Clearly then it becomes of some importance to assess the likelihood of publication bias in any meta-analysis. A well-known, informal method of assessing publication bias is the so-called *funnel plot*. This assumes that the results from smaller studies will be more widely spread around the mean effect because of larger random error; a plot of a measure of the precision (such as inverse standard error or sample size) of the studies versus treatment effect from individual studies in a meta-analysis, should therefore be shaped like a funnel if there is no publication bias. If the chance of publication is greater

```
R> summary(BCG_mod)

Call:
lm(formula = y ~ Latitude + Year, data = BCG,
  weights = studyweights)

Residuals:
     Min       1Q   Median       3Q      Max
-1.66012 -0.36910 -0.02937  0.31565  1.26040

Coefficients:
              Estimate Std. Error t value Pr(>|t|)
(Intercept) -16.199115  37.605403  -0.431   0.6758
Latitude     -0.025808   0.013680  -1.887   0.0886
Year          0.008279   0.018972   0.436   0.6718

Residual standard error: 0.7992 on 10 degrees of freedom
Multiple R-squared: 0.4387,      Adjusted R-squared: 0.3265
F-statistic: 3.909 on 2 and 10 DF,  p-value: 0.05569
```

**Figure 15.5** R output of the `summary` method for `BCG_mod`.

for studies with statistically significant results, the shape of the funnel may become skewed.

Example funnel plots, inspired by those shown in Duval and Tweedie (2000), are displayed in Figure 15.7. In the first of these plots, there is little evidence of publication bias, while in the second, there is definite asymmetry with a clear lack of studies in the bottom left hand corner of the plot.

We can construct a funnel plot for the nicotine gum data using the R code depicted with Figure 15.8. There does not appear to be any strong evidence of publication bias here.

## 15.7 Summary

It is probably fair to say that the majority of statisticians and clinicians are largely enthusiastic about the advantages of meta-analysis over the classical review, although there remain sceptics who feel that the conclusions from meta-analyses often go beyond what the techniques and the data justify. Some of their concerns are echoed in the following quotation from Oakes (1993):

> The term meta-analysis refers to the quantitative combination of data from independent trials. Where the result of such combination is a descriptive summary of the weight of the available evidence, the exercise is of undoubted value. Attempts to apply inferential methods, however, are subject to considerable methodological and logical difficulties. The selection and quality of trials included, population bias and the specification of the population to which inference may properly be made are problems to which no satisfactory solutions have been proposed.

```
R> plot(y ~ Latitude, data = BCG, ylab = "Estimated log-OR")
R> abline(lm(y ~ Latitude, data = BCG, weights = studyweights))
```

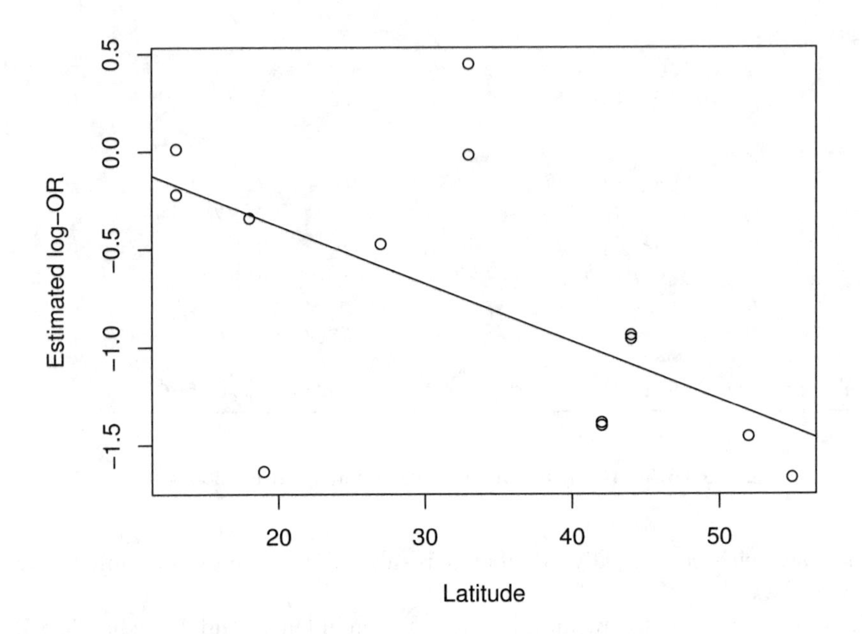


**Figure 15.6** Plot of observed effect size for the BCG vaccine data against latitude, with a weighted least squares regression fit shown in addition.

But despite such concerns the systematic review, in particular its quantitative component, meta-analysis, has had a major impact on medical science in the past ten years, and has been largely responsible for the development of evidence-based medical practise. One of the principal reasons that meta-analysis has been so successful is the large number of clinical trials that are now conducted. For example, the number of randomised clinical trials is now of the order of 10,000 per year. Synthesising results from many studies can be difficult, confusing and ultimately misleading. Meta-analysis has the potential to demonstrate treatment effects with a high degree of precision, possibly revealing small, but clinically important effects. But as with an individual clinical trial, careful planning, comprehensive data collection and a formal approach to statistical methods are necessary in order to achieve an acceptable and convincing meta-analysis.

A more comprehensive treatment of this subject will be available soon from the book *Meta-analysis with R* (Schwarzer et al., 2009), the associated R pack-

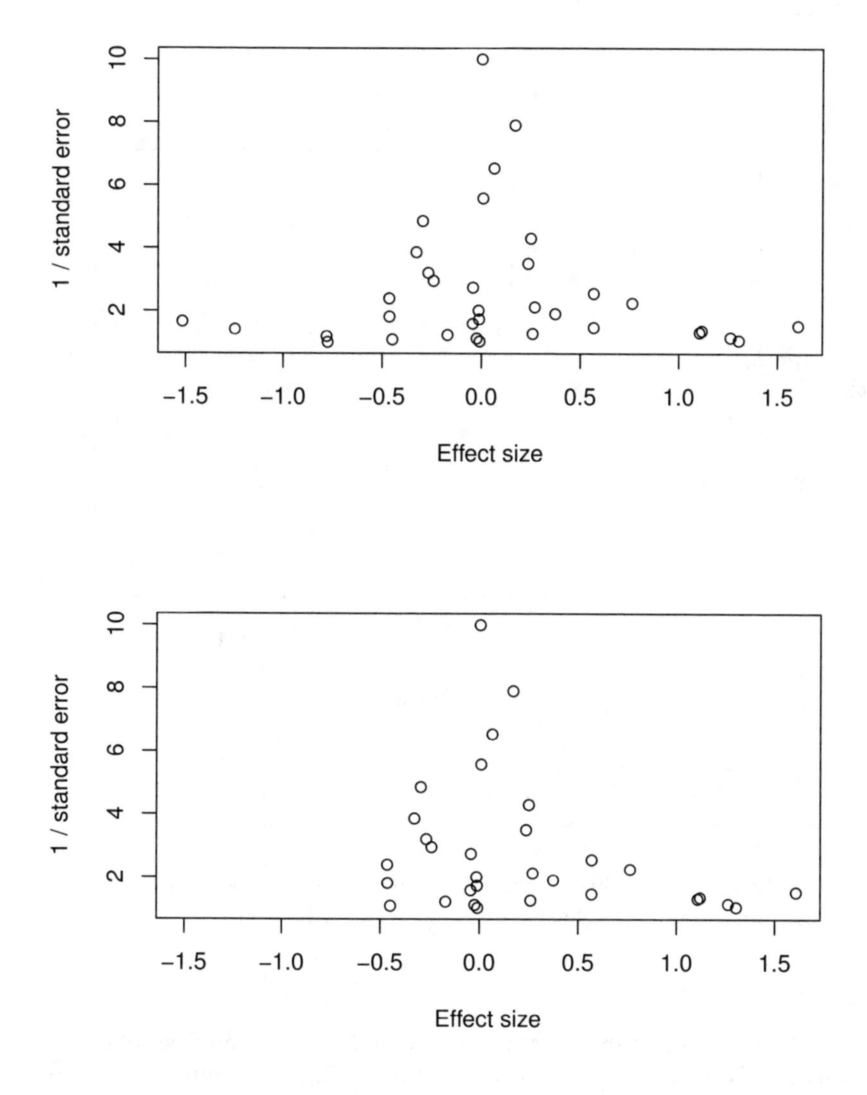


**Figure 15.7** Example funnel plots from simulated data. The asymmetry in the lower plot is a hint that a publication bias might be a problem.

```
R> funnelplot(smokingDSL$logs, smokingDSL$selogs,
+             summ = smokingDSL$logDSL, xlim = c(-1.7, 1.7))
R> abline(v = 0, lty = 2)
```

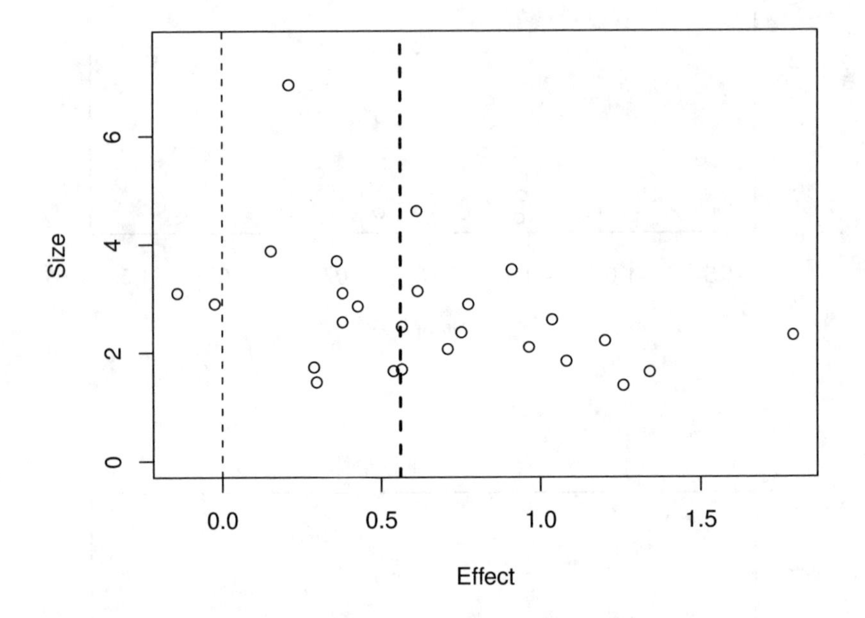


**Figure 15.8** Funnel plot for nicotine gum data.

age **meta** (Schwarzer, 2009), which for example offers functionality for testing on funnel plot asymmetry, has already been published on CRAN.

## Exercises

Ex. 15.1 The data in Table 15.4 were collected for a meta-analysis of the effectiveness of aspirin (versus placebo) in preventing death after a myocardial infarction (Fleiss, 1993). Calculate the log-odds ratio for each study and its variance, and then fit both a fixed effects and random effects model. Investigate the effect of possible publication bias.

**Table 15.4**: `aspirin` data. Meta-analysis on aspirin and myocardial infarct, the table shows the number of deaths after placebo (`dp`), the total number subjects treated with placebo (`tp`) as well as the number of deaths after aspirin (`da`) and the total number of subjects treated with aspirin (`ta`).

| | dp | tp | da | ta |
|---|---|---|---|---|
| Elwood et al. (1974) | 67 | 624 | 49 | 615 |
| Coronary Drug Project Group (1976) | 64 | 77 | 44 | 757 |
| Elwood and Sweetman (1979) | 126 | 850 | 102 | 832 |
| Breddin et al. (1979) | 38 | 309 | 32 | 317 |
| Persantine-Aspirin Reinfarction Study Research Group (1980) | 52 | 406 | 85 | 810 |
| Aspirin Myocardial Infarction Study Research Group (1980) | 219 | 2257 | 346 | 2267 |
| ISIS-2 (Second International Study of Infarct Survival) Collaborative Group (1988) | 1720 | 8600 | 1570 | 8587 |

Ex. 15.2 The data in Table 15.5 show the results of nine randomised trials comparing two different toothpastes for the prevention of caries development (see Everitt and Pickles, 2000). The outcomes in each trial was the change from baseline, in the decayed, missing (due to caries) and filled surface dental index (DMFS). Calculate an appropriate measure of effect size for each study and then carry out a meta-analysis of the results. What conclusions do you draw from the results?

**Table 15.5:** `toothpaste` data. Meta-analysis on trials comparing two toothpastes, the number of individuals in the study, the mean and the standard deviation for each study `A` and `B` are shown.

| Study | nA | meanA | sdA | nB | meanB | sdB |
|---|---|---|---|---|---|---|
| 1 | 134 | 5.96 | 4.24 | 113 | 4.72 | 4.72 |
| 2 | 175 | 4.74 | 4.64 | 151 | 5.07 | 5.38 |
| 3 | 137 | 2.04 | 2.59 | 140 | 2.51 | 3.22 |
| 4 | 184 | 2.70 | 2.32 | 179 | 3.20 | 2.46 |
| 5 | 174 | 6.09 | 4.86 | 169 | 5.81 | 5.14 |
| 6 | 754 | 4.72 | 5.33 | 736 | 4.76 | 5.29 |
| 7 | 209 | 10.10 | 8.10 | 209 | 10.90 | 7.90 |
| 8 | 1151 | 2.82 | 3.05 | 1122 | 3.01 | 3.32 |
| 9 | 679 | 3.88 | 4.85 | 673 | 4.37 | 5.37 |

Ex. 15.3 As an exercise in writing R code write your own meta-analysis function that allows the plotting of observed effect sizes and their associated confidence intervals (*forest plot*), estimates the overall effect size and its standard error by both the fixed effects and random effect models, and shows both on the constructed forest plot.

CHAPTER 16

# Principal Component Analysis: The Olympic Heptathlon

## 16.1 Introduction

The pentathlon for women was first held in Germany in 1928. Initially this consisted of the shot put, long jump, 100m, high jump and javelin events held over two days. In the 1964 Olympic Games the pentathlon became the first combined Olympic event for women, consisting now of the 80m hurdles, shot, high jump, long jump and 200m. In 1977 the 200m was replaced by the 800m and from 1981 the IAAF brought in the seven-event heptathlon in place of the pentathlon, with day one containing the events 100m hurdles, shot, high jump, 200m and day two, the long jump, javelin and 800m. A scoring system is used to assign points to the results from each event and the winner is the woman who accumulates the most points over the two days. The event made its first Olympic appearance in 1984.

In the 1988 Olympics held in Seoul, the heptathlon was won by one of the stars of women's athletics in the USA, Jackie Joyner-Kersee. The results for all 25 competitors in all seven disciplines are given in Table 16.1 (from Hand et al., 1994). We shall analyse these data using *principal component analysis* with a view to exploring the structure of the data and assessing how the derived principal component scores (see later) relate to the scores assigned by the official scoring system.

## 16.2 Principal Component Analysis

The basic aim of principal component analysis is to describe variation in a set of correlated variables, $x_1, x_2, \dots, x_q$, in terms of a new set of uncorrelated variables, $y_1, y_2, \dots, y_q$, each of which is a linear combination of the $x$ variables. The new variables are derived in decreasing order of 'importance' in the sense that $y_1$ accounts for as much of the variation in the original data amongst all linear combinations of $x_1, x_2, \dots, x_q$. Then $y_2$ is chosen to account for as much as possible of the remaining variation, subject to being uncorrelated with $y_1$ – and so on, i.e., forming an orthogonal coordinate system. The new variables defined by this process, $y_1, y_2, \dots, y_q$, are the principal components.

The general hope of principal component analysis is that the first few components will account for a substantial proportion of the variation in the original variables, $x_1, x_2, \dots, x_q$, and can, consequently, be used to provide a conve-

**Table 16.1**: `heptathlon` data. Results Olympic heptathlon, Seoul, 1988.

| | `hurdles` | `highjump` | `shot` | `run200m` | `longjump` | `javelin` | `run800m` | `score` |
|---|---|---|---|---|---|---|---|---|
| Joyner-Kersee (USA) | 12.69 | 1.86 | 15.80 | 22.56 | 7.27 | 45.66 | 128.51 | 7291 |
| John (GDR) | 12.85 | 1.80 | 16.23 | 23.65 | 6.71 | 42.56 | 126.12 | 6897 |
| Behmer (GDR) | 13.20 | 1.83 | 14.20 | 23.10 | 6.68 | 44.54 | 124.20 | 6858 |
| Sablovskaite (URS) | 13.61 | 1.80 | 15.23 | 23.92 | 6.25 | 42.78 | 132.24 | 6540 |
| Choubenkova (URS) | 13.51 | 1.74 | 14.76 | 23.93 | 6.32 | 47.46 | 127.90 | 6540 |
| Schulz (GDR) | 13.75 | 1.83 | 13.50 | 24.65 | 6.33 | 42.82 | 125.79 | 6411 |
| Fleming (AUS) | 13.38 | 1.80 | 12.88 | 23.59 | 6.37 | 40.28 | 132.54 | 6351 |
| Greiner (USA) | 13.55 | 1.80 | 14.13 | 24.48 | 6.47 | 38.00 | 133.65 | 6297 |
| Lajbnerova (CZE) | 13.63 | 1.83 | 14.28 | 24.86 | 6.11 | 42.20 | 136.05 | 6252 |
| Bouraga (URS) | 13.25 | 1.77 | 12.62 | 23.59 | 6.28 | 39.06 | 134.74 | 6252 |
| Wijnsma (HOL) | 13.75 | 1.86 | 13.01 | 25.03 | 6.34 | 37.86 | 131.49 | 6205 |
| Dimitrova (BUL) | 13.24 | 1.80 | 12.88 | 23.59 | 6.37 | 40.28 | 132.54 | 6171 |
| Scheider (SWI) | 13.85 | 1.86 | 11.58 | 24.87 | 6.05 | 47.50 | 134.93 | 6137 |
| Braun (FRG) | 13.71 | 1.83 | 13.16 | 24.78 | 6.12 | 44.58 | 142.82 | 6109 |
| Ruotsalainen (FIN) | 13.79 | 1.80 | 12.32 | 24.61 | 6.08 | 45.44 | 137.06 | 6101 |
| Yuping (CHN) | 13.93 | 1.86 | 14.21 | 25.00 | 6.40 | 38.60 | 146.67 | 6087 |
| Hagger (GB) | 13.47 | 1.80 | 12.75 | 25.47 | 6.34 | 35.76 | 138.48 | 5975 |
| Brown (USA) | 14.07 | 1.83 | 12.69 | 24.83 | 6.13 | 44.34 | 146.43 | 5972 |
| Mulliner (GB) | 14.39 | 1.71 | 12.68 | 24.92 | 6.10 | 37.76 | 138.02 | 5746 |
| Hautenauve (BEL) | 14.04 | 1.77 | 11.81 | 25.61 | 5.99 | 35.68 | 133.90 | 5734 |
| Kytola (FIN) | 14.31 | 1.77 | 11.66 | 25.69 | 5.75 | 39.48 | 133.35 | 5686 |
| Geremias (BRA) | 14.23 | 1.71 | 12.95 | 25.50 | 5.50 | 39.64 | 144.02 | 5508 |
| Hui-Ing (TAI) | 14.85 | 1.68 | 10.00 | 25.23 | 5.47 | 39.14 | 137.30 | 5290 |
| Jeong-Mi (KOR) | 14.53 | 1.71 | 10.83 | 26.61 | 5.50 | 39.26 | 139.17 | 5289 |
| Launa (PNG) | 16.42 | 1.50 | 11.78 | 26.16 | 4.88 | 46.38 | 163.43 | 4566 |

nient lower-dimensional summary of these variables that might prove useful for a variety of reasons.

In some applications, the principal components may be an end in themselves and might be amenable to interpretation in a similar fashion as the factors in an *exploratory factor analysis* (see Everitt and Dunn, 2001). More often they are obtained for use as a means of constructing a low-dimensional informative graphical representation of the data, or as input to some other analysis.

The low-dimensional representation produced by principal component analysis is such that

$$\sum_{r=1}^{n}\sum_{s=1}^{n}\left(d_{rs}^2 - \hat{d}_{rs}^2\right)$$

is minimised with respect to $\hat{d}_{rs}^2$. In this expression, $d_{rs}$ is the Euclidean distance (see Chapter 17) between observations $r$ and $s$ in the original $q$ dimensional space, and $\hat{d}_{rs}$ is the corresponding distance in the space of the first $m$ components.

As stated previously, the first principal component of the observations is that linear combination of the original variables whose sample variance is greatest amongst all possible such linear combinations. The second principal component is defined as that linear combination of the original variables that accounts for a maximal proportion of the remaining variance subject to being uncorrelated with the first principal component. Subsequent components are defined similarly. The question now arises as to how the coefficients specifying the linear combinations of the original variables defining each component are found? The algebra of *sample* principal components is summarised briefly.

The first principal component of the observations, $y_1$, is the linear combination

$$y_1 = a_{11}x_1 + a_{12}x_2 + \dots, a_{1q}x_q$$

whose sample variance is greatest among all such linear combinations. Since the variance of $y_1$ could be increased without limit simply by increasing the coefficients $\mathbf{a}_1^\top = (a_{11}, a_{12}, \dots, a_{1q})$ (here written in form of a vector for convenience), a restriction must be placed on these coefficients. As we shall see later, a sensible constraint is to require that the sum of squares of the coefficients, $\mathbf{a}_1^\top\mathbf{a}_1$, should take the value one, although other constraints are possible.

The second principal component $y_2 = \mathbf{a}_2^\top\mathbf{x}$ with $\mathbf{x} = (x_1, \dots, x_q)$ is the linear combination with greatest variance subject to the two conditions $\mathbf{a}_2^\top\mathbf{a}_2 = 1$ and $\mathbf{a}_2^\top\mathbf{a}_1 = 0$. The second condition ensures that $y_1$ and $y_2$ are uncorrelated. Similarly, the $j$th principal component is that linear combination $y_j = \mathbf{a}_j^\top\mathbf{x}$ which has the greatest variance subject to the conditions $\mathbf{a}_j^\top\mathbf{a}_j = 1$ and $\mathbf{a}_j^\top\mathbf{a}_i = 0$ for $(i < j)$.

To find the coefficients defining the first principal component we need to choose the elements of the vector $\mathbf{a}_1$ so as to maximise the variance of $y_1$ subject to the constraint $\mathbf{a}_1^\top\mathbf{a}_1 = 1$.

To maximise a function of several variables subject to one or more constraints, the method of *Lagrange multipliers* is used. In this case this leads to the solution that $\mathbf{a}_1$ is the eigenvector of the sample covariance matrix, $\mathbf{S}$, corresponding to its largest eigenvalue – full details are given in Morrison (2005).

The other components are derived in similar fashion, with $\mathbf{a}_j$ being the eigenvector of $\mathbf{S}$ associated with its $j$th largest eigenvalue. If the eigenvalues of $\mathbf{S}$ are $\lambda_1, \lambda_2, \dots, \lambda_q$, then since $\mathbf{a}_j^\top \mathbf{a}_j = 1$, the variance of the $j$th component is given by $\lambda_j$.

The total variance of the $q$ principal components will equal the total variance of the original variables so that

$$\sum_{j=1}^{q} \lambda_j = s_1^2 + s_2^2 + \cdots + s_q^2$$

where $s_j^2$ is the sample variance of $x_j$. We can write this more concisely as

$$\sum_{j=1}^{q} \lambda_j = \text{trace}(\mathbf{S}).$$

Consequently, the $j$th principal component accounts for a proportion $P_j$ of the total variation of the original data, where

$$P_j = \frac{\lambda_j}{\text{trace}(\mathbf{S})}.$$

The first $m$ principal components, where $m < q$, account for a proportion

$$P^{(m)} = \frac{\sum_{j=1}^{m} \lambda_j}{\text{trace}(\mathbf{S})}.$$

When the variables are on very different scales principal component analysis is usally carried out on the correlation matrix rather than the covariance matrix.

## 16.3 Analysis Using R

To begin it will help to score all seven events in the same direction, so that 'large' values are 'good'. We will recode the running events to achieve this;

```
R> data("heptathlon", package = "HSAUR2")
R> heptathlon$hurdles <- max(heptathlon$hurdles) -
+     heptathlon$hurdles
R> heptathlon$run200m <- max(heptathlon$run200m) -
+     heptathlon$run200m
R> heptathlon$run800m <- max(heptathlon$run800m) -
+     heptathlon$run800m
```

Figure 16.1 shows a scatterplot matrix of the results from all 25 competitors for the seven events. Most of the scatterplots in the diagram suggest that there

```
R> score <- which(colnames(heptathlon) == "score")
R> plot(heptathlon[,-score])
```

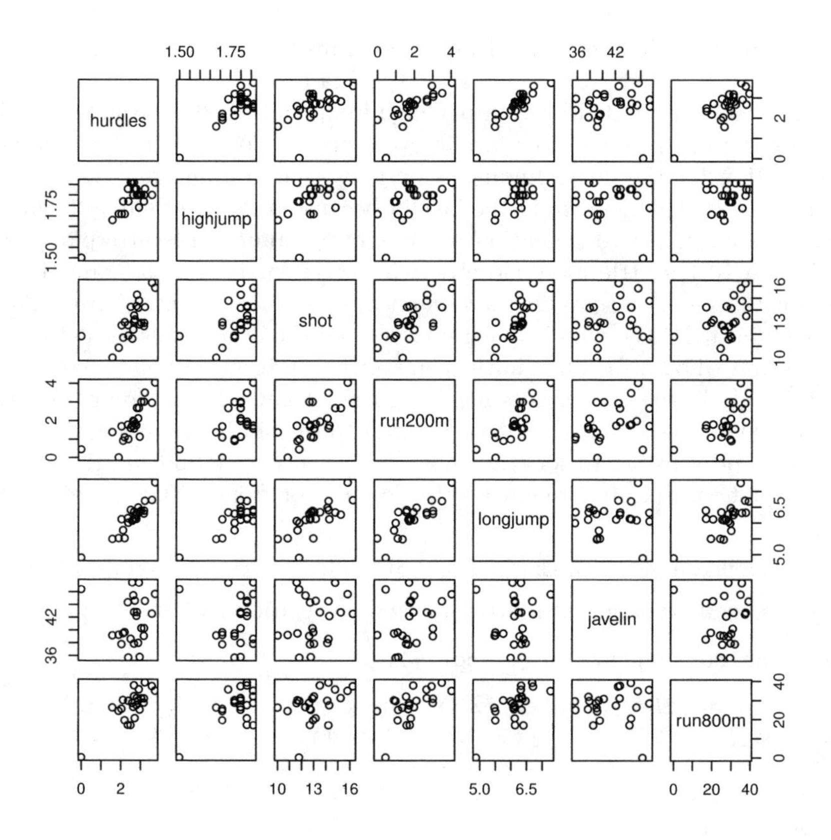


**Figure 16.1** Scatterplot matrix for the **heptathlon** data (all countries).

is a positive relationship between the results for each pairs of events. The exception are the plots involving the javelin event which give little evidence of any relationship between the result for this event and the results from the other six events; we will suggest possible reasons for this below, but first we will examine the numerical values of the between pairs events correlations by applying the `cor` function

```
R> round(cor(heptathlon[,-score]), 2)
```

```
         hurdles highjump shot run200m longjump javelin run800m
hurdles     1.00     0.81 0.65    0.77     0.91    0.01    0.78
highjump    0.81     1.00 0.44    0.49     0.78    0.00    0.59
shot        0.65     0.44 1.00    0.68     0.74    0.27    0.42
run200m     0.77     0.49 0.68    1.00     0.82    0.33    0.62
```

```
longjump     0.91      0.78 0.74    0.82      1.00     0.07    0.70
javelin      0.01      0.00 0.27    0.33      0.07     1.00   -0.02
run800m      0.78      0.59 0.42    0.62      0.70    -0.02    1.00
```

Examination of these numerical values confirms that most pairs of events are positively correlated, some moderately (for example, high jump and shot) and others relatively highly (for example, high jump and hurdles). And we see that the correlations involving the javelin event are all close to zero. One possible explanation for the latter finding is perhaps that training for the other six events does not help much in the javelin because it is essentially a 'technical' event. An alternative explanation is found if we examine the scatterplot matrix in Figure 16.1 a little more closely. It is very clear in this diagram that for all events except the javelin there is an outlier, the competitor from Papua New Guinea (PNG), who is much poorer than the other athletes at these six events and who finished last in the competition in terms of points scored. But surprisingly in the scatterplots involving the javelin it is this competitor who again stands out but because she has the third highest value for the event. It might be sensible to look again at both the correlation matrix and the scatterplot matrix after removing the competitor from PNG; the relevant R code is

```
R> heptathlon <- heptathlon[-grep("PNG", rownames(heptathlon)),]
```

Now, we again look at the scatterplot and correlation matrix;

```
R> round(cor(heptathlon[,-score]), 2)
```

```
         hurdles highjump shot run200m longjump javelin run800m
hurdles     1.00     0.58 0.77    0.83     0.89    0.33    0.56
highjump    0.58     1.00 0.46    0.39     0.66    0.35    0.15
shot        0.77     0.46 1.00    0.67     0.78    0.34    0.41
run200m     0.83     0.39 0.67    1.00     0.81    0.47    0.57
longjump    0.89     0.66 0.78    0.81     1.00    0.29    0.52
javelin     0.33     0.35 0.34    0.47     0.29    1.00    0.26
run800m     0.56     0.15 0.41    0.57     0.52    0.26    1.00
```

The correlations change quite substantially and the new scatterplot matrix in Figure 16.2 does not point us to any further extreme observations. In the remainder of this chapter we analyse the heptathlon data with the observations of the competitor from Papua New Guinea removed.

Because the results for the seven heptathlon events are on different scales we shall extract the principal components from the correlation matrix. A principal component analysis of the data can be applied using the prcomp function with the scale argument set to TRUE to ensure the analysis is carried out on the correlation matrix. The result is a list containing the coefficients defining each component (sometimes referred to as *loadings*), the principal component scores, etc. The required code is (omitting the score variable)

```
R> heptathlon_pca <- prcomp(heptathlon[, -score], scale = TRUE)
R> print(heptathlon_pca)
```

```
R> score <- which(colnames(heptathlon) == "score")
R> plot(heptathlon[,-score])
```

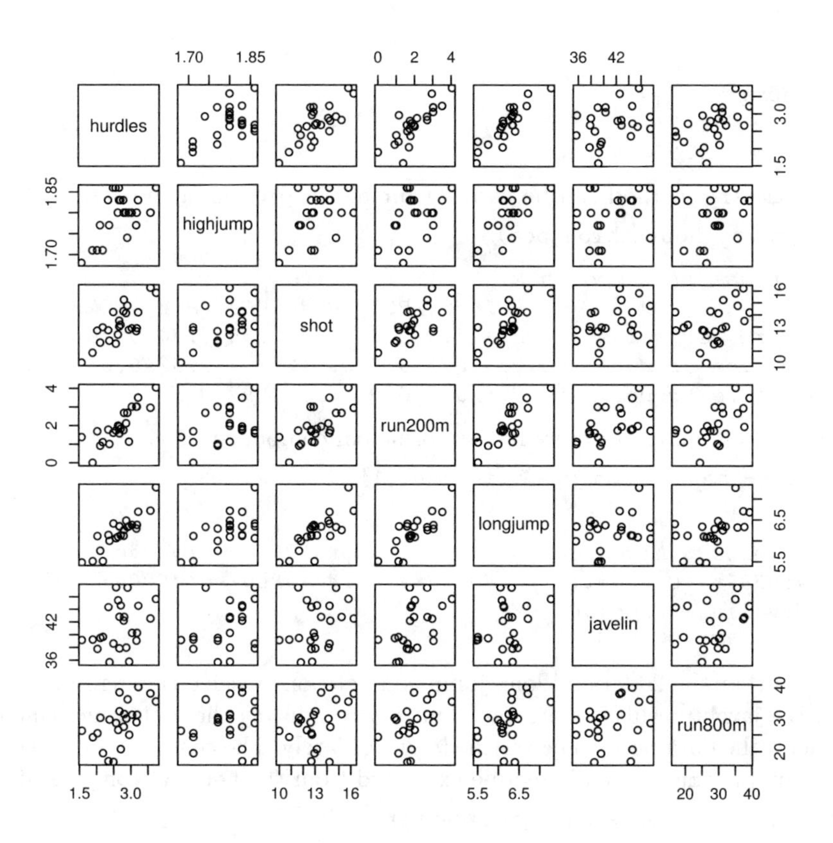


**Figure 16.2** Scatterplot matrix for the `heptathlon` data after removing observations of the PNG competitor.

```
Standard deviations:
[1] 2.0793 0.9482 0.9109 0.6832 0.5462 0.3375 0.2620

Rotation:
             PC1       PC2     PC3      PC4      PC5      PC6
hurdles  -0.4504  0.05772 -0.1739  0.04841 -0.19889  0.84665
highjump -0.3145 -0.65133 -0.2088 -0.55695  0.07076 -0.09008
shot     -0.4025 -0.02202 -0.1535  0.54827  0.67166 -0.09886
run200m  -0.4271  0.18503  0.1301  0.23096 -0.61782 -0.33279
longjump -0.4510 -0.02492 -0.2698 -0.01468 -0.12152 -0.38294
javelin  -0.2423 -0.32572  0.8807  0.06025  0.07874  0.07193
run800m  -0.3029  0.65651  0.1930 -0.57418  0.31880 -0.05218
```

```
              PC7
hurdles  -0.06962
highjump  0.33156
shot      0.22904
run200m   0.46972
longjump -0.74941
javelin  -0.21108
run800m   0.07719
```

The `summary` method can be used for further inspection of the details:

```
R> summary(heptathlon_pca)
```

```
Importance of components:
                       PC1 PC2 PC3  PC4  PC5  PC6  PC7
Standard deviation     2.1 0.9 0.9 0.68 0.55 0.34 0.26
Proportion of Variance 0.6 0.1 0.1 0.07 0.04 0.02 0.01
Cumulative Proportion  0.6 0.7 0.9 0.93 0.97 0.99 1.00
```

The linear combination for the first principal component is

```
R> a1 <- heptathlon_pca$rotation[,1]
R> a1
```

```
   hurdles   highjump       shot    run200m   longjump 
-0.4503876 -0.3145115 -0.4024884 -0.4270860 -0.4509639 
   javelin     run800m 
-0.2423079 -0.3029068 
```

We see that the 200m and long jump competitions receive the highest weight but the javelin result is less important. For computing the first principal component, the data need to be rescaled appropriately. The center and the scaling used by `prcomp` internally can be extracted from the `heptathlon_pca` via

```
R> center <- heptathlon_pca$center
R> scale <- heptathlon_pca$scale
```

Now, we can apply the `scale` function to the data and multiply with the loadings matrix in order to compute the first principal component score for each competitor

```
R> hm <- as.matrix(heptathlon[,-score])
R> drop(scale(hm, center = center, scale = scale) %*%
+       heptathlon_pca$rotation[,1])
```

```
Joyner-Kersee (USA)          John (GDR)         Behmer (GDR)
       -4.757530189        -3.147943402         -2.926184760
 Sablovskaite (URS)   Choubenkova (URS)         Schulz (GDR)
       -1.288135516        -1.503450994         -0.958467101
     Fleming (AUS)       Greiner (USA)     Lajbnerova (CZE)
       -0.953445060        -0.633239267         -0.381571974
      Bouraga (URS)       Wijnsma (HOL)      Dimitrova (BUL)
       -0.522322004        -0.217701500         -1.075984276
     Scheider (SWI)         Braun (FRG)   Ruotsalainen (FIN)
        0.003014986         0.109183759          0.208868056
```

```
        Yuping (CHN)          Hagger (GB)          Brown (USA)
         0.232507119          0.659520046          0.756854602
       Mulliner (GB)    Hautenauve (BEL)          Kytola (FIN)
         1.880932819          1.828170404          2.118203163
      Geremias (BRA)       Hui-Ing (TAI)       Jeong-Mi (KOR)
         2.770706272          3.901166920          3.896847898
```

or, more conveniently, by extracting the first from all precomputed principal components

```
R> predict(heptathlon_pca)[,1]
```

```
Joyner-Kersee (USA)          John (GDR)         Behmer (GDR)
        -4.757530189        -3.147943402         -2.926184760
 Sablovskaite (URS)  Choubenkova (URS)          Schulz (GDR)
        -1.288135516        -1.503450994         -0.958467101
       Fleming (AUS)       Greiner (USA)     Lajbnerova (CZE)
        -0.953445060        -0.633239267         -0.381571974
       Bouraga (URS)       Wijnsma (HOL)      Dimitrova (BUL)
        -0.522322004        -0.217701500         -1.075984276
      Scheider (SWI)        Braun (FRG)   Ruotsalainen (FIN)
         0.003014986         0.109183759          0.208868056
        Yuping (CHN)         Hagger (GB)          Brown (USA)
         0.232507119         0.659520046          0.756854602
       Mulliner (GB)    Hautenauve (BEL)          Kytola (FIN)
         1.880932819         1.828170404          2.118203163
      Geremias (BRA)       Hui-Ing (TAI)       Jeong-Mi (KOR)
         2.770706272         3.901166920          3.896847898
```

The first two components account for 75% of the variance. A barplot of each component's variance (see Figure 16.3) shows how the first two components dominate. A plot of the data in the space of the first two principal components, with the points labelled by the name of the corresponding competitor, can be produced as shown with Figure 16.4. In addition, the first two loadings for the events are given in a second coordinate system, also illustrating the special role of the javelin event. This graphical representation is known as *biplot* (Gabriel, 1971). A biplot is a graphical representation of the information in an $n \times p$ data matrix. The "bi" is a reflection that the technique produces a diagram that gives variance and covariance information about the variables and information about generalised distances between individuals. The coordinates used to produce the biplot can all be obtained directly from the principal components analysis of the covariance matrix of the data and so the plots can be viewed as an alternative representation of the results of such an analysis. Full details of the technical details of the biplot are given in Gabriel (1981) and in Gower and Hand (1996). Here we simply construct the biplot for the heptathlon data (without PNG); the result is shown in Figure 16.4. The plot clearly shows that the winner of the gold medal, Jackie Joyner-Kersee, accumulates the majority of her points from the three events long jump, hurdles, and 200m.

```
R> plot(heptathlon_pca)
```

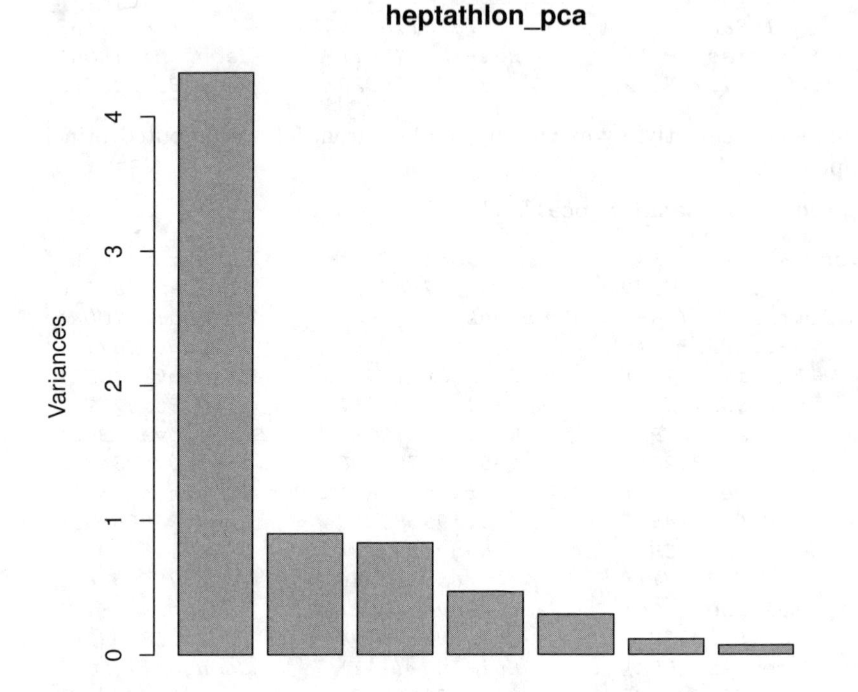


**Figure 16.3** Barplot of the variances explained by the principal components. (with observations for PNG removed).

The correlation between the score given to each athlete by the standard scoring system used for the heptathlon and the first principal component score can be found from

```
R> cor(heptathlon$score, heptathlon_pca$x[,1])

[1] -0.9931168
```

This implies that the first principal component is in good agreement with the score assigned to the athletes by official Olympic rules; a scatterplot of the official score and the first principal component is given in Figure 16.5.

```
R> biplot(heptathlon_pca, col = c("gray", "black"))
```

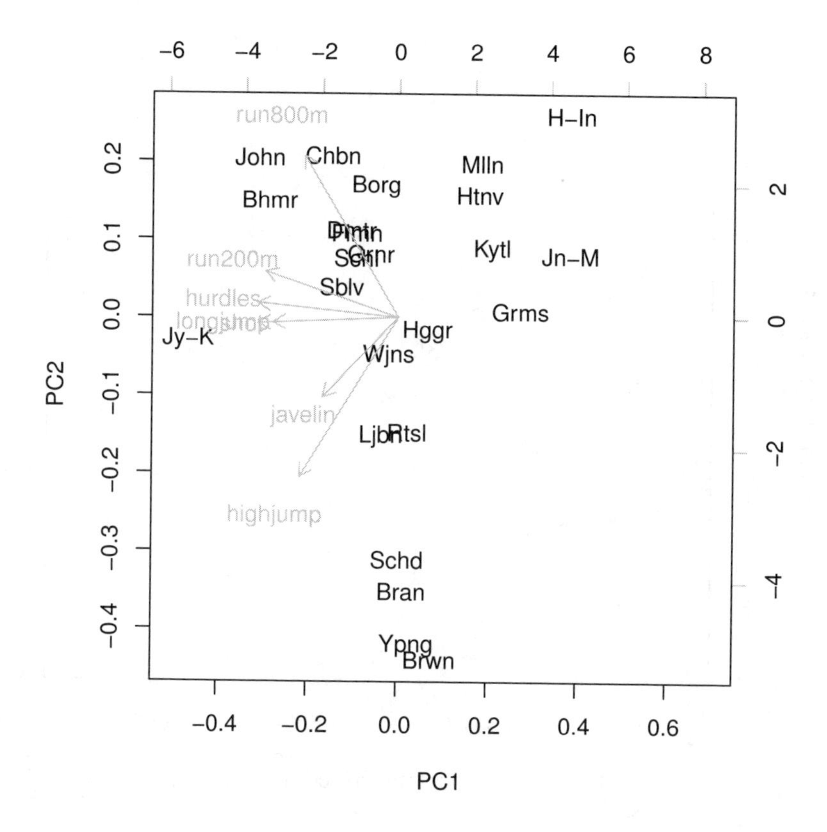


**Figure 16.4** Biplot of the (scaled) first two principal components (with observations for PNG removed).

## 16.4 Summary

Principal components look for a few linear combinations of the original variables that can be used to summarise a data set, losing in the process as little information as possible. The derived variables might be used in a variety of ways, in particular for simplifying later analyses and providing informative plots of the data. The method consists of transforming a set of correlated variables to a new set of variables that are uncorrelated. Consequently it should be noted that if the original variables are themselves almost uncorrelated there is little point in carrying out a principal components analysis, since it will merely find components that are close to the original variables but arranged in decreasing order of variance.

```
R> plot(heptathlon$score, heptathlon_pca$x[,1])
```

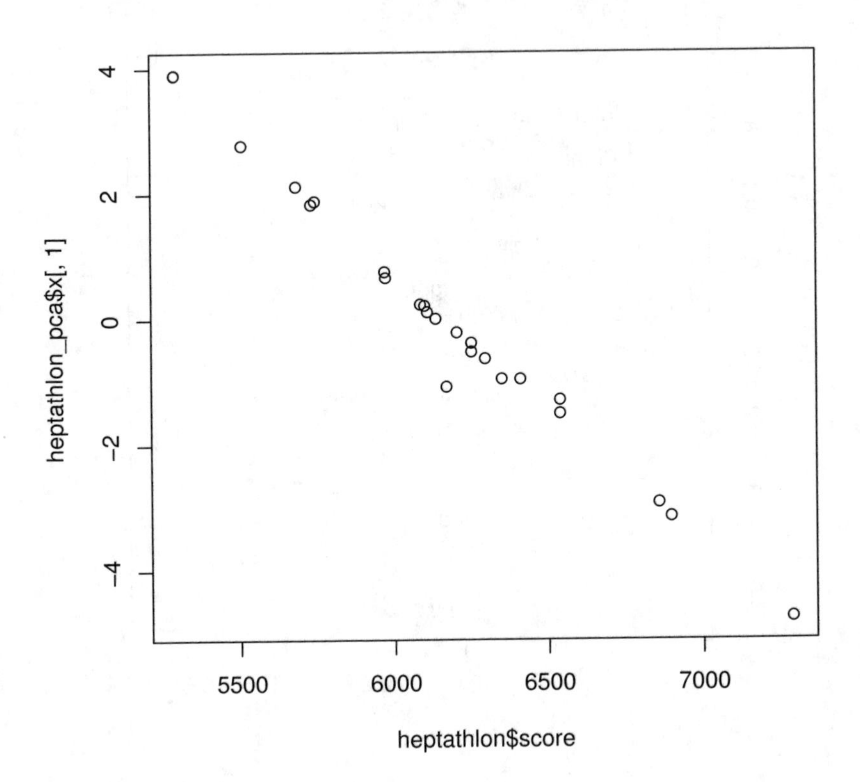


**Figure 16.5** Scatterplot of the score assigned to each athlete in 1988 and the first principal component.

## Exercises

Ex. 16.1 Apply principal components analysis to the covariance matrix of the heptathlon data (excluding the score variable) and compare your results with those given in the text, derived from the correlation matrix of the data. Which results do you think are more appropriate for these data?

Ex. 16.2 The data in Table 16.2 give measurements on five meteorological variables over an 11-year period (taken from Everitt and Dunn, 2001). The variables are

`year`: the corresponding year,

`rainNovDec`: rainfall in November and December (mm),

`temp`: average July temperature,

`rainJuly`: rainfall in July (mm),

`radiation`: radiation in July (curies), and

`yield`: average harvest yield (quintals per hectare).

Carry out a principal components analysis of both the covariance matrix and the correlation matrix of the data and compare the results. Which set of components leads to the most meaningful interpretation?

**Table 16.2**: `meteo` data. Meteorological measurements in an 11-year period.

| year | rainNovDec | temp | rainJuly | radiation | yield |
|---|---|---|---|---|---|
| 1920-21 | 87.9 | 19.6 | 1.0 | 1661 | 28.37 |
| 1921-22 | 89.9 | 15.2 | 90.1 | 968 | 23.77 |
| 1922-23 | 153.0 | 19.7 | 56.6 | 1353 | 26.04 |
| 1923-24 | 132.1 | 17.0 | 91.0 | 1293 | 25.74 |
| 1924-25 | 88.8 | 18.3 | 93.7 | 1153 | 26.68 |
| 1925-26 | 220.9 | 17.8 | 106.9 | 1286 | 24.29 |
| 1926-27 | 117.7 | 17.8 | 65.5 | 1104 | 28.00 |
| 1927-28 | 109.0 | 18.3 | 41.8 | 1574 | 28.37 |
| 1928-29 | 156.1 | 17.8 | 57.4 | 1222 | 24.96 |
| 1929-30 | 181.5 | 16.8 | 140.6 | 902 | 21.66 |
| 1930-31 | 181.4 | 17.0 | 74.3 | 1150 | 24.37 |

*Source*: From Everitt, B. S. and Dunn, G., *Applied Multivariate Data Analysis, 2nd Edition*, Arnold, London, 2001. With permission.

Ex. 16.3 The correlations below are for the calculus measurements for the six anterior mandibular teeth. Find all six principal components of the data and use a screeplot to suggest how many components are needed to adequately account for the observed correlations. Can you interpret the components?

**Table 16.3**: Correlations for calculus measurements for the six anterior mandibular teeth.

| | | | | | |
|---|---|---|---|---|---|
| 1.00 | | | | | |
| 0.54 | 1.00 | | | | |
| 0.34 | 0.65 | 1.00 | | | |
| 0.37 | 0.65 | 0.84 | 1.00 | | |
| 0.36 | 0.59 | 0.67 | 0.80 | 1.00 | |
| 0.62 | 0.49 | 0.43 | 0.42 | 0.55 | 1.00 |

CHAPTER 17

# Multidimensional Scaling: British Water Voles and Voting in US Congress

## 17.1 Introduction

Corbet et al. (1970) report a study of water voles (genus *Arvicola*) in which the aim was to compare British populations of these animals with those in Europe, to investigate whether more than one species might be present in Britain. The original data consisted of observations of the presence or absence of 13 characteristics in about 300 water vole skulls arising from six British populations and eight populations from the rest of Europe. Table 17.1 gives a distance matrix derived from the data as described in Corbet et al. (1970).

Romesburg (1984) gives a set of data that shows the number of times 15 congressmen from New Jersey voted differently in the House of Representatives on 19 environmental bills. Abstentions are not recorded, but two congressmen abstained more frequently than the others, these being Sandman (nine abstentions) and Thompson (six abstentions). The data are available in Table 17.2 and of interest is if party affiliations can be detected.

## 17.2 Multidimensional Scaling

The data in Tables 17.1 and 17.2 are both examples of *proximity matrices*. The elements of such matrices attempt to quantify how similar are stimuli, objects, individuals, etc. In Table 17.1 the values measure the 'distance' between populations of water voles; in Table 17.2 it is the similarity of the voting behaviour of the congressmen that is measured. Models are fitted to proximities in order to clarify, display and possibly explain any structure or pattern not readily apparent in the collection of numerical values. In some areas, particularly psychology, the ultimate goal in the analysis of a set of proximities is more specifically theories for explaining similarity judgements, or in other words, finding an answer to the question "what makes things seem alike or seem different?". Here though we will concentrate on how proximity data can be best displayed to aid in uncovering any interesting structure.

The class of techniques we shall consider here, generally collected under the label *multidimensional scaling* (MDS), has the unifying feature that they seek to represent an observed proximity matrix by a simple geometrical model or map. Such a model consists of a series of say $q$-dimensional coordinate values,

**Table 17.1**: `watervoles` data. Water voles data – dissimilarity matrix.

| | Srry | Shrp | Yrks | Prth | Abrd | ElnG | Alps | Ygsl | Grmn | Nrwy | PyrI | PyII | NrtS | SthS |
|---|---|---|---|---|---|---|---|---|---|---|---|---|---|---|
| Surrey | 0.000 | | | | | | | | | | | | | |
| Shropshire | 0.099 | 0.000 | | | | | | | | | | | | |
| Yorkshire | 0.033 | 0.022 | 0.000 | | | | | | | | | | | |
| Perthshire | 0.183 | 0.114 | 0.042 | 0.000 | | | | | | | | | | |
| Aberdeen | 0.148 | 0.224 | 0.059 | 0.068 | 0.000 | | | | | | | | | |
| Elean Gamhna | 0.198 | 0.039 | 0.053 | 0.085 | 0.051 | 0.000 | | | | | | | | |
| Alps | 0.462 | 0.266 | 0.322 | 0.435 | 0.268 | 0.025 | 0.000 | | | | | | | |
| Yugoslavia | 0.628 | 0.442 | 0.444 | 0.406 | 0.240 | 0.129 | 0.014 | 0.000 | | | | | | |
| Germany | 0.113 | 0.070 | 0.046 | 0.047 | 0.034 | 0.002 | 0.106 | 0.129 | 0.000 | | | | | |
| Norway | 0.173 | 0.119 | 0.162 | 0.331 | 0.177 | 0.039 | 0.089 | 0.237 | 0.071 | 0.000 | | | | |
| Pyrenees I | 0.434 | 0.419 | 0.339 | 0.505 | 0.469 | 0.390 | 0.315 | 0.349 | 0.151 | 0.430 | 0.000 | | | |
| Pyrenees II | 0.762 | 0.633 | 0.781 | 0.700 | 0.758 | 0.625 | 0.469 | 0.618 | 0.440 | 0.538 | 0.607 | 0.000 | | |
| North Spain | 0.530 | 0.389 | 0.482 | 0.579 | 0.597 | 0.498 | 0.374 | 0.562 | 0.247 | 0.383 | 0.387 | 0.084 | 0.000 | |
| South Spain | 0.586 | 0.435 | 0.550 | 0.530 | 0.552 | 0.509 | 0.369 | 0.471 | 0.234 | 0.346 | 0.456 | 0.090 | 0.038 | 0.000 |

**Table 17.2**: `voting` data. House of Representatives voting data.

| | Hnt | Snd | Hwr | Thm | Fry | Frs | Wdn | Roe | Hlt | Rdn | Mns | Rnl | Mrz | Dnl | Ptt |
|---|---|---|---|---|---|---|---|---|---|---|---|---|---|---|---|
| Hunt(R) | 0 | | | | | | | | | | | | | | |
| Sandman(R) | 8 | 0 | | | | | | | | | | | | | |
| Howard(D) | 15 | 17 | 0 | | | | | | | | | | | | |
| Thompson(D) | 15 | 12 | 9 | 0 | | | | | | | | | | | |
| Freylinghuysen(R) | 10 | 13 | 16 | 14 | 0 | | | | | | | | | | |
| Forsythe(R) | 9 | 13 | 12 | 12 | 8 | 0 | | | | | | | | | |
| Widnall(R) | 7 | 12 | 15 | 13 | 9 | 7 | 0 | | | | | | | | |
| Roe(D) | 15 | 16 | 5 | 10 | 13 | 12 | 17 | 0 | | | | | | | |
| Heltoski(D) | 16 | 17 | 5 | 8 | 14 | 11 | 16 | 4 | 0 | | | | | | |
| Rodino(D) | 14 | 15 | 6 | 8 | 12 | 10 | 15 | 5 | 3 | 0 | | | | | |
| Minish(D) | 15 | 16 | 5 | 8 | 12 | 9 | 14 | 5 | 2 | 1 | 0 | | | | |
| Rinaldo(R) | 16 | 17 | 4 | 6 | 12 | 10 | 15 | 3 | 1 | 2 | 1 | 0 | | | |
| Maraziti(R) | 7 | 13 | 11 | 15 | 10 | 6 | 10 | 12 | 13 | 11 | 12 | 12 | 0 | | |
| Daniels(D) | 11 | 12 | 10 | 10 | 11 | 6 | 11 | 7 | 7 | 4 | 5 | 6 | 9 | 0 | |
| Patten(D) | 13 | 16 | 7 | 7 | 11 | 10 | 13 | 6 | 5 | 6 | 5 | 4 | 13 | 9 | 0 |

$n$ in number, where $n$ is the number of rows (and columns) of the proximity matrix, and an associated measure of distance between pairs of points. Each point is used to represent one of the stimuli in the resulting spatial model for the proximities and the objective of a multidimensional approach is to determine both the dimensionality of the model (i.e., the value of $q$) that provides an adequate 'fit', and the positions of the points in the resulting $q$-dimensional space. Fit is judged by some numerical index of the correspondence between the observed proximities and the inter-point distances. In simple terms this means that the larger the perceived distance or dissimilarity between two stimuli (or the smaller their similarity), the further apart should be the points representing them in the final geometrical model.

A number of inter-point distance measures might be used, but by far the most common is *Euclidean distance.* For two points, $i$ and $j$, with $q$-dimensional coordinate values, $\mathbf{x}_i = (x_{i1}, x_{i2}, \dots, x_{iq})$ and $\mathbf{x}_j = (x_{j1}, x_{j2}, \dots, x_{jq})$ the Euclidean distance is defined as

$$d_{ij} = \sqrt{\sum_{k=1}^{q} (x_{ik} - x_{jk})^2}.$$

Having decided on a suitable distance measure the problem now becomes one of estimating the coordinate values to represent the stimuli, and this is achieved by optimising the chosen goodness of fit index measuring how well the fitted distances match the observed proximities. A variety of optimisation schemes combined with a variety of goodness of fit indices leads to a variety of MDS methods. For details see, for example, Everitt and Rabe-Hesketh (1997). Here we give a brief account of two methods, *classical scaling* and *non-metric scaling*, which will then be used to analyse the two data sets described earlier.

### *17.2.1 Classical Multidimensional Scaling*

Classical scaling provides one answer to how we estimate $q$, and the $n$, $q$-dimensional, coordinate values $\mathbf{x}_1, \mathbf{x}_2, \dots, \mathbf{x}_n$, from the observed proximity matrix, based on the work of Young and Householder (1938). To begin we must note that there is no unique set of coordinate values since the Euclidean distances involved are unchanged by shifting the whole configuration of points from one place to another, or by rotation or reflection of the configuration. In other words, we cannot uniquely determine either the location or the orientation of the configuration. The location problem is usually overcome by placing the mean vector of the configuration at the origin. The orientation problem means that any configuration derived can be subjected to an arbitrary *orthogonal transformation.* Such transformations can often be used to facilitate the interpretation of solutions as will be seen later.

To begin our account of the method we shall assume that the proximity matrix we are dealing with is a matrix of Euclidean distances $\mathbf{D}$ derived from a raw data matrix, $\mathbf{X}$. Previously we saw how to calculate Euclidean distances from $\mathbf{X}$; multidimensional scaling is essentially concerned with the reverse problem, given the distances how do we find $\mathbf{X}$?

An $n \times n$ inner products matrix $\mathbf{B}$ is first calculated as $\mathbf{B} = \mathbf{X}\mathbf{X}^\top$, the elements of $\mathbf{B}$ are given by

$$b_{ij} = \sum_{k=1}^{q} x_{ik} x_{jk}. \tag{17.1}$$

It is easy to see that the squared Euclidean distances between the rows of $\mathbf{X}$ can be written in terms of the elements of $\mathbf{B}$ as

$$d_{ij}^2 = b_{ii} + b_{jj} - 2b_{ij}. \tag{17.2}$$

If the $b$s could be found in terms of the $d$s as in the equation above, then the required coordinate value could be derived by factoring $\mathbf{B} = \mathbf{X}\mathbf{X}^\top$.

No unique solution exists unless a location constraint is introduced; usually the centre of the points $\bar{\mathbf{x}}$ is set at the origin, so that $\sum_{i=1}^{n} x_{ik} = 0$ for all $k$.

These constraints and the relationship given in (17.1) imply that the sum of the terms in any row of $\mathbf{B}$ must be zero.

Consequently, summing the relationship given in (17.2) over $i$, over $j$ and finally over both $i$ and $j$, leads to the following series of equations:

$$\begin{aligned}
\sum_{i=1}^{n} d_{ij}^2 &= \text{trace}(\mathbf{B}) + n b_{jj} \\
\sum_{j=1}^{n} d_{ij}^2 &= \text{trace}(\mathbf{B}) + n b_{ii} \\
\sum_{i=1}^{n} \sum_{j=1}^{n} d_{ij}^2 &= 2n \times \text{trace}(\mathbf{B})
\end{aligned}$$

where trace($\mathbf{B}$) is the trace of the matrix $\mathbf{B}$. The elements of $\mathbf{B}$ can now be found in terms of squared Euclidean distances as

$$b_{ij} = -\frac{1}{2}\left( d_{ij}^2 - n^{-1}\sum_{j=1}^{n} d_{ij}^2 - n^{-1}\sum_{i=1}^{n} d_{ij}^2 + n^{-2}\sum_{i=1}^{n}\sum_{j=1}^{n} d_{ij}^2 \right).$$

Having now derived the elements of $\mathbf{B}$ in terms of Euclidean distances, it remains to factor it to give the coordinate values. In terms of its singular value decomposition $\mathbf{B}$ can be written as

$$\mathbf{B} = \mathbf{V}\Lambda\mathbf{V}^\top$$

where $\Lambda = \text{diag}(\lambda_1, \dots, \lambda_n)$ is the diagonal matrix of eigenvalues of $\mathbf{B}$ and $\mathbf{V}$ the corresponding matrix of eigenvectors, normalised so that the sum of squares of their elements is unity, that is, $\mathbf{V}^\top\mathbf{V} = \mathbf{I}_n$. The eigenvalues are assumed labeled such that $\lambda_1 \geq \lambda_2 \geq \cdots \geq \lambda_n$.

When the matrix of Euclidian distances $\mathbf{D}$ arises from an $n \times k$ matrix of full column rank, then the rank of $\mathbf{B}$ is $k$, so that the last $n - k$ of its eigenvalues will be zero. So $\mathbf{B}$ can be written as $\mathbf{B} = \mathbf{V}_1\Lambda_1\mathbf{V}_1^\top$, where $\mathbf{V}_1$ contains the first $k$ eigenvectors and $\Lambda_1$ the $q$ non-zero eigenvalues. The required coordinate values are thus $\mathbf{X} = \mathbf{V}_1\Lambda_1^{1/2}$, where $\Lambda_1^{1/2} = \text{diag}(\sqrt{\lambda_1}, \dots, \sqrt{\lambda_k})$.

The best fitting $k$-dimensional representation is given by the $k$ eigenvectors of $\mathbf{B}$ corresponding to the $k$ largest eigenvalues. The adequacy of the $k$-dimensional representation can be judged by the size of the criterion

$$P_k = \frac{\sum_{i=1}^{k} \lambda_i}{\sum_{i=1}^{n-1} \lambda_i}.$$

Values of $P_k$ of the order of 0.8 suggest a reasonable fit.

When the observed dissimilarity matrix is not Euclidean, the matrix $\mathbf{B}$ is not positive-definite. In such cases some of the eigenvalues of $\mathbf{B}$ will be negative; corresponding, some coordinate values will be complex numbers. If, however, $\mathbf{B}$ has only a small number of small negative eigenvalues, a useful representation of the proximity matrix may still be possible using the eigenvectors associated with the $k$ largest positive eigenvalues.

The adequacy of the resulting solution might be assessed using one of the following two criteria suggested by Mardia et al. (1979); namely

$$\frac{\sum_{i=1}^{k} |\lambda_i|}{\sum_{i=1}^{n} |\lambda_i|} \quad \text{or} \quad \frac{\sum_{i=1}^{k} \lambda_i^2}{\sum_{i=1}^{n} \lambda_i^2}.$$

Alternatively, Sibson (1979) recommends the following:

1. *Trace criterion*: Choose the number of coordinates so that the sum of their positive eigenvalues is approximately equal to the sum of all the eigenvalues.
2. *Magnitude criterion*: Accept as genuinely positive only those eigenvalues whose magnitude substantially exceeds that of the largest negative eigenvalue.

### *17.2.2 Non-metric Multidimensional Scaling*

In classical scaling the goodness-of-fit measure is based on a direct numerical comparison of observed proximities and fitted distances. In many situations however, it might be believed that the observed proximities contain little reliable information beyond that implied by their rank order. In psychological experiments, for example, proximity matrices frequently arise from asking subjects to make judgements about the similarity or dissimilarity of the stimuli of interest; in many such experiments the investigator may feel that, realistically, subjects can give only 'ordinal' judgements. For example, in comparing a range of colours they might be able to specify that one was say 'brighter' than another without being able to attach any realistic value to the extent that they differed. For such situations, what is needed is a method of multidimensional scaling, the solutions from which depend only on the rank order of the proximities, rather than their actual numerical values. In other words the solution should be invariant under monotonic transformations of the proximities. Such a method was originally suggested by Shepard (1962a,b) and Kruskal (1964a). The quintessential component of the method is the use of *monotonic regression* (see Barlow et al., 1972). In essence the aim is to represent the fitted distances, $d_{ij}$, as $d_{ij} = \hat{d}_{ij} + \varepsilon_{ij}$ where the *disparities* $\hat{d}_{ij}$ are monotonic with the observed proximities and, subject to this constraint, resemble the $d_{ij}$ as closely as possible. Algorithms to achieve this are described in Kruskal (1964b). For a given set of disparities the required coordinates can be found by minimising some function of the squared differences between the observed proximities and the derived disparities (generally known as *stress* in this context). The procedure then iterates until some convergence criterion is satisfied. Again for details see Kruskal (1964b).

## 17.3 Analysis Using R

We can apply classical scaling to the distance matrix for populations of water voles using the R function `cmdscale`. The following code finds the classical scaling solution and computes the two criteria for assessing the required number of dimensions as described above.

```
R> data("watervoles", package = "HSAUR2")
R> voles_mds <- cmdscale(watervoles, k = 13, eig = TRUE)
R> voles_mds$eig
```

```
 [1]  7.359910e-01  2.626003e-01  1.492622e-01  6.990457e-02
 [5]  2.956972e-02  1.931184e-02  8.326673e-17 -1.139451e-02
 [9] -1.279569e-02 -2.849924e-02 -4.251502e-02 -5.255450e-02
[13] -7.406143e-02
```

Note that some of the eigenvalues are negative. The criterion $P_2$ can be computed by

```
R> sum(abs(voles_mds$eig[1:2]))/sum(abs(voles_mds$eig))
```

```
[1] 0.6708889
```

and the criterion suggested by Mardia et al. (1979) is

```
R> sum((voles_mds$eig[1:2])^2)/sum((voles_mds$eig)^2)
```

```
[1] 0.9391378
```

The two criteria for judging number of dimensions differ considerably, but both values are reasonably large, suggesting that the original distances between the water vole populations can be represented adequately in two dimensions. The two-dimensional solution can be plotted by extracting the coordinates from the `points` element of the `voles_mds` object; the plot is shown in Figure 17.1.

It appears that the six British populations are close to populations living in the Alps, Yugoslavia, Germany, Norway and Pyrenees I (consisting of the

```
R> x <- voles_mds$points[,1]
R> y <- voles_mds$points[,2]
R> plot(x, y, xlab = "Coordinate 1", ylab = "Coordinate 2",
+      xlim = range(x)*1.2, type = "n")
R> text(x, y, labels = colnames(watervoles))
```

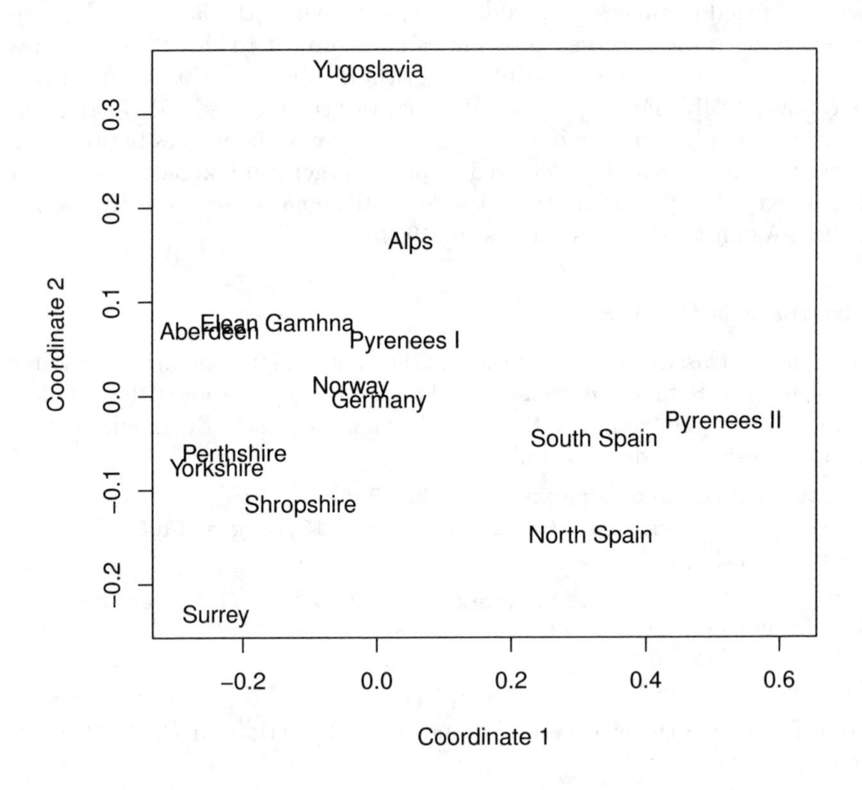


**Figure 17.1** Two-dimensional solution from classical multidimensional scaling of distance matrix for water vole populations.

species *Arvicola terrestris*) but rather distant from the populations in Pyrenees II, North Spain and South Spain (species *Arvicola sapidus*). This result would seem to imply that *Arvicola terrestris* might be present in Britain but it is less likely that this is so for *Arvicola sapidus*.

A useful graphic for highlighting possible distortions in a multidimensional scaling solution is the *minimum spanning tree*, which is defined as follows. Suppose $n$ points are given (possibly in many dimensions), then a tree spanning these points, i.e., a spanning tree, is any set of straight line segments joining pairs of points such that

- No closed loops occur,
- Every point is visited at least one time,
- The tree is connected, i.e., it has paths between any pairs of points.

The length of the tree is defined to be the sum of the length of its segments, and when a set of $n$ points and the length of all $\binom{n}{2}$ segments are given, then the minimum spanning tree is defined as the spanning tree with minimum length. Algorithms to find the minimum spanning tree of a set of $n$ points given the distances between them are given in Prim (1957) and Gower and Ross (1969).

The links of the minimum spanning tree (of the spanning tree) of the proximity matrix of interest may be plotted onto the two-dimensional scaling representation in order to identify possible distortions produced by the scaling solutions. Such distortions are indicated when nearby points on the plot are not linked by an edge of the tree.

To find the minimum spanning tree of the water vole proximity matrix, the function `mst` from package **ape** (Paradis et al., 2009) can be used and we can plot the minimum spanning tree on the two-dimensional scaling solution as shown in Figure 17.2.

The plot indicates, for example, that the apparent closeness of the populations in Germany and Norway, suggested by the points representing them in the MDS solution, does not reflect accurately their calculated dissimilarity; the links of the minimum spanning tree show that the Aberdeen and Elean Gamhna populations are actually more similar to the German water voles than those from Norway.

We shall now apply non-metric scaling to the voting behaviour shown in Table 17.2. Non-metric scaling is available with function `isoMDS` from package **MASS** (Venables and Ripley, 2002):

```
R> library("MASS")
R> data("voting", package = "HSAUR2")
R> voting_mds <- isoMDS(voting)
```

and we again depict the two-dimensional solution (Figure 17.3). The Figure suggests that voting behaviour is essentially along party lines, although there is more variation among Republicans. The voting behaviour of one of the Republicans (Rinaldo) seems to be closer to his democratic colleagues rather than to the voting behaviour of other Republicans.

The quality of a multidimensional scaling can be assessed informally by plotting the original dissimilarities and the distances obtained from a multidimensional scaling in a scatterplot, a so-called Shepard diagram. For the `voting` data, such a plot is shown in Figure 17.4. In an ideal situation, the points fall on the bisecting line; in our case, some deviations are observable.

```
R> library("ape")
R> st <- mst(watervoles)
R> plot(x, y, xlab = "Coordinate 1", ylab = "Coordinate 2",
+      xlim = range(x)*1.2, type = "n")
R> for (i in 1:nrow(watervoles)) {
+     w1 <- which(st[i, ] == 1)
+     segments(x[i], y[i], x[w1], y[w1])
+ }
R> text(x, y, labels = colnames(watervoles))
```

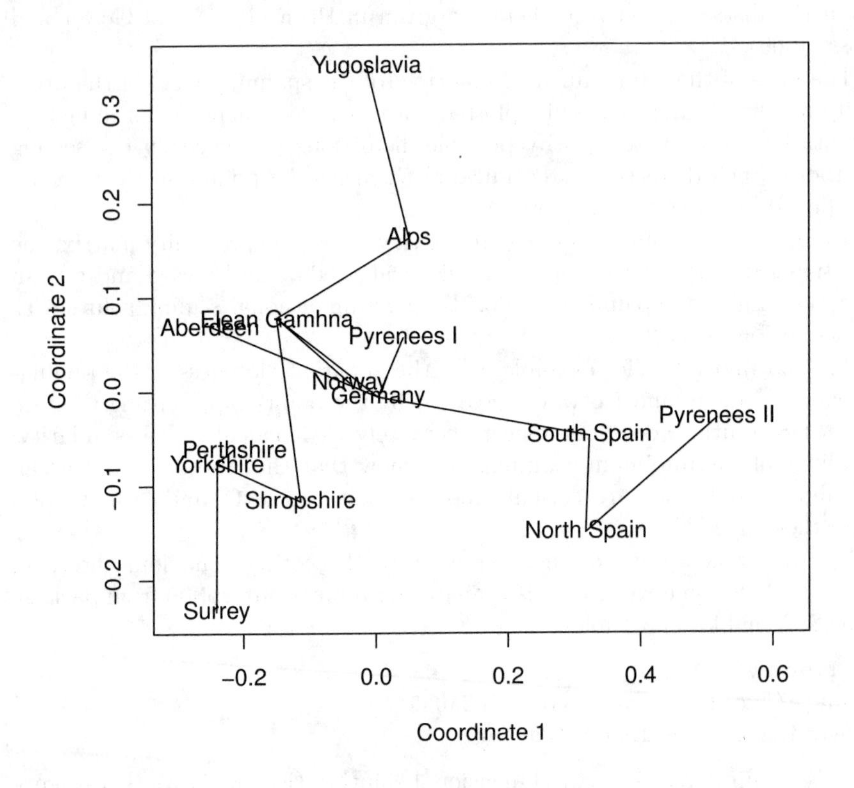


**Figure 17.2** Minimum spanning tree for the `watervoles` data.

```
R> x <- voting_mds$points[,1]
R> y <- voting_mds$points[,2]
R> plot(x, y, xlab = "Coordinate 1", ylab = "Coordinate 2",
+      xlim = range(voting_mds$points[,1])*1.2, type = "n")
R> text(x, y, labels = colnames(voting))
R> voting_sh <- Shepard(voting[lower.tri(voting)],
+                       voting_mds$points)
```

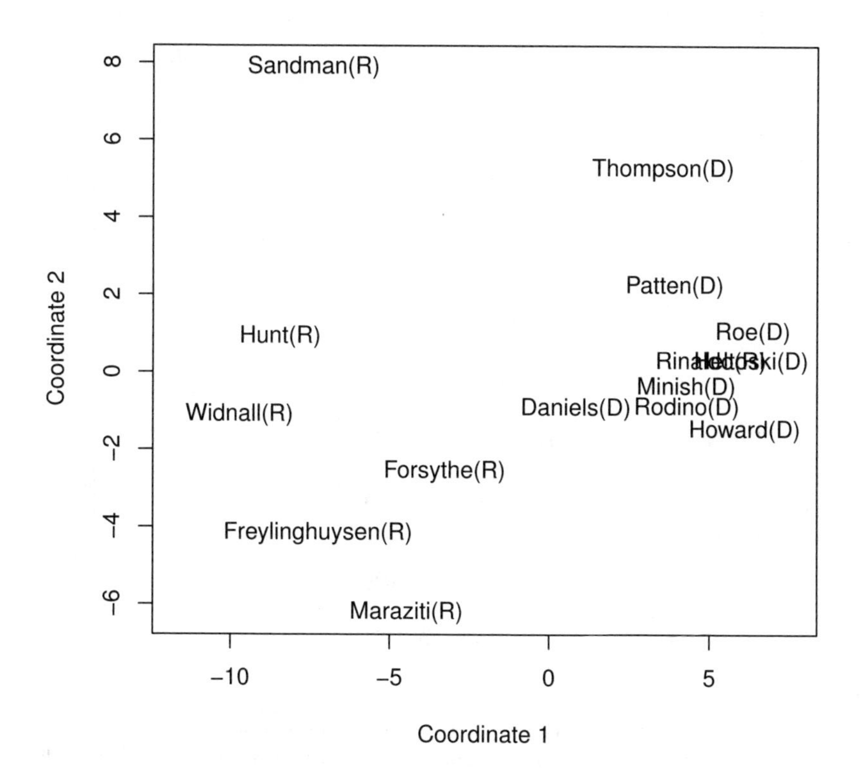


**Figure 17.3** Two-dimensional solution from non-metric multidimensional scaling of distance matrix for voting matrix.

```
R> plot(voting_sh, pch = ".", xlab = "Dissimilarity",
+       ylab = "Distance", xlim = range(voting_sh$x),
+       ylim = range(voting_sh$x))
R> lines(voting_sh$x, voting_sh$yf, type = "S")
```

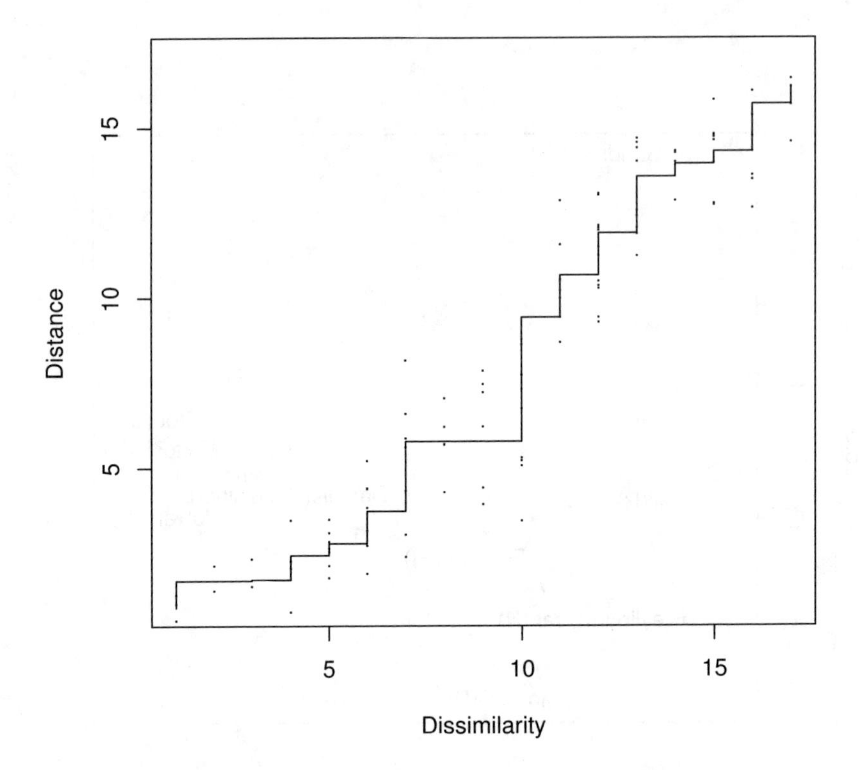


**Figure 17.4** The Shepard diagram for the `voting` data shows some discrepancies between the original dissimilarities and the multidimensional scaling solution.

## 17.4 Summary

Multidimensional scaling provides a powerful approach to extracting the structure in observed proximity matrices. Uncovering the pattern in this type of data may be important for a number of reasons, in particular for discovering the dimensions on which similarity judgements have been made.

**Exercises**

Ex. 17.1 The data in Table 17.3 shows road distances between 21 European cities. Apply classical scaling to the matrix and compare the plotted two-dimensional solution with a map of Europe.

Ex. 17.2 In Table 17.4 (from Kaufman and Rousseeuw, 1990), the dissimilarity matrix of 18 species of garden flowers is shown. Use some form of multidimensional scaling to investigate which species share common properties.

Ex. 17.3 Consider 51 objects $O_1, \ldots, O_{51}$ assumed to be arranged along a straight line with the $j$th object being located at a point with coordinate $j$. Define the similarity $s_{ij}$ between object $i$ and object $j$ as

$$s_{ij} = \begin{cases} 9 & \text{if} \quad i = j \\ 8 & \text{if} \quad 1 \leq |i - j| \leq 3 \\ 7 & \text{if} \quad 4 \leq |i - j| \leq 6 \\ & \cdots \\ 1 & \text{if} \quad 22 \leq |i - j| \leq 24 \\ 0 & \text{if} \quad |i - j| \geq 25 \end{cases}$$

Convert these similarities into dissimilarities ($\delta_{ij}$) by using

$$\delta_{ij} = \sqrt{s_{ii} + s_{jj} - 2s_{ij}}$$

and then apply classical multidimensional scaling to the resulting dissimilaritiy matrix. Explain the shape of the derived two-dimensional solution.

**Table 17.3**: eurodist data (package datasets). Distances between European cities, in km.

| | Athn | Brcl | Brss | Cals | Chrb | Clgn | Cpnh | Genv | Gbrl | Hmbr | HkoH | Lsbn | Lyns | Mdrd | Mrsl | Miln | Mnch | Pars | Rome | Stck | Vinn |
|---|---|---|---|---|---|---|---|---|---|---|---|---|---|---|---|---|---|---|---|---|---|
| Athens | 0 | | | | | | | | | | | | | | | | | | | | |
| Barcelona | 3313 | 0 | | | | | | | | | | | | | | | | | | | |
| Brussels | 2963 | 1318 | 0 | | | | | | | | | | | | | | | | | | |
| Calais | 3175 | 1326 | 204 | 0 | | | | | | | | | | | | | | | | | |
| Cherbourg | 3339 | 1294 | 583 | 460 | 0 | | | | | | | | | | | | | | | | |
| Cologne | 2762 | 1498 | 206 | 409 | 785 | 0 | | | | | | | | | | | | | | | |
| Copenhagen | 3276 | 2218 | 966 | 1136 | 1545 | 760 | 0 | | | | | | | | | | | | | | |
| Geneva | 2610 | 803 | 677 | 747 | 853 | 1662 | 1418 | 0 | | | | | | | | | | | | | |
| Gibraltar | 4485 | 1172 | 2256 | 2224 | 2047 | 2436 | 3196 | 1975 | 0 | | | | | | | | | | | | |
| Hamburg | 2977 | 2018 | 597 | 714 | 1115 | 460 | 460 | 1118 | 2897 | 0 | | | | | | | | | | | |
| Hook of Holland | 3030 | 1490 | 172 | 330 | 731 | 269 | 269 | 895 | 2428 | 550 | 0 | | | | | | | | | | |
| Lisbon | 4532 | 1305 | 2084 | 2052 | 1827 | 2290 | 2971 | 1936 | 676 | 2671 | 2280 | 0 | | | | | | | | | |
| Lyons | 2753 | 645 | 690 | 739 | 789 | 714 | 1458 | 158 | 1817 | 1159 | 863 | 1178 | 0 | | | | | | | | |
| Madrid | 3949 | 636 | 1558 | 1550 | 1347 | 1764 | 2498 | 1439 | 698 | 2198 | 1730 | 668 | 1281 | 0 | | | | | | | |
| Marseilles | 2865 | 521 | 1011 | 1059 | 1101 | 1035 | 1778 | 425 | 1693 | 1479 | 1183 | 1762 | 320 | 1157 | 0 | | | | | | |
| Milan | 2282 | 1014 | 925 | 1077 | 1209 | 911 | 1537 | 328 | 2185 | 1238 | 1098 | 2250 | 328 | 1724 | 618 | 0 | | | | | |
| Munich | 2179 | 1365 | 747 | 977 | 1160 | 583 | 1104 | 591 | 2565 | 805 | 851 | 2507 | 724 | 2010 | 1109 | 331 | 0 | | | | |
| Paris | 3000 | 1033 | 285 | 280 | 340 | 465 | 1176 | 513 | 1971 | 877 | 457 | 1799 | 471 | 1273 | 792 | 856 | 821 | 0 | | | |
| Rome | 817 | 1460 | 1511 | 1662 | 1794 | 1497 | 2050 | 995 | 2631 | 1751 | 1683 | 2700 | 1048 | 2097 | 1011 | 586 | 946 | 1476 | 0 | | |
| Stockholm | 3927 | 2868 | 1616 | 1786 | 2196 | 1403 | 650 | 2068 | 3886 | 949 | 1500 | 3231 | 2108 | 3188 | 2428 | 2187 | 1754 | 1827 | 2707 | 0 | |
| Vienna | 1991 | 1802 | 1175 | 1381 | 1588 | 937 | 1455 | 1019 | 2974 | 1155 | 1205 | 2937 | 1157 | 2409 | 1363 | 898 | 428 | 1249 | 1209 | 2105 | 0 |

**Table 17.4**: `gardenflowers` data. Dissimilarity matrix of 18 species of gardenflowers.

| | Bgn | Brm | Cml | Dhl | F- | Fch | Grn | Gld | Hth | Hyd | Irs | Lly | L- | Pny | Pnc | Rdr | Scr | Tlp |
|---|---|---|---|---|---|---|---|---|---|---|---|---|---|---|---|---|---|---|
| Begonia | 0.00 | | | | | | | | | | | | | | | | | |
| Broom | 0.91 | 0.00 | | | | | | | | | | | | | | | | |
| Camellia | 0.49 | 0.67 | 0.00 | | | | | | | | | | | | | | | |
| Dahlia | 0.47 | 0.59 | 0.59 | 0.00 | | | | | | | | | | | | | | |
| Forget-me-not | 0.43 | 0.90 | 0.57 | 0.61 | 0.00 | | | | | | | | | | | | | |
| Fuchsia | 0.23 | 0.79 | 0.29 | 0.52 | 0.44 | 0.00 | | | | | | | | | | | | |
| Geranium | 0.31 | 0.70 | 0.54 | 0.44 | 0.54 | 0.24 | 0.00 | | | | | | | | | | | |
| Gladiolus | 0.49 | 0.57 | 0.71 | 0.26 | 0.49 | 0.68 | 0.49 | 0.00 | | | | | | | | | | |
| Heather | 0.57 | 0.57 | 0.57 | 0.89 | 0.50 | 0.61 | 0.70 | 0.77 | 0.00 | | | | | | | | | |
| Hydrangae | 0.76 | 0.58 | 0.58 | 0.62 | 0.39 | 0.61 | 0.86 | 0.70 | 0.55 | 0.00 | | | | | | | | |
| Iris | 0.32 | 0.77 | 0.63 | 0.75 | 0.46 | 0.52 | 0.60 | 0.63 | 0.46 | 0.47 | 0.00 | | | | | | | |
| Lily | 0.51 | 0.69 | 0.69 | 0.53 | 0.51 | 0.65 | 0.77 | 0.47 | 0.51 | 0.39 | 0.36 | 0.00 | | | | | | |
| Lily-of-the-valley | 0.59 | 0.75 | 0.75 | 0.77 | 0.35 | 0.63 | 0.72 | 0.65 | 0.35 | 0.41 | 0.45 | 0.24 | 0.00 | | | | | |
| Peony | 0.37 | 0.68 | 0.68 | 0.38 | 0.52 | 0.48 | 0.63 | 0.49 | 0.52 | 0.39 | 0.37 | 0.17 | 0.39 | 0.00 | | | | |
| Pink carnation | 0.74 | 0.54 | 0.70 | 0.58 | 0.54 | 0.74 | 0.50 | 0.49 | 0.36 | 0.52 | 0.60 | 0.48 | 0.39 | 0.49 | 0.00 | | | |
| Red rose | 0.84 | 0.41 | 0.75 | 0.37 | 0.82 | 0.71 | 0.61 | 0.64 | 0.81 | 0.43 | 0.84 | 0.62 | 0.67 | 0.47 | 0.45 | 0.00 | | |
| Scotch rose | 0.94 | 0.20 | 0.70 | 0.48 | 0.77 | 0.83 | 0.74 | 0.45 | 0.77 | 0.38 | 0.80 | 0.58 | 0.62 | 0.57 | 0.40 | 0.21 | 0.00 | |
| Tulip | 0.44 | 0.50 | 0.79 | 0.48 | 0.59 | 0.68 | 0.47 | 0.22 | 0.59 | 0.92 | 0.59 | 0.67 | 0.72 | 0.67 | 0.61 | 0.85 | 0.67 | 0.00 |

CHAPTER 18

# Cluster Analysis: Classifying Romano-British Pottery and Exoplanets

## 18.1 Introduction

The data shown in Table 18.1 give the chemical composition of 48 specimens of Romano-British pottery, determined by atomic absorption spectrophotometry, for nine oxides (Tubb et al., 1980). In addition to the chemical composition of the pots, the kiln site at which the pottery was found is known for these data. For these data, interest centres on whether, on the basis of their chemical compositions, the pots can be divided into distinct groups, and how these groups relate to the kiln site.

**Table 18.1**: pottery data. Romano-British pottery data.

| Al2O3 | Fe2O3 | MgO | CaO | Na2O | K2O | TiO2 | MnO | BaO | kiln |
|---|---|---|---|---|---|---|---|---|---|
| 18.8 | 9.52 | 2.00 | 0.79 | 0.40 | 3.20 | 1.01 | 0.077 | 0.015 | 1 |
| 16.9 | 7.33 | 1.65 | 0.84 | 0.40 | 3.05 | 0.99 | 0.067 | 0.018 | 1 |
| 18.2 | 7.64 | 1.82 | 0.77 | 0.40 | 3.07 | 0.98 | 0.087 | 0.014 | 1 |
| 16.9 | 7.29 | 1.56 | 0.76 | 0.40 | 3.05 | 1.00 | 0.063 | 0.019 | 1 |
| 17.8 | 7.24 | 1.83 | 0.92 | 0.43 | 3.12 | 0.93 | 0.061 | 0.019 | 1 |
| 18.8 | 7.45 | 2.06 | 0.87 | 0.25 | 3.26 | 0.98 | 0.072 | 0.017 | 1 |
| 16.5 | 7.05 | 1.81 | 1.73 | 0.33 | 3.20 | 0.95 | 0.066 | 0.019 | 1 |
| 18.0 | 7.42 | 2.06 | 1.00 | 0.28 | 3.37 | 0.96 | 0.072 | 0.017 | 1 |
| 15.8 | 7.15 | 1.62 | 0.71 | 0.38 | 3.25 | 0.93 | 0.062 | 0.017 | 1 |
| 14.6 | 6.87 | 1.67 | 0.76 | 0.33 | 3.06 | 0.91 | 0.055 | 0.012 | 1 |
| 13.7 | 5.83 | 1.50 | 0.66 | 0.13 | 2.25 | 0.75 | 0.034 | 0.012 | 1 |
| 14.6 | 6.76 | 1.63 | 1.48 | 0.20 | 3.02 | 0.87 | 0.055 | 0.016 | 1 |
| 14.8 | 7.07 | 1.62 | 1.44 | 0.24 | 3.03 | 0.86 | 0.080 | 0.016 | 1 |
| 17.1 | 7.79 | 1.99 | 0.83 | 0.46 | 3.13 | 0.93 | 0.090 | 0.020 | 1 |
| 16.8 | 7.86 | 1.86 | 0.84 | 0.46 | 2.93 | 0.94 | 0.094 | 0.020 | 1 |
| 15.8 | 7.65 | 1.94 | 0.81 | 0.83 | 3.33 | 0.96 | 0.112 | 0.019 | 1 |
| 18.6 | 7.85 | 2.33 | 0.87 | 0.38 | 3.17 | 0.98 | 0.081 | 0.018 | 1 |
| 16.9 | 7.87 | 1.83 | 1.31 | 0.53 | 3.09 | 0.95 | 0.092 | 0.023 | 1 |
| 18.9 | 7.58 | 2.05 | 0.83 | 0.13 | 3.29 | 0.98 | 0.072 | 0.015 | 1 |
| 18.0 | 7.50 | 1.94 | 0.69 | 0.12 | 3.14 | 0.93 | 0.035 | 0.017 | 1 |
| 17.8 | 7.28 | 1.92 | 0.81 | 0.18 | 3.15 | 0.90 | 0.067 | 0.017 | 1 |

**Table 18.1**: pottery data (continued).

| Al2O3 | Fe2O3 | MgO | CaO | Na2O | K2O | TiO2 | MnO | BaO | kiln |
|---|---|---|---|---|---|---|---|---|---|
| 14.4 | 7.00 | 4.30 | 0.15 | 0.51 | 4.25 | 0.79 | 0.160 | 0.019 | 2 |
| 13.8 | 7.08 | 3.43 | 0.12 | 0.17 | 4.14 | 0.77 | 0.144 | 0.020 | 2 |
| 14.6 | 7.09 | 3.88 | 0.13 | 0.20 | 4.36 | 0.81 | 0.124 | 0.019 | 2 |
| 11.5 | 6.37 | 5.64 | 0.16 | 0.14 | 3.89 | 0.69 | 0.087 | 0.009 | 2 |
| 13.8 | 7.06 | 5.34 | 0.20 | 0.20 | 4.31 | 0.71 | 0.101 | 0.021 | 2 |
| 10.9 | 6.26 | 3.47 | 0.17 | 0.22 | 3.40 | 0.66 | 0.109 | 0.010 | 2 |
| 10.1 | 4.26 | 4.26 | 0.20 | 0.18 | 3.32 | 0.59 | 0.149 | 0.017 | 2 |
| 11.6 | 5.78 | 5.91 | 0.18 | 0.16 | 3.70 | 0.65 | 0.082 | 0.015 | 2 |
| 11.1 | 5.49 | 4.52 | 0.29 | 0.30 | 4.03 | 0.63 | 0.080 | 0.016 | 2 |
| 13.4 | 6.92 | 7.23 | 0.28 | 0.20 | 4.54 | 0.69 | 0.163 | 0.017 | 2 |
| 12.4 | 6.13 | 5.69 | 0.22 | 0.54 | 4.65 | 0.70 | 0.159 | 0.015 | 2 |
| 13.1 | 6.64 | 5.51 | 0.31 | 0.24 | 4.89 | 0.72 | 0.094 | 0.017 | 2 |
| 11.6 | 5.39 | 3.77 | 0.29 | 0.06 | 4.51 | 0.56 | 0.110 | 0.015 | 3 |
| 11.8 | 5.44 | 3.94 | 0.30 | 0.04 | 4.64 | 0.59 | 0.085 | 0.013 | 3 |
| 18.3 | 1.28 | 0.67 | 0.03 | 0.03 | 1.96 | 0.65 | 0.001 | 0.014 | 4 |
| 15.8 | 2.39 | 0.63 | 0.01 | 0.04 | 1.94 | 1.29 | 0.001 | 0.014 | 4 |
| 18.0 | 1.50 | 0.67 | 0.01 | 0.06 | 2.11 | 0.92 | 0.001 | 0.016 | 4 |
| 18.0 | 1.88 | 0.68 | 0.01 | 0.04 | 2.00 | 1.11 | 0.006 | 0.022 | 4 |
| 20.8 | 1.51 | 0.72 | 0.07 | 0.10 | 2.37 | 1.26 | 0.002 | 0.016 | 4 |
| 17.7 | 1.12 | 0.56 | 0.06 | 0.06 | 2.06 | 0.79 | 0.001 | 0.013 | 5 |
| 18.3 | 1.14 | 0.67 | 0.06 | 0.05 | 2.11 | 0.89 | 0.006 | 0.019 | 5 |
| 16.7 | 0.92 | 0.53 | 0.01 | 0.05 | 1.76 | 0.91 | 0.004 | 0.013 | 5 |
| 14.8 | 2.74 | 0.67 | 0.03 | 0.05 | 2.15 | 1.34 | 0.003 | 0.015 | 5 |
| 19.1 | 1.64 | 0.60 | 0.10 | 0.03 | 1.75 | 1.04 | 0.007 | 0.018 | 5 |

*Source*: Tubb, A., et al., *Archaeometry*, 22, 153–171, 1980. With permission.

Exoplanets are planets outside the Solar System. The first such planet was discovered in 1995 by Mayor and Queloz (1995). The planet, similar in mass to Jupiter, was found orbiting a relatively ordinary star, 51 Pegasus. In the intervening period over a hundred exoplanets have been discovered, nearly all detected indirectly, using the gravitational influence they exert on their associated central stars. A fascinating account of exoplanets and their discovery is given in Mayor and Frei (2003).

From the properties of the exoplanets found up to now it appears that the theory of planetary development constructed for the planets of the Solar System may need to be reformulated. The exoplanets are not at all like the nine local planets that we know so well. A first step in the process of understanding the exoplanets might be to try to classify them with respect to their known properties and this will be the aim in this chapter. The data in Table 18.2 (taken with permission from Mayor and Frei, 2003) give the mass (in Jupiter mass, `mass`), the period (in earth days, `period`) and the eccentricity (`eccent`) of the exoplanets discovered up until October 2002.

We shall investigate the structure of both the pottery data and the exoplanets data using a number of methods of cluster analysis.

**Table 18.2:** `planets` data. Jupiter mass, period and eccentricity of exoplanets.

| mass | period | eccen | mass | period | eccen |
|---|---|---|---|---|---|
| 0.120 | 4.950000 | 0.0000 | 1.890 | 61.020000 | 0.1000 |
| 0.197 | 3.971000 | 0.0000 | 1.900 | 6.276000 | 0.1500 |
| 0.210 | 44.280000 | 0.3400 | 1.990 | 743.000000 | 0.6200 |
| 0.220 | 75.800000 | 0.2800 | 2.050 | 241.300000 | 0.2400 |
| 0.230 | 6.403000 | 0.0800 | 0.050 | 1119.000000 | 0.1700 |
| 0.250 | 3.024000 | 0.0200 | 2.080 | 228.520000 | 0.3040 |
| 0.340 | 2.985000 | 0.0800 | 2.240 | 311.300000 | 0.2200 |
| 0.400 | 10.901000 | 0.4980 | 2.540 | 1089.000000 | 0.0600 |
| 0.420 | 3.509700 | 0.0000 | 2.540 | 627.340000 | 0.0600 |
| 0.470 | 4.229000 | 0.0000 | 2.550 | 2185.000000 | 0.1800 |
| 0.480 | 3.487000 | 0.0500 | 2.630 | 414.000000 | 0.2100 |
| 0.480 | 22.090000 | 0.3000 | 2.840 | 250.500000 | 0.1900 |
| 0.540 | 3.097000 | 0.0100 | 2.940 | 229.900000 | 0.3500 |
| 0.560 | 30.120000 | 0.2700 | 3.030 | 186.900000 | 0.4100 |
| 0.680 | 4.617000 | 0.0200 | 3.320 | 267.200000 | 0.2300 |
| 0.685 | 3.524330 | 0.0000 | 3.360 | 1098.000000 | 0.2200 |
| 0.760 | 2594.000000 | 0.1000 | 3.370 | 133.710000 | 0.5110 |
| 0.770 | 14.310000 | 0.2700 | 3.440 | 1112.000000 | 0.5200 |
| 0.810 | 828.950000 | 0.0400 | 3.550 | 18.200000 | 0.0100 |
| 0.880 | 221.600000 | 0.5400 | 3.810 | 340.000000 | 0.3600 |
| 0.880 | 2518.000000 | 0.6000 | 3.900 | 111.810000 | 0.9270 |
| 0.890 | 64.620000 | 0.1300 | 4.000 | 15.780000 | 0.0460 |
| 0.900 | 1136.000000 | 0.3300 | 4.000 | 5360.000000 | 0.1600 |
| 0.930 | 3.092000 | 0.0000 | 4.120 | 1209.900000 | 0.6500 |
| 0.930 | 14.660000 | 0.0300 | 4.140 | 3.313000 | 0.0200 |
| 0.990 | 39.810000 | 0.0700 | 4.270 | 1764.000000 | 0.3530 |
| 0.990 | 500.730000 | 0.1000 | 4.290 | 1308.500000 | 0.3100 |
| 0.990 | 872.300000 | 0.2800 | 4.500 | 951.000000 | 0.4500 |
| 1.000 | 337.110000 | 0.3800 | 4.800 | 1237.000000 | 0.5150 |
| 1.000 | 264.900000 | 0.3800 | 5.180 | 576.000000 | 0.7100 |
| 1.010 | 540.400000 | 0.5200 | 5.700 | 383.000000 | 0.0700 |
| 1.010 | 1942.000000 | 0.4000 | 6.080 | 1074.000000 | 0.0110 |
| 1.020 | 10.720000 | 0.0440 | 6.292 | 71.487000 | 0.1243 |
| 1.050 | 119.600000 | 0.3500 | 7.170 | 256.000000 | 0.7000 |
| 1.120 | 500.000000 | 0.2300 | 7.390 | 1582.000000 | 0.4780 |
| 1.130 | 154.800000 | 0.3100 | 7.420 | 116.700000 | 0.4000 |

**Table 18.2**: `planets` data (continued).

| mass | period | eccen | mass | period | eccen |
|---|---|---|---|---|---|
| 1.150 | 2614.000000 | 0.0000 | 7.500 | 2300.000000 | 0.3950 |
| 1.230 | 1326.000000 | 0.1400 | 7.700 | 58.116000 | 0.5290 |
| 1.240 | 391.000000 | 0.4000 | 7.950 | 1620.000000 | 0.2200 |
| 1.240 | 435.600000 | 0.4500 | 8.000 | 1558.000000 | 0.3140 |
| 1.282 | 7.126200 | 0.1340 | 8.640 | 550.650000 | 0.7100 |
| 1.420 | 426.000000 | 0.0200 | 9.700 | 653.220000 | 0.4100 |
| 1.550 | 51.610000 | 0.6490 | 10.000 | 3030.000000 | 0.5600 |
| 1.560 | 1444.500000 | 0.2000 | 10.370 | 2115.200000 | 0.6200 |
| 1.580 | 260.000000 | 0.2400 | 10.960 | 84.030000 | 0.3300 |
| 1.630 | 444.600000 | 0.4100 | 11.300 | 2189.000000 | 0.3400 |
| 1.640 | 406.000000 | 0.5300 | 11.980 | 1209.000000 | 0.3700 |
| 1.650 | 401.100000 | 0.3600 | 14.400 | 8.428198 | 0.2770 |
| 1.680 | 796.700000 | 0.6800 | 16.900 | 1739.500000 | 0.2280 |
| 1.760 | 903.000000 | 0.2000 | 17.500 | 256.030000 | 0.4290 |
| 1.830 | 454.000000 | 0.2000 | | | |

*Source*: From Mayor, M., Frei, P.-Y., and Roukema, B., *New Worlds in the Cosmos*, Cambridge University Press, Cambridge, England, 2003. With permission.

## 18.2 Cluster Analysis

Cluster analysis is a generic term for a wide range of numerical methods for examining multivariate data with a view to uncovering or discovering groups or clusters of observations that are homogeneous and separated from other groups. In medicine, for example, discovering that a sample of patients with measurements on a variety of characteristics and symptoms actually consists of a small number of groups within which these characteristics are relatively similar, and between which they are different, might have important implications both in terms of future treatment and for investigating the aetiology of a condition. More recently cluster analysis techniques have been applied to microarray data (Alon et al., 1999, among many others), image analysis (Everitt and Bullmore, 1999) or in marketing science (Dolnicar and Leisch, 2003).

Clustering techniques essentially try to formalise what human observers do so well in two or three dimensions. Consider, for example, the scatterplot shown in Figure 18.1. The conclusion that there are three natural groups or clusters of dots is reached with no conscious effort or thought. Clusters are identified by the assessment of the relative distances between points and in this example, the relative homogeneity of each cluster and the degree of their separation makes the task relatively simple.

Detailed accounts of clustering techniques are available in Everitt et al. (2001) and Gordon (1999). Here we concentrate on three types of cluster-

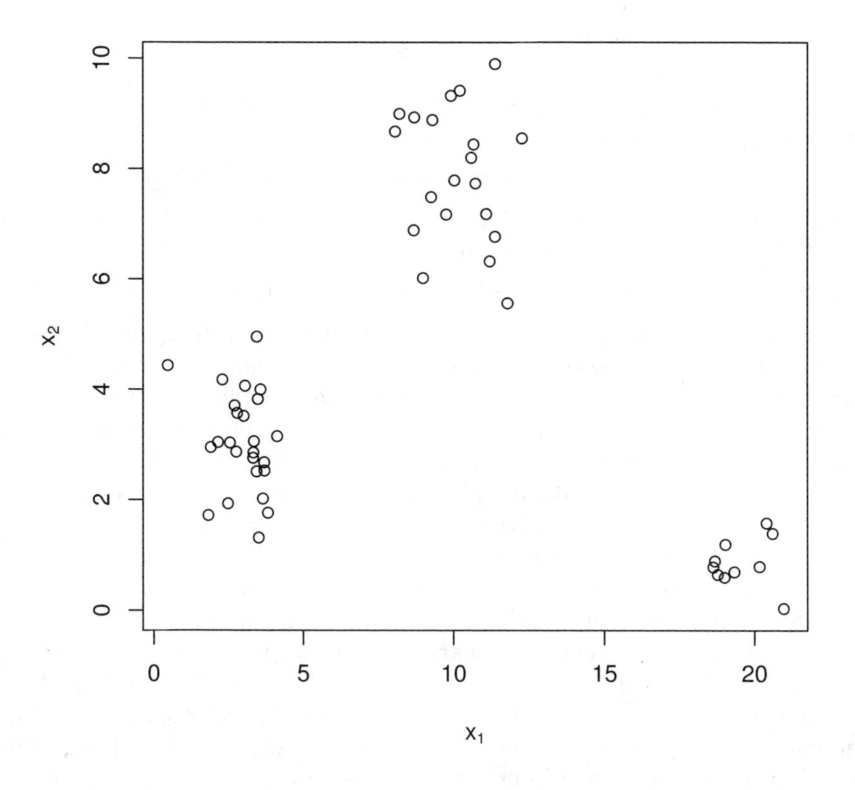


**Figure 18.1** Bivariate data showing the presence of three clusters.

ing procedures: agglomerative hierarchical clustering, $k$-means clustering and classification maximum likelihood methods for clustering.

### *18.2.1 Agglomerative Hierarchical Clustering*

In a hierarchical classification the data are not partitioned into a particular number of classes or clusters at a single step. Instead the classification consists of a series of partitions that may run from a single 'cluster' containing all individuals, to $n$ clusters each containing a single individual. Agglomerative hierarchical clustering techniques produce partitions by a series of successive fusions of the $n$ individuals into groups. With such methods, fusions, once made, are irreversible, so that when an agglomerative algorithm has placed two individuals in the same group they cannot subsequently appear in different groups. Since all agglomerative hierarchical techniques ultimately reduce the data to a single cluster containing all the individuals, the investigator seeking the solution with the 'best' fitting number of clusters will need to decide which division to choose. The problem of deciding on the 'correct' number of clusters will be taken up later.

An agglomerative hierarchical clustering procedure produces a series of partitions of the data, $P_n, P_{n-1}, \ldots, P_1$. The first, $P_n$, consists of $n$ single-member clusters, and the last, $P_1$, consists of a single group containing all $n$ individuals. The basic operation of all methods is similar:

**Start** Clusters $C_1, C_2, \ldots, C_n$ each containing a single individual.

**Step 1** Find the nearest pair of distinct clusters, say $C_i$ and $C_j$, merge $C_i$ and $C_j$, delete $C_j$ and decrease the number of clusters by one.

**Step 2** If number of clusters equals one then stop; else return to Step 1.

At each stage in the process the methods fuse individuals or groups of individuals that are closest (or most similar). The methods begin with an inter-individual distance matrix (for example, one containing Euclidean distances), but as groups are formed, distance between an individual and a group containing several individuals or between two groups of individuals will need to be calculated. How such distances are defined leads to a variety of different techniques; see the next sub-section.

Hierarchic classifications may be represented by a two-dimensional diagram known as a dendrogram, which illustrates the fusions made at each stage of the analysis. An example of such a diagram is given in Figure 18.2. The structure of Figure 18.2 resembles an evolutionary tree, a concept introduced by Darwin under the term "Tree of Life" in his book *On the Origin of Species by Natural Selection* in 1859 (see Figure 18.3), and it is in biological applications that hierarchical classifications are most relevant and most justified (although this type of clustering has also been used in many other areas). According to Rohlf (1970), a biologist, all things being equal, aims for a system of nested clusters. Hawkins et al. (1982), however, issue the following caveat: "users should be very wary of using hierarchic methods if they are not clearly necessary".

### *18.2.2 Measuring Inter-cluster Dissimilarity*

Agglomerative hierarchical clustering techniques differ primarily in how they measure the distance between or similarity of two clusters (where a cluster may, at times, consist of only a single individual). Two simple inter-group measures are

$$\begin{aligned} d_{\min}(A, B) &= \min_{i \in A, j \in B} d_{ij} \\ d_{\max}(A, B) &= \max_{i \in A, j \in B} d_{ij} \end{aligned}$$

where $d(A, B)$ is the distance between two clusters $A$ and $B$, and $d_{ij}$ is the distance between individuals $i$ and $j$. This could be Euclidean distance or one of a variety of other distance measures (see Everitt et al., 2001, for details).

The inter-group dissimilarity measure $d_{\min}(A, B)$ is the basis of *single link-*

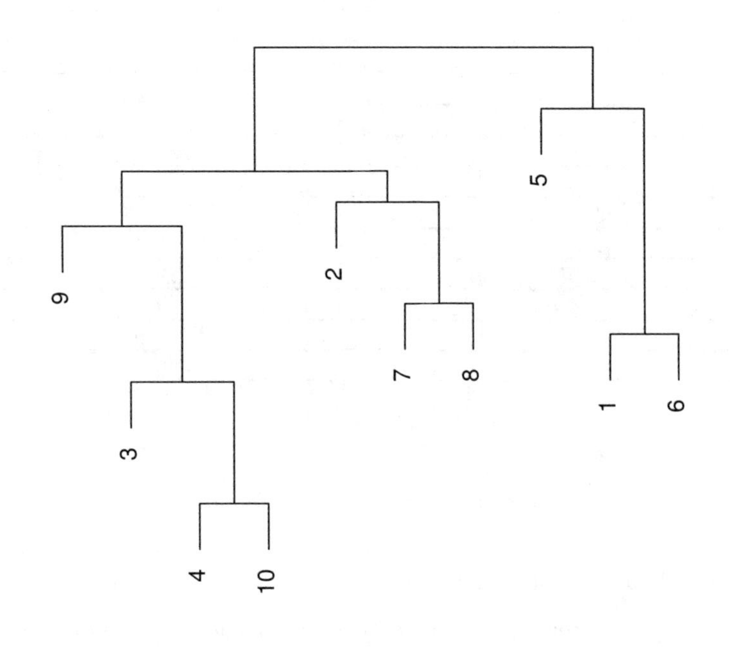


**Figure 18.2** Example of a dendrogram.

*age clustering*, $d_{\text{max}}(A, B)$ that of *complete linkage clustering*. Both these techniques have the desirable property that they are invariant under monotone transformations of the original inter-individual dissimilarities or distances. A further possibility for measuring inter-cluster distance or dissimilarity is

$$d_{\text{mean}}(A, B) = \frac{1}{|A| \cdot |B|} \sum_{i \in A, j \in B} d_{ij}$$

where $|A|$ and $|B|$ are the number of individuals in clusters $A$ and $B$. This measure is the basis of a commonly used procedure known as *average linkage clustering*.

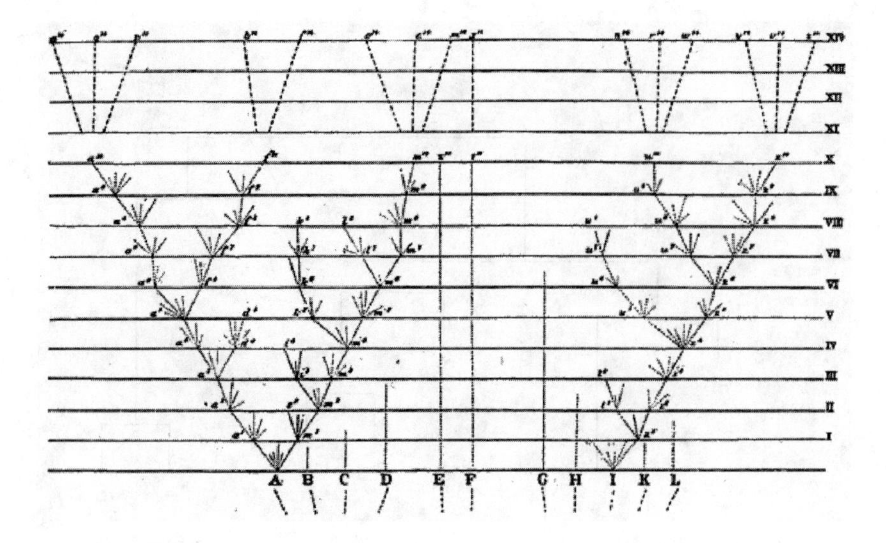


**Figure 18.3** Darwin's Tree of Life.

*18.2.3 k-Means Clustering*

The $k$-means clustering technique seeks to partition a set of data into a specified number of groups, $k$, by minimising some numerical criterion, low values of which are considered indicative of a 'good' solution. The most commonly used approach, for example, is to try to find the partition of the $n$ individuals into $k$ groups, which minimises the within-group sum of squares over all variables. The problem then appears relatively simple; namely, consider every possible partition of the $n$ individuals into $k$ groups, and select the one with the lowest within-group sum of squares. Unfortunately, the problem in practise is not so straightforward. The numbers involved are so vast that complete enumeration of *every* possible partition remains impossible even with the fastest computer. The scale of the problem is illustrated by the numbers in Table 18.3.

**Table 18.3**: Number of possible partitions depending on the sample size $n$ and number of clusters $k$.

| $n$ | $k$ | Number of possible partitions |
|---|---|---|
| 15 | 3 | $2,375,101$ |
| 20 | 4 | $45,232,115,901$ |
| 25 | 8 | $690,223,721,118,368,580$ |
| 100 | 5 | $10^{68}$ |

The impracticability of examining every possible partition has led to the development of algorithms designed to search for the minimum values of the clustering criterion by rearranging existing partitions and keeping the new one only if it provides an improvement. Such algorithms do not, of course, guarantee finding the global minimum of the criterion. The essential steps in these algorithms are as follows:

1. Find some initial partition of the individuals into the required number of groups. Such an initial partition could be provided by a solution from one of the hierarchical clustering techniques described in the previous section.
2. Calculate the change in the clustering criterion produced by 'moving' each individual from its own to another cluster.
3. Make the change that leads to the greatest improvement in the value of the clustering criterion.
4. Repeat steps 2 and 3 until no move of an individual causes the clustering criterion to improve.

When variables are on very different scales (as they are for the exoplanets data) some form of standardisation will be needed before applying $k$-means clustering (for a detailed discussion of this problem see Everitt et al., 2001).

### *18.2.4 Model-based Clustering*

The $k$-means clustering method described in the previous section is based largely in heuristic but intuitively reasonable procedures. But it is not based on formal models thus making problems such as deciding on a particular method, estimating the number of clusters, etc., particularly difficult. And, of course, without a reasonable model, formal inference is precluded. In practise these may not be insurmountable objections to the use of the technique since cluster analysis is essentially an 'exploratory' tool. But model-based cluster methods do have some advantages, and a variety of possibilities have been proposed. The most successful approach has been that proposed by Scott and Symons (1971) and extended by Banfield and Raftery (1993) and Fraley and Raftery (1999, 2002), in which it is assumed that the population from which the observations arise consists of $c$ subpopulations each corresponding to a cluster, and that the density of a $q$-dimensional observation $\mathbf{x}^\top = (x_1, \dots, x_q)$ from the $j$th subpopulation is $f_j(\mathbf{x}, \vartheta_j), j = 1, \dots, c$, for some unknown vector of parameters, $\vartheta_j$. They also introduce a vector $\gamma = (\gamma_1, \dots, \gamma_n)$, where $\gamma_i = j$ of $\mathbf{x}_i$ is from the $j$ subpopulation. The $\gamma_i$ label the subpopulation for each observation $i = 1, \dots, n$. The clustering problem now becomes that of choosing $\vartheta = (\vartheta_1, \dots, \vartheta_c)$ and $\gamma$ to maximise the likelihood function associated with such assumptions. This classification maximum likelihood procedure is described briefly in the sequel.

### *18.2.5 Classification Maximum Likelihood*

Assume the population consists of $c$ subpopulations, each corresponding to a cluster of observations, and that the density function of a $q$-dimensional observation from the $j$th subpopulation is $f_j(\mathbf{x}, \vartheta_j)$ for some unknown vector of parameters, $\vartheta_j$. Also, assume that $\gamma = (\gamma_1, \dots, \gamma_n)$ gives the labels of the subpopulation to which the observation belongs: so $\gamma_i = j$ if $\mathbf{x}_i$ is from the $j$th population.

The clustering problem becomes that of choosing $\vartheta = (\vartheta_1, \dots, \vartheta_c)$ and $\gamma$ to maximise the likelihood

$$L(\vartheta, \gamma) = \prod_{i=1}^{n} f_{\gamma_i}(\mathbf{x}_i, \vartheta_{\gamma_i}). \qquad (18.1)$$

If $f_j(\mathbf{x}, \vartheta_j)$ is taken as the multivariate normal density with mean vector $\mu_j$ and covariance matrix $\Sigma_j$, this likelihood has the form

$$L(\vartheta, \gamma) = \prod_{j=1}^{c} \prod_{i:\gamma_i = j} |\Sigma_j|^{-1/2} \exp\left(-\frac{1}{2}(\mathbf{x}_i - \mu_j)^\top \Sigma_j^{-1} (\mathbf{x}_i - \mu_j)\right). \qquad (18.2)$$

The maximum likelihood estimator of $\mu_j$ is $\hat{\mu}_j = n_j^{-1} \sum_{i:\gamma_i = j} \mathbf{x}_i$ where the number of observations in each subpopulation is $n_j = \sum_{i=1}^{n} I(\gamma_i = j)$. Replacing $\mu_j$ in (18.2) yields the following log-likelihood

$$l(\vartheta, \gamma) = -\frac{1}{2} \sum_{j=1}^{c} \text{trace}(\mathbf{W}_j \Sigma_j^{-1}) + n_j \log |\Sigma_j|$$

where $\mathbf{W}_j$ is the $q \times q$ matrix of sums of squares and cross-products of the variables for subpopulation $j$.

Banfield and Raftery (1993) demonstrate the following: If the covariance matrix $\Sigma_j$ is $\sigma^2$ times the identity matrix for all populations $j = 1, \dots, c$, then the likelihood is maximised by choosing $\gamma$ to minimise trace($\mathbf{W}$), where $\mathbf{W} = \sum_{j=1}^{c} \mathbf{W}_j$, i.e., minimisation of the written group sum of squares. Use of this criterion in a cluster analysis will tend to produce spherical clusters of largely equal sizes which may or may not match the 'real' clusters in the data.

If $\Sigma_j = \Sigma$ for $j = 1, \dots, c$, then the likelihood is maximised by choosing $\gamma$ to minimise $|\mathbf{W}|$, a clustering criterion discussed by Friedman and Rubin (1967) and Marriott (1982). Use of this criterion in a cluster analysis will tend to produce clusters with the same elliptical shape, which again may not necessarily match the actual clusters in the data.

If $\Sigma_j$ is not constrained, the likelihood is maximised by choosing $\gamma$ to minimise $\sum_{j=1}^{c} n_j \log |\mathbf{W}_j / n_j|$, a criterion that allows for different shaped clusters in the data.

Banfield and Raftery (1993) also consider criteria that allow the shape of clusters to be less constrained than with the minimisation of trace($\mathbf{W}$) and $|\mathbf{W}|$ criteria, but to remain more parsimonious than the completely unconstrained model. For example, constraining clusters to be spherical but not to have the same volume, or constraining clusters to have diagonal covariance matrices but allowing their shapes, sizes and orientations to vary.

The EM algorithm (see Dempster et al., 1977) is used for maximum likelihood estimation – details are given in Fraley and Raftery (1999). Model selection is a combination of choosing the appropriate clustering model and the optimal number of clusters. A Bayesian approach is used (see Fraley and Raftery, 1999), using what is known as the Bayesian Information Criterion (BIC).

## 18.3 Analysis Using R

### *18.3.1 Classifying Romano-British Pottery*

We start our analysis with computing the dissimilarity matrix containing the Euclidean distance of the chemical measurements on all 45 pots. The resulting $45 \times 45$ matrix can be inspected by an *image plot*, here obtained from function `levelplot` available in package **lattice** (Sarkar, 2009, 2008). Such a plot associates each cell of the dissimilarity matrix with a color or a grey value. We choose a very dark grey for cells with distance zero (i.e., the diagonal elements of the dissimilarity matrix) and pale values for cells with greater Euclidean distance. Figure 18.4 leads to the impression that there are at least three distinct groups with small inter-cluster differences (the dark rectangles) whereas much larger distances can be observed for all other cells.

We now construct three series of partitions using single, complete, and average linkage hierarchical clustering as introduced in subsections 18.2.1 and 18.2.2. The function `hclust` performs all three procedures based on the dissimilarity matrix of the data; its `method` argument is used to specify how the distance between two clusters is assessed. The corresponding `plot` method draws a dendrogram; the code and results are given in Figure 18.5. Again, all three dendrograms lead to the impression that three clusters fit the data best (although this judgement is very informal).

From the `pottery_average` object representing the average linkage hierarchical clustering, we derive the three-cluster solution by cutting the dendrogram at a height of four (which, based on the right display in Figure 18.5 leads to a partition of the data into three groups). Our interest is now a comparison with the kiln sites at which the pottery was found.

```
R> pottery_cluster <- cutree(pottery_average, h = 4)
R> xtabs(~ pottery_cluster + kiln, data = pottery)
```

```
               kiln
pottery_cluster  1  2  3  4  5
              1 21  0  0  0  0
              2  0 12  2  0  0
              3  0  0  0  5  5
```

The contingency table shows that cluster 1 contains all pots found at kiln site number one, cluster 2 contains all pots from kiln sites number two and three, and cluster three collects the ten pots from kiln sites four and five. In fact, the five kiln sites are from three different regions defined by one, two and three, and four and five, so the clusters actually correspond to pots from three different regions.

### *18.3.2 Classifying Exoplanets*

Prior to a cluster analysis we present a graphical representation of the three-dimensional `planets` data by means of the **scatterplot3d** package (Ligges and

```
R> pottery_dist <- dist(pottery[, colnames(pottery) != "kiln"])
R> library("lattice")
R> levelplot(as.matrix(pottery_dist), xlab = "Pot Number",
+            ylab = "Pot Number")
```

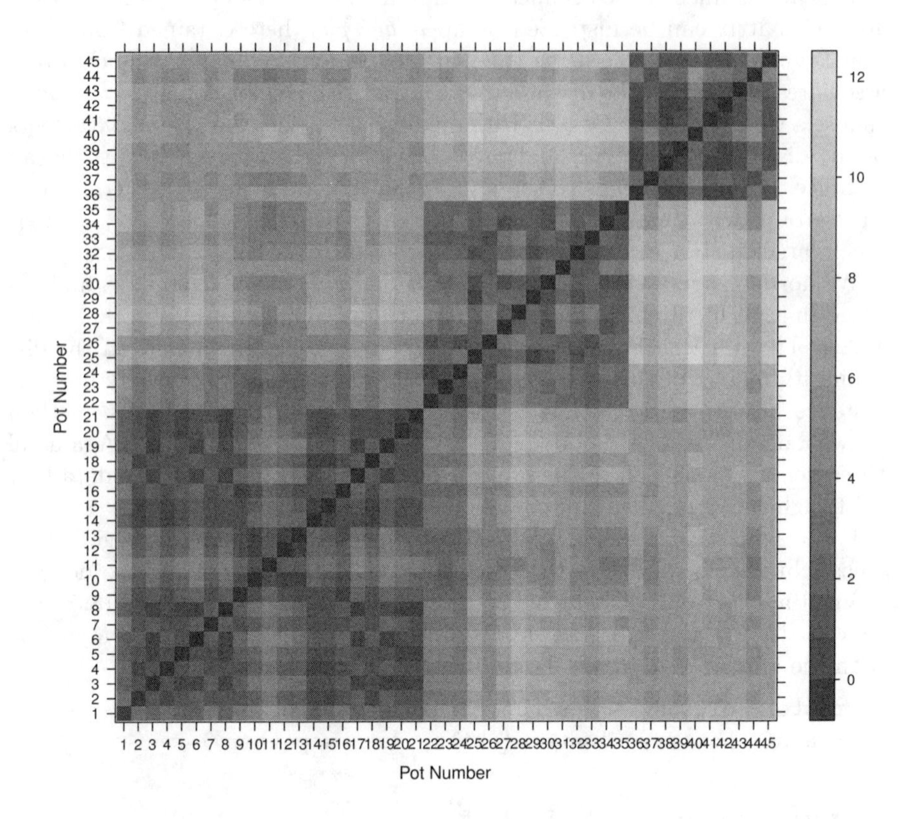


**Figure 18.4** Image plot of the dissimilarity matrix of the `pottery` data.

Mächler, 2003). The logarithms of the mass, period and eccentricity measurements are shown in a scatterplot in Figure 18.6. The diagram gives no clear indication of distinct clusters in the data but nevertheless we shall continue to investigate this possibility by applying $k$-means clustering with the `kmeans` function in R. In essence this method finds a partition of the observations for a particular number of clusters by minimising the total within-group sum of squares over all variables. Deciding on the 'optimal' number of groups is often difficult and there is no method that can be recommended in all circumstances (see Everitt et al., 2001). An informal approach to the number of groups problem is to plot the within-group sum of squares for each par-

```
R> pottery_single <- hclust(pottery_dist, method = "single")
R> pottery_complete <- hclust(pottery_dist, method = "complete")
R> pottery_average <- hclust(pottery_dist, method = "average")
R> layout(matrix(1:3, ncol = 3))
R> plot(pottery_single, main = "Single Linkage",
+       sub = "", xlab = "")
R> plot(pottery_complete, main = "Complete Linkage",
+       sub = "", xlab = "")
R> plot(pottery_average, main = "Average Linkage",
+       sub = "", xlab = "")
```

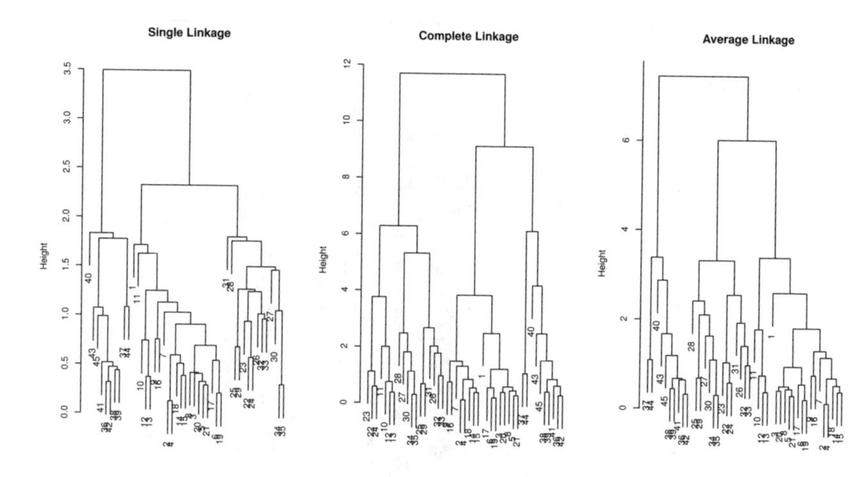


**Figure 18.5** Hierarchical clustering of `pottery` data and resulting dendrograms.

tition given by applying the `kmeans` procedure and looking for an 'elbow' in the resulting curve (cf. scree plots in factor analysis). Such a plot can be constructed in R for the `planets` data using the code displayed with Figure 18.7 (note that since the three variables are on very different scales they first need to be standardised in some way – here we use the range of each).

Sadly Figure 18.7 gives no completely convincing verdict on the number of groups we should consider, but using a little imagination 'little elbows' can be spotted at the three and five group solutions. We can find the number of planets in each group using

```
R> planet_kmeans3 <- kmeans(planet.dat, centers = 3)
R> table(planet_kmeans3$cluster)

 1  2  3
34 53 14
```

The centres of the clusters for the untransformed data can be computed using a small convenience function

```
R> data("planets", package = "HSAUR2")
R> library("scatterplot3d")
R> scatterplot3d(log(planets$mass), log(planets$period),
+       log(planets$eccen), type = "h", angle = 55,
+       pch = 16, y.ticklabs = seq(0, 10, by = 2),
+       y.margin.add = 0.1, scale.y = 0.7)
```

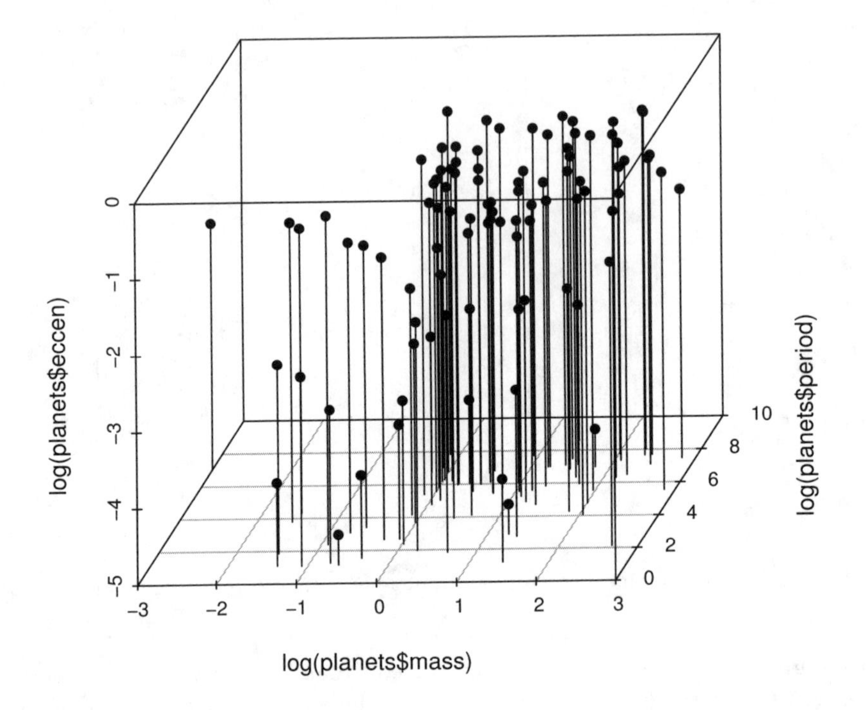


**Figure 18.6** 3D scatterplot of the logarithms of the three variables available for each of the exoplanets.

```
R> ccent <- function(cl) {
+       f <- function(i) colMeans(planets[cl == i,])
+       x <- sapply(sort(unique(cl)), f)
+       colnames(x) <- sort(unique(cl))
+       return(x)
+ }
```

which, applied to the three-cluster solution obtained by $k$-means gets

```
R> rge <- apply(planets, 2, max) - apply(planets, 2, min)
R> planet.dat <- sweep(planets, 2, rge, FUN = "/")
R> n <- nrow(planet.dat)
R> wss <- rep(0, 10)
R> wss[1] <- (n - 1) * sum(apply(planet.dat, 2, var))
R> for (i in 2:10)
+     wss[i] <- sum(kmeans(planet.dat,
+                          centers = i)$withinss)
R> plot(1:10, wss, type = "b", xlab = "Number of groups",
+       ylab = "Within groups sum of squares")
```

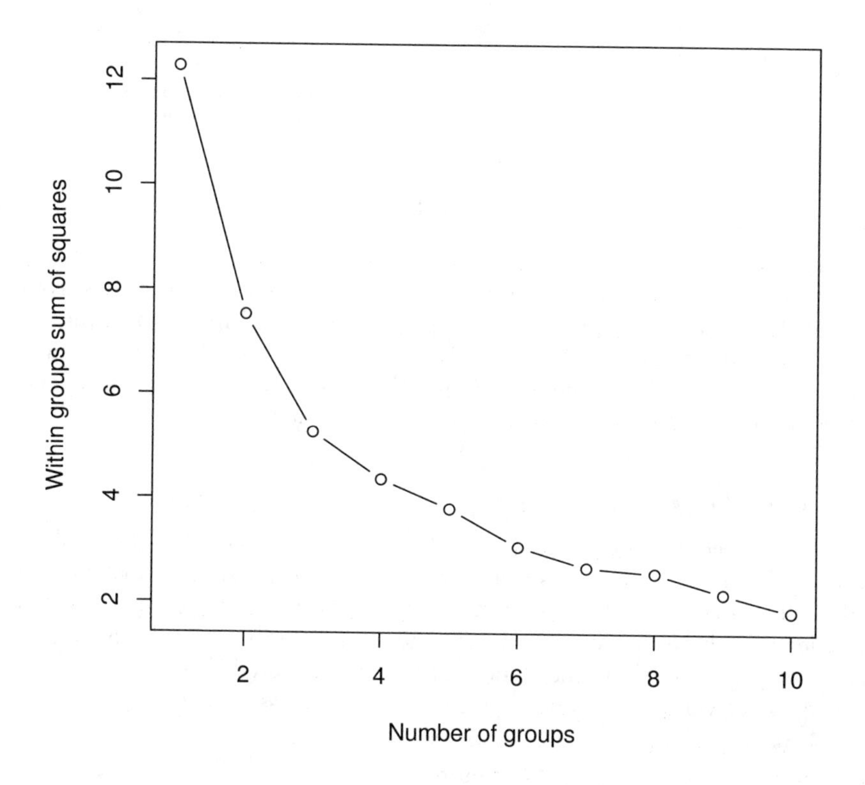


**Figure 18.7** Within-cluster sum of squares for different numbers of clusters for the exoplanet data.

```
R> ccent(planet_kmeans3$cluster)
```

```
               1           2          3
mass    2.9276471   1.6710566   10.56786
period 616.0760882 427.7105892 1693.17201
eccen   0.4953529   0.1219491    0.36650
```

for the three-cluster solution and, for the five cluster solution using

```
R> planet_kmeans5 <- kmeans(planet.dat, centers = 5)
R> table(planet_kmeans5$cluster)
```

```
 1  2  3  4  5
18 35 14 30  4
```

```
R> ccent(planet_kmeans5$cluster)
```

```
               1           2            3          4
mass    3.4916667   1.7448571   10.8121429   1.743533
period 638.0220556 552.3494286 1318.6505856 176.297374
eccen   0.6032778   0.2939143    0.3836429   0.049310
               5
mass       2.115
period  3188.250
eccen      0.110
```

Interpretation of both the three- and five-cluster solutions clearly requires a detailed knowledge of astronomy. But the mean vectors of the three-group solution, for example, imply a relatively large class of Jupiter-sized planets with small periods and small eccentricities, a smaller class of massive planets with moderate periods and large eccentricities, and a very small class of large planets with extreme periods and moderate eccentricities.

### *18.3.3 Model-based Clustering in* R

We now proceed to apply model-based clustering to the `planets` data. R functions for model-based clustering are available in package **mclust** (Fraley et al., 2009, Fraley and Raftery, 2002). Here we use the `Mclust` function since this selects both the most appropriate model for the data *and* the optimal number of groups based on the values of the BIC computed over several models and a range of values for number of groups. The necessary code is:

```
R> library("mclust")
R> planet_mclust <- Mclust(planet.dat)
```

and we first examine a plot of BIC values using the R code that is displayed on top of Figure 18.8. In this diagram the different plotting symbols refer to different model assumptions about the shape of clusters:

EII: spherical, equal volume,

VII: spherical, unequal volume,

EEI: diagonal, equal volume and shape,

VEI: diagonal, varying volume, equal shape,

```
R> plot(planet_mclust, planet.dat, what = "BIC", col = "black",
+       ylab = "-BIC", ylim = c(0, 350))
```

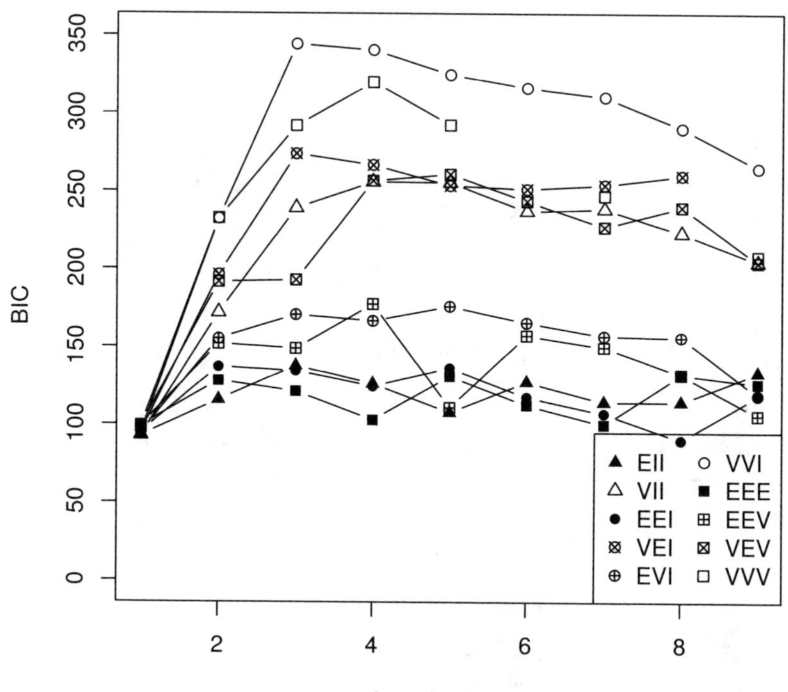


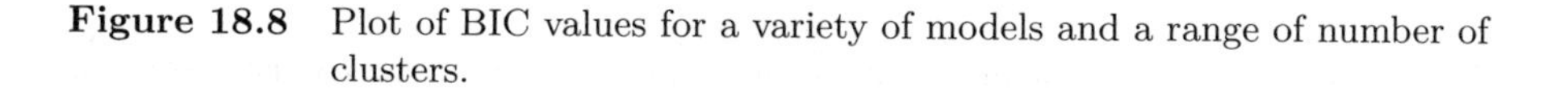

**Figure 18.8** Plot of BIC values for a variety of models and a range of number of clusters.

EVI: diagonal, equal volume, varying shape,

VVI: diagonal, varying volume and shape,

EEE: ellipsoidal, equal volume, shape, and orientation,

EEV: ellipsoidal, equal volume and equal shape,

VEV: ellipsoidal, equal shape,

VVV: ellipsoidal, varying volume, shape, and orientation

The BIC selects model VVI (diagonal varying volume and varying shape) with three clusters as the best solution as can be seen from the `print` output:

```
R> print(planet_mclust)
best model: diagonal, varying volume and shape with 3 components
```

```
R> clPairs(planet.dat,
+       classification = planet_mclust$classification,
+       symbols = 1:3, col = "black")
```

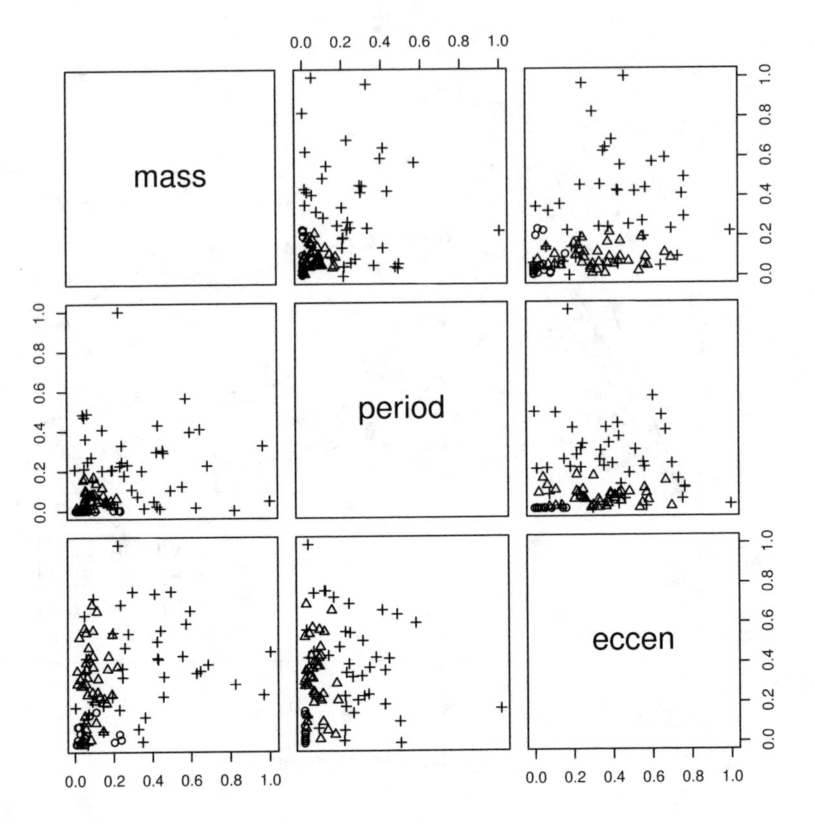


**Figure 18.9** Scatterplot matrix of planets data showing a three-cluster solution from `Mclust`.

This solution can be shown graphically as a scatterplot matrix. The plot is shown in Figure 18.9. Figure 18.10 depicts the clustering solution in the three-dimensional space.

The number of planets in each cluster and the mean vectors of the three clusters for the untransformed data can now be inspected by using

```
R> table(planet_mclust$classification)

 1  2  3
19 41 41

R> ccent(planet_mclust$classification)
```

```
R> scatterplot3d(log(planets$mass), log(planets$period),
+       log(planets$eccen), type = "h", angle = 55,
+       scale.y = 0.7, pch = planet_mclust$classification,
+       y.ticklabs = seq(0, 10, by = 2), y.margin.add = 0.1)
```

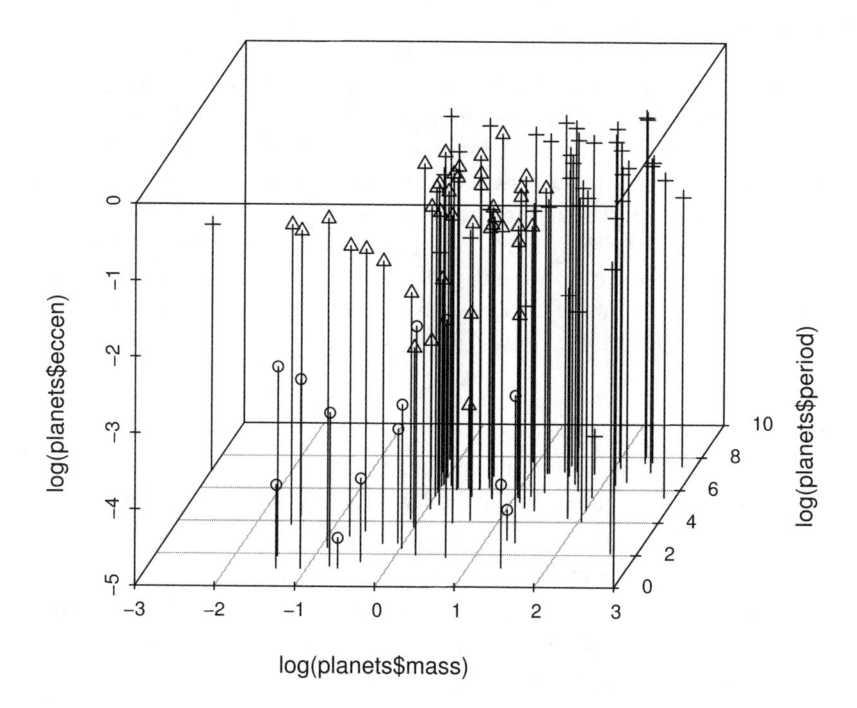


**Figure 18.10** 3D scatterplot of planets data showing a three-cluster solution from Mclust.

```
                 1           2            3
mass   1.16652632   1.5797561    6.0761463
period 6.47180158 313.4127073 1325.5310048
eccen  0.03652632   0.3061463    0.3704951
```

Cluster 1 consists of planets about the same size as Jupiter with very short periods and eccentricities (similar to the first cluster of the $k$-means solution). Cluster 2 consists of slightly larger planets with moderate periods and large eccentricities, and cluster 3 contains the very large planets with very large periods. These two clusters do not match those found by the $k$-means approach.

## 18.4 Summary

Cluster analysis techniques provide a rich source of possible strategies for exploring complex multivariate data. But the use of cluster analysis in practise does not involve simply the application of one particular technique to the data under investigation, but rather necessitates a series of steps, each of which may be dependent on the results of the preceding one. It is generally impossible a priori to anticipate what combination of variables, similarity measures and clustering technique is likely to lead to interesting and informative classifications. Consequently, the analysis proceeds through several stages, with the researcher intervening if necessary to alter variables, choose a different similarity measure, concentrate on a particular subset of individuals, and so on. The final, extremely important, stage concerns the evaluation of the clustering solutions obtained. Are the clusters 'real' or merely artefacts of the algorithms? Do other solutions exist that are better in some sense? Can the clusters be given a convincing interpretation? A long list of such questions might be posed, and readers intending to apply clustering to their data are recommended to read the detailed accounts of cluster evaluation given in Dubes and Jain (1979) and in Everitt et al. (2001).

## Exercises

Ex. 18.1 Construct a three-dimensional drop-line scatterplot of the planets data in which the points are labelled with a suitable cluster label.

Ex. 18.2 Write an R function to fit a mixture of $k$ normal densities to a data set using maximum likelihood.

Ex. 18.3 Apply complete linkage and average linkage hierarchical clustering to the `planets` data. Compare the results with those given in the text.

Ex. 18.4 Write a general R function that will display a particular partition from the $k$-means cluster method on both a scatterplot matrix of the original data and a scatterplot or scatterplot matrix of a selected number of principal components of the data.

# Bibliography


Adler, D. and Murdoch, D. (2009), ***rgl**: 3D Visualization Device System (OpenGL)*, URL `http://rgl.neoscientists.org`, R package version 0.84.

Agresti, A. (1996), *An Introduction to Categorical Data Analysis*, New York, USA: John Wiley & Sons.

Agresti, A. (2002), *Categorical Data Analysis*, Hoboken, New Jersey, USA: John Wiley & Sons, 2nd edition.

Aitkin, M. (1978), "The analysis of unbalanced cross-classifications," *Journal of the Royal Statistical Society, Series A*, 141, 195–223, with discussion.

Alon, U., Barkai, N., Notternam, D. A., Gish, K., Ybarra, S., Mack, D., and Levine, A. J. (1999), "Broad patterns of gene expressions revealed by clustering analysis of tumour and normal colon tissues probed by oligonucleotide arrays," *Cell Biology*, 99, 6754–6760.

Ambler, G. and Benner, A. (2009), ***mfp**: Multivariable Fractional Polynomials*, URL `http://CRAN.R-project.org/package=mfp`, R package version 1.4.6.

Aspirin Myocardial Infarction Study Research Group (1980), "A randomized, controlled trial of aspirin in persons recovered from myocardial infarction," *Journal of the American Medical Association*, 243, 661–669.

Bailey, K. R. (1987), "Inter-study differences: how should they influence the interpretation of results?" *Statistics in Medicine*, 6, 351–360.

Banfield, J. D. and Raftery, A. E. (1993), "Model-based Gaussian and non-Gaussian clustering," *Biometrics*, 49, 803–821.

Barlow, R. E., Bartholomew, D. J., Bremner, J. M., and Brunk, H. D. (1972), *Statistical Inference under Order Restrictions*, New York, USA: John Wiley & Sons.

Bates, D. (2005), "Fitting linear mixed models in R," *R News*, 5, 27–30, URL `http://CRAN.R-project.org/doc/Rnews/`.

Bates, D. and Sarkar, D. (2008), ***lme4**: Linear Mixed-Effects Models Using S4 Classes*, URL `http://CRAN.R-project.org/package=lme4`, R package version 0.999375-28.

Beck, A., Steer, R., and Brown, G. (1996), *BDI-II Manual*, The Psychological Corporation, San Antonio, 2nd edition.

Becker, R. A., Chambers, J. M., and Wilks, A. R. (1988), *The New S Language*, London, UK: Chapman & Hall.

Bönsch, D., Lederer, T., Reulbach, U., Hothorn, T., Kornhuber, J., and Bleich, S. (2005), "Joint analysis of the NACP-REP1 marker within the alpha synuclein gene concludes association with alcohol dependence," *Human Molecular Genetics*, 14, 967–971.

Breddin, K., Loew, D., Lechner, K., Überla, K., and Walter, E. (1979), "Secondary prevention of myocardial infarction. Comparison of acetylsalicylic acid, phenprocoumon and placebo. A multicenter two-year prospective study," *Thrombosis and Haemostasis*, 41, 225–236.

Breiman, L. (1996), "Bagging predictors," *Machine Learning*, 24, 123–140.

Breiman, L. (2001a), "Random forests," *Machine Learning*, 45, 5–32.

Breiman, L. (2001b), "Statistical modeling: The two cultures," *Statistical Science*, 16, 199–231, with discussion.

Breiman, L., Cutler, A., Liaw, A., and Wiener, M. (2009), ***randomForest**: Breiman and Cutler's Random Forests for Classification and Regression*, URL `http://stat-www.berkeley.edu/users/breiman/RandomForests`, R package version 4.5-30.

Breiman, L., Friedman, J. H., Olshen, R. A., and Stone, C. J. (1984), *Classification and Regression Trees*, California, USA: Wadsworth.

Bühlmann, P. (2004), "Bagging, boosting and ensemble methods," in *Handbook of Computational Statistics*, eds. J. E. Gentle, W. Härdle, and Y. Mori, Berlin, Heidelberg: Springer-Verlag, pp. 877–907.

Bühlmann, P. and Hothorn, T. (2007), "Boosting algorithms: Regularization, prediction and model fitting," *Statistical Science*, 22, 477–505.

Canty, A. and Ripley, B. D. (2009), ***boot**: Bootstrap R (S-PLUS) Functions*, URL `http://CRAN.R-project.org/package=boot`, R package version 1.2-36.

Carey, V. J., Lumley, T., and Ripley, B. D. (2008), ***gee**: Generalized Estimation Equation Solver*, URL `http://CRAN.R-project.org/package=gee`, R package version 4.13-13.

Carlin, J. B., Ryan, L. M., Harvey, E. A., and Holmes, L. B. (2000), "Anticonvulsant teratogenesis 4: Inter-rater agreement in assessing minor physical features related to anticonvulsant therapy," *Teratology*, 62, 406–412.

Carpenter, J., Pocock, S., and Lamm, C. J. (2002), "Coping with missing data in clinical trials: A model-based approach applied to asthma trials," *Statistics in Medicine*, 21, 1043–1066.

Chalmers, T. C. and Lau, J. (1993), "Meta-analytic stimulus for changes in clinical trials," *Statistical Methods in Medical Research*, 2, 161–172.

Chambers, J. M. (1998), *Programming with Data*, New York, USA: Springer-Verlag.

Chambers, J. M., Cleveland, W. S., Kleiner, B., and Tukey, P. A. (1983), *Graphical Methods for Data Analysis*, London: Chapman & Hall/CRC.

Chambers, J. M. and Hastie, T. J. (1992), *Statistical Models in S*, London, UK: Chapman & Hall.

Chen, C., Härdle, W., and Unwin, A., eds. (2008), *Handbook of Data Visualization*, Berlin, Heidelberg: Springer-Verlag.

Cleveland, W. S. (1979), "Robust locally weighted regression and smoothing scatterplots," *Journal of the American Statistical Association*, 74, 829–836.

Colditz, G. A., Brewer, T. F., Berkey, C. S., Wilson, M. E., Burdick, E., Fineberg, H. V., and Mosteller, F. (1994), "Efficacy of BCG vaccine in the prevention of tuberculosis. Meta-analysis of the published literature," *Journal of the American Medical Association*, 271, 698–702.

Collett, D. (2003), *Modelling Binary Data*, London, UK: Chapman & Hall/CRC, 2nd edition.

Collett, D. and Jemain, A. A. (1985), "Residuals, outliers and influential observations in regression analysis," *Sains Malaysiana*, 4, 493–511.

Cook, R. D. and Weisberg, S. (1982), *Residuals and Influence in Regression*, London, UK: Chapman & Hall/CRC.

Cook, R. J. (1998), "Generalized linear model," in *Encyclopedia of Biostatistics*, eds. P. Armitage and T. Colton, Chichester, UK: John Wiley & Sons.

Corbet, G. B., Cummins, J., Hedges, S. R., and Krzanowski, W. J. (1970), "The taxonomic structure of British water voles, genus *Arvicola*," *Journal of Zoology*, 61, 301–316.

Coronary Drug Project Group (1976), "Asprin in coronary heart disease," *Journal of Chronic Diseases*, 29, 625–642.

Cox, D. R. (1972), "Regression models and life-tables," *Journal of the Royal Statistical Society, Series B*, 34, 187–202, with discussion.

Dalgaard, P. (2002), *Introductory Statistics with R*, New York, USA: Springer-Verlag.

Davis, C. S. (1991), "Semi-parametric and non-parametric methods for the analysis of repeated measurements with applications to clinical trials," *Statistics in Medicine*, 10, 1959–1980.

Davis, C. S. (2002), *Statistical Methods for the Analysis of Repeated Measurements*, New York, USA: Springer-Verlag.

DeMets, D. L. (1987), "Methods for combining randomized clinical trials: strengths and limitations," *Statistics in Medicine*, 6, 341–350.

Dempster, A. P., Laird, N. M., and Rubin, D. B. (1977), "Maximum likelihood from incomplete data via the EM algorithm (C/R: p22-37)," *Journal of the Royal Statistical Society, Series B*, 39, 1–22.

DerSimonian, R. and Laird, N. (1986), "Meta-analysis in clinical trials," *Controlled Clinical Trials*, 7, 177–188.

Diggle, P. J. (1998), "Dealing with missing values in longitudinal studies," in *Statistical Analysis of Medical Data*, eds. B. S. Everitt and G. Dunn, London, UK: Arnold.

Diggle, P. J., Heagerty, P. J., Liang, K. Y., and Zeger, S. L. (2003), *Analysis of Longitudinal Data*, Oxford, UK: Oxford University Press.

Diggle, P. J. and Kenward, M. G. (1994), "Informative dropout in longitudinal data analysis," *Journal of the Royal Statistical Society, Series C*, 43, 49–93.

Dolnicar, S. and Leisch, F. (2003), "Winter tourist segments in Austria: Identifying stable vacation styles using bagged clustering techniques," *Journal of Travel Research*, 41, 281–292.

Dubes, R. and Jain, A. K. (1979), "Validity studies in clustering methodologies," *Pattern Recognition*, 8, 247–260.

Duval, S. and Tweedie, R. L. (2000), "A nonparametric 'trim and fill' method of accounting for publication bias in meta-analysis," *Journal of the American Statistical Association*, 95, 89–98.

Easterbrook, P. J., Berlin, J. A., Gopalan, R., and Matthews, D. R. (1991), "Publication bias in research," *Lancet*, 337, 867–872.

Edgington, E. S. (1987), *Randomization Tests*, New York, USA: Marcel Dekker.

Efron, B. and Tibshirani, R. J. (1993), *An Introduction to the Bootstrap*, London, UK: Chapman & Hall/CRC.

Elwood, P. C., Cochrane, A. L., Burr, M. L., Sweetman, P. M., Williams, G., Welsby, E., Hughes, S. J., and Renton, R. (1974), "A randomized controlled trial of acetyl salicilic acid in the secondary prevention of mortality from myocardial infarction," *British Medical Journal*, 1, 436–440.

Elwood, P. C. and Sweetman, P. M. (1979), "Asprin and secondary mortality after myocardial infarction," *Lancet*, 2, 1313–1315.

Everitt, B. S. (1992), *The Analysis of Contingency Tables*, London, UK: Chapman & Hall/CRC, 2nd edition.

Everitt, B. S. (1996), *Making Sense of Statistics in Psychology: A Second-Level Course*, Oxford, UK: Oxford University Press.

Everitt, B. S. (2001), *Statistics for Psychologists*, Mahwah, New Jersey, USA: Lawrence Erlbaum.

Everitt, B. S. (2002a), *Cambridge Dictionary of Statistics in the Medical Sciences*, Cambridge, UK: Cambridge University Press.

Everitt, B. S. (2002b), *Modern Medical Statistics*, London, UK: Arnold.

Everitt, B. S. and Bullmore, E. T. (1999), "Mixture model mapping of brain activation in functional magnetic resonance images," *Human Brain Mapping*, 7, 1–14.

Everitt, B. S. and Dunn, G. (2001), *Applied Multivariate Data Analysis*, London, UK: Arnold, 2nd edition.

Everitt, B. S., Landau, S., and Leese, M. (2001), *Cluster Analysis*, London, UK: Arnold, 4th edition.

Everitt, B. S. and Pickles, A. (2000), *Statistical Aspects of the Design and Analysis of Clinical Trials*, London, UK: Imperial College Press.

Everitt, B. S. and Rabe-Hesketh, S. (1997), *The Analysis of Proximity Data*, London, UK: Arnold.

Everitt, B. S. and Rabe-Hesketh, S. (2001), *Analysing Medical Data Using S-Plus*, New York, USA: Springer-Verlag.

Fisher, L. D. and Belle, G. V. (1993), *Biostatistics. A Methodology for the Health Sciences*, New York, USA: John Wiley & Sons.

Fisher, R. A. (1935), *The Design of Experiments*, Edinburgh, UK: Oliver and Boyd.

Fleiss, J. L. (1993), "The statistical basis of meta-analysis," *Statistical Methods in Medical Research*, 2, 121–145.

Flury, B. and Riedwyl, H. (1988), *Multivariate Statistics: A Practical Approach*, London, UK: Chapman & Hall.

Fraley, C. and Raftery, A. E. (2002), "Model-based clustering, discriminant analysis, and density estimation," *Journal of the American Statistical Association*, 97, 611–631.

Fraley, C., Raftery, A. E., and Wehrens, R. (2009), ***mclust**: Model-based Cluster Analysis*, URL `http://www.stat.washington.edu/mclust`, R package version 3.1-10.3.

Fraley, G. and Raftery, A. E. (1999), "MCLUST: Software for model-based cluster analysis," *Journal of Classification*, 16, 297–306.

Freedman, W. L., Madore, B. F., Gibson, B. K., Ferrarese, L., Kelson, D. D., Sakai, S., Mould, J. R., Kennicutt, R. C., Ford, H. C., Graham, J. A., Huchra, J. P., Hughes, S. M. G., Illingworth, G. D., Macri, L. M., and Stetson, P. B. (2001), "Final results from the Hubble Space Telescope key project to measure the Hubble constant," *The Astrophysical Journal*, 553, 47–72.

Freeman, G. H. and Halton, J. H. (1951), "Note on an exact treatment of contingency, goodness of fit and other problems of significance," *Biometrika*, 38, 141–149.

Friedman, H. P. and Rubin, J. (1967), "On some invariant criteria for grouping data," *Journal of the American Statistical Association*, 62, 1159–1178.

Friendly, M. (1994), "Mosaic displays for multi-way contingency tables," *Journal of the American Statistical Association*, 89, 190–200.

Gabriel, K. R. (1971), "The biplot graphical display of matrices with application to principal component analysis," *Biometrika*, 58, 453–467.

Gabriel, K. R. (1981), "Biplot display of multivariate matrices for inspection of data and diagnosis," in *Interpreting Multivariate Data*, ed. V. Barnett, Chichester, UK: John Wiley & Sons.

Garcia, A. L., Wagner, K., Hothorn, T., Koebnick, C., Zunft, H. J., and Trippo, U. (2005), "Improved prediction of body fat by measuring skinfold thickness, circumferences, and bone breadths," *Obesity Research*, 13, 626–634.

Garczarek, U. M. and Weihs, C. (2003), "Standardizing the comparison of partitions," *Computational Statistics*, 18, 143–162.

Gentleman, R. (2005), "Reproducible research: A bioinformatics case study," *Statistical Applications in Genetics and Molecular Biology*, 4, URL `http://www.bepress.com/sagmb/vol4/iss1/art2`, Article 2.

Giardiello, F. M., Hamilton, S. R., Krush, A. J., Piantadosi, S., Hylind, L. M., Celano, P., Booker, S. V., Robinson, C. R., and Offerhaus, G. J. A. (1993), "Treatment of colonic and rectal adenomas with sulindac in familial adenomatous polyposis," *New England Journal of Medicine*, 328, 1313–1316.

Gordon, A. (1999), *Classification*, Boca Raton, Florida, USA: Chapman & Hall/CRC, 2nd edition.

Gower, J. C. and Hand, D. J. (1996), *Biplots*, London, UK: Chapman & Hall/CRC.

Gower, J. C. and Ross, G. J. S. (1969), "Minimum spanning trees and single linkage cluster analysis," *Applied Statistics*, 18, 54–64.

Grana, C., Chinol, M., Robertson, C., Mazzetta, C., Bartolomei, M., Cicco, C. D., Fiorenza, M., Gatti, M., Caliceti, P., and Paganelli1, G. (2002), "Pretargeted adjuvant radioimmunotherapy with Yttrium-90-biotin in malignant glioma patients: A pilot study," *British Journal of Cancer*, 86, 207–212.

Greenwald, A. G. (1975), "Consequences of prejudice against the null hypothesis," *Psychological Bulletin*, 82, 1–20.

Greenwood, M. and Yule, G. U. (1920), "An inquiry into the nature of frequency distribution of multiple happenings with particular reference of multiple attacks of disease or of repeated accidents," *Journal of the Royal Statistical Society*, 83, 255–279.

Haberman, S. J. (1973), "The analysis of residuals in cross-classified tables," *Biometrics*, 29, 205–220.

Hand, D. J., Daly, F., Lunn, A. D., McConway, K. J., and Ostrowski, E. (1994), *A Handbook of Small Datasets*, London, UK: Chapman & Hall/CRC.

Harrison, D. and Rubinfeld, D. L. (1978), "Hedonic prices and the demand for clean air," *Journal of Environmental Economics & Management*, 5, 81–102.

Hartigan, J. A. (1975), *Clustering Algorithms*, New York, USA: John Wiley & Sons.

Hastie, T. and Tibshirani, R. (1990), *Generalized Additive Models*, Boca Raton, Florida: Chapman & Hall.

Hawkins, D. M., Muller, M. W., and ten Krooden, J. A. (1982), "Cluster analysis," in *Topics in Applied Multivariate Analysis*, ed. D. M. Hawkins, Cambridge, UK: Cambridge University Press.

Heitjan, D. F. (1997), "Annotation: What can be done about missing data? Approaches to imputation," *American Journal of Public Health*, 87, 548–550.

Hochberg, Y. and Tamhane, A. C. (1987), *Multiple Comparison Procedures*, New York, USA: John Wiley & Sons.

Hofmann, H. and Theus, M. (2005), "Interactive graphics for visualizing conditional distributions," Unpublished Manuscript.

Hothorn, T., Bretz, F., and Westfall, P. (2008a), "Simultaneous inference in general parametric models," *Biometrical Journal*, 50, 346–363.

Hothorn, T., Bretz, F., and Westfall, P. (2009a), ***multcomp**: Simultaneous Inference for General Linear Hypotheses*, URL `http://CRAN.R-project.org/package=multcomp`, R package version 1.0-7.

Hothorn, T., Bühlmann, P., Kneib, T., Schmid, M., and Hofner, B. (2009b), ***mboost**: Model-Based Boosting*, URL `http://CRAN.R-project.org/package=mboost`, R package version 1.1-1.

Hothorn, T., Hornik, K., Strobl, C., and Zeileis, A. (2009c), ***party**: A Laboratory for Recursive Partytioning*, URL `http://CRAN.R-project.org/package=party`, R package version 0.9-996.

Hothorn, T., Hornik, K., van de Wiel, M., and Zeileis, A. (2008b), ***coin**: Conditional Inference Procedures in a Permutation Test Framework*, URL `http://CRAN.R-project.org/package=coin`, R package version 1.0-3.

Hothorn, T., Hornik, K., van de Wiel, M. A., and Zeileis, A. (2006a), "A Lego system for conditional inference," *The American Statistician*, 60, 257–263.

Hothorn, T., Hornik, K., and Zeileis, A. (2006b), "Unbiased recursive partitioning: A conditional inference framework," *Journal of Computational and Graphical Statistics*, 15, 651–674.

Hothorn, T. and Zeileis, A. (2009), ***partykit**: A Toolkit for Recursive Partytioning*, URL `http://R-forge.R-project.org/projects/partykit/`, R package version 0.0-1.

Hsu, J. C. (1996), *Multiple Comparisons: Theory and Methods*, London: CRC Press, Chapman & Hall.

ISIS-2 (Second International Study of Infarct Survival) Collaborative Group (1988), "Randomised trial of intravenous streptokinase, oral aspirin, both, or neither among 17,187 cases of suspected acute myocardial infarction: ISIS-2," *Lancet*, 13, 349–360.

Kalbfleisch, J. D. and Prentice, R. L. (1980), *The Statistical Analysis of Failure Time Data*, New York, USA: John Wiley & Sons.

Kaplan, E. L. and Meier, P. (1958), "Nonparametric estimation from incomplete observations," *Journal of the American Statistical Association*, 53, 457–481.

Kaufman, L. and Rousseeuw, P. J. (1990), *Finding Groups in Data: An Introduction to Cluster Analysis*, New York, USA: John Wiley & Sons.

Keele, L. J. (2008), *Semiparametric Regression for the Social Sciences*, New York, USA: John Wiley & Sons.

Kelsey, J. L. and Hardy, R. J. (1975), "Driving of motor vehicles as a risk factor for acute herniated lumbar intervertebral disc," *American Journal of Epidemiology*, 102, 63–73.

Kraepelin, E. (1919), *Dementia Praecox and Paraphrenia*, Edinburgh, UK: Livingstone.

Kruskal, J. B. (1964a), "Multidimensional scaling by optimizing goodness-of-fit to a nonmetric hypothesis," *Psychometrika*, 29, 1–27.

Kruskal, J. B. (1964b), "Nonmetric multidimensional scaling: A numerical method," *Psychometrika*, 29, 115–129.

Lanza, F. L. (1987), "A double-blind study of prophylactic effect of misoprostol on lesions of gastric and duodenal mucosa induced by oral administration of tolmetin in healthy subjects," *British Journal of Clinical Practice*, 40, 91–101.

Lanza, F. L., Aspinall, R. L., Swabb, E. A., Davis, R. E., Rack, M. F., and Rubin, A. (1988a), "Double-blind, placebo-controlled endoscopic comparison of the mucosal protective effects of misoprostol versus cimetidine on tolmetin-induced mucosal injury to the stomach and duodenum," *Gastroenterology*, 95, 289–294.

Lanza, F. L., Fakouhi, D., Rubin, A., Davis, R. E., Rack, M. F., Nissen, C., and Geis, S. (1989), "A double-blind placebo-controlled comparison of the efficacy and safety of 50, 100, and 200 micrograms of misoprostol QID in the prevention of Ibuprofen-induced gastric and duodenal mucosal lesions and symptoms," *American Journal of Gastroenterology*, 84, 633–636.

Lanza, F. L., Peace, K., Gustitus, L., Rack, M. F., and Dickson, B. (1988b), "A blinded endoscopic comparative study of misoprostol versus sucralfate and placebo in the prevention of aspirin-induced gastric and duodenal ulceration," *American Journal of Gastroenterology*, 83, 143–146.

Leisch, F. (2002a), "Sweave: Dynamic generation of statistical reports using literate data analysis," in *Compstat 2002 — Proceedings in Computational Statistics*, eds. W. Härdle and B. Rönz, Physica Verlag, Heidelberg, pp. 575–580, ISBN 3-7908-1517-9.

Leisch, F. (2002b), "Sweave, Part I: Mixing R and LaTeX," *R News*, 2, 28–31, URL `http://CRAN.R-project.org/doc/Rnews/`.

Leisch, F. (2003), "Sweave, Part II: Package vignettes," *R News*, 3, 21–24, URL `http://CRAN.R-project.org/doc/Rnews/`.

Leisch, F. (2004), "FlexMix: A general framework for finite mixture models and latent class regression in R," *Journal of Statistical Software*, 11, URL `http://www.jstatsoft.org/v11/i08/`.

Leisch, F. and Dimitriadou, E. (2009), ***mlbench****: Machine Learning Benchmark Problems*, URL `http://CRAN.R-project.org/package=mlbench`, R package version 1.1-6.

Leisch, F. and Rossini, A. J. (2003), "Reproducible statistical research," *Chance*, 16, 46–50.

Liang, K. and Zeger, S. L. (1986), "Longitudinal data analysis using generalized linear models," *Biometrika*, 73, 13–22.

Ligges, U. and Mächler, M. (2003), "Scatterplot3d – An R package for visualizing multivariate data," *Journal of Statistical Software*, 8, 1–20, URL `http://www.jstatsoft.org/v08/i11`.

Longford, N. T. (1993), *Random Coefficient Models*, Oxford, UK: Oxford University Press.

Lumley, T. (2009), ***rmeta**: Meta-Analysis*, URL `http://CRAN.R-project.org/package=rmeta`, R package version 2.15.

Lumley, T. and Miller, A. (2009), ***leaps**: Regression Subset Selection*, URL `http://CRAN.R-project.org/package=leaps`, R package version 2.8.

Mann, L. (1981), "The baiting crowd in episodes of threatened suicide," *Journal of Personality and Social Psychology*, 41, 703–709.

Mardia, K. V., Kent, J. T., and Bibby, J. M. (1979), *Multivariate Analysis*, London, UK: Academic Press.

Mardin, C. Y., Hothorn, T., Peters, A., Jünemann, A. G., Nguyen, N. X., and Lausen, B. (2003), "New glaucoma classification method based on standard HRT parameters by bagging classification trees," *Journal of Glaucoma*, 12, 340–346.

Marriott, F. H. C. (1982), "Optimization methods of cluster analysis," *Biometrika*, 69, 417–421.

Mayor, M. and Frei, P. (2003), *New Worlds in the Cosmos: The Discovery of Exoplanets*, Cambridge, UK: Cambridge University Press.

Mayor, M. and Queloz, D. (1995), "A Jupiter-mass companion to a solar-type star," *Nature*, 378, 355.

McCullagh, P. and Nelder, J. A. (1989), *Generalized Linear Models*, London, UK: Chapman & Hall/CRC.

McLachlan, G. and Peel, D. (2000), *Finite Mixture Models*, New York, USA: John Wiley & Sons.

Mehta, C. R. and Patel, N. R. (2003), *StatXact-6: Statistical Software for Exact Nonparametric Inference*, Cytel Software Corporation, Cambridge, MA, USA.

Meyer, D., Zeileis, A., Karatzoglou, A., and Hornik, K. (2009), ***vcd**: Visualizing Categorical Data*, URL `http://CRAN.R-project.org/package=vcd`, R package version 1.2-3.

Miller, A. (2002), *Subset Selection in Regression*, New York, USA: Chapman & Hall, 2nd edition.

Morrison, D. F. (2005), "Multivariate analysis of variance," in *Encyclopedia of Biostatistics*, eds. P. Armitage and T. Colton, Chichester, UK: John Wiley & Sons, 2nd edition.

Murray, G. D. and Findlay, J. G. (1988), "Correcting for bias caused by dropouts in hypertension trials," *Statistics in Medicine*, 7, 941–946.

Murrell, P. (2005), *R Graphics*, Boca Raton, Florida, USA: Chapman & Hall/CRC.

Murthy, S. K. (1998), "Automatic construction of decision trees from data: A multi-disciplinary survey," *Data Mining and Knowledge Discovery*, 2, 345–389.

Nelder, J. A. (1977), "A reformulation of linear models," *Journal of the Royal Statistical Society, Series A*, 140, 48–76, with commentary.

Nelder, J. A. and Wedderburn, R. W. M. (1972), "Generalized linear models," *Journal of the Royal Statistical Society, Series A*, 135, 370–384.

Oakes, M. (1993), "The logic and role of meta-analysis in clinical research," *Statistical Methods in Medical Research*, 2, 147–160.

Paradis, E., Strimmer, K., Claude, J., Jobb, G., Opgen-Rhein, R., Dutheil, J., Noel, Y., and Bolker, B. (2009), ***ape**: Analyses of Phylogenetics and Evolution*, URL `http://CRAN.R-project.org/package=ape`, R package version 2.3.

Pearson, K. (1894), "Contributions to the mathematical theory of evolution," *Philosophical Transactions A*, 185, 71–110.

Persantine-Aspirin Reinfarction Study Research Group (1980), "Persantine and Aspirin in coronary heart disease," *Circulation*, 62, 449–461.

Pesarin, F. (2001), *Multivariate Permutation Tests: With Applications to Biostatistics*, Chichester, UK: John Wiley & Sons.

Peters, A., Hothorn, T., and Lausen, B. (2002), "ipred: Improved predictors," *R News*, 2, 33–36, URL `http://CRAN.R-project.org/doc/Rnews/`, ISSN 1609-3631.

Petitti, D. B. (2000), *Meta-Analysis, Decision Analysis and Cost-Effectiveness Analysis*, New York, USA: Oxford University Press.

Piantadosi, S. (1997), *Clinical Trials: A Methodologic Perspective*, New York, USA: John Wiley & Sons.

Pinheiro, J. C. and Bates, D. M. (2000), *Mixed-Effects Models in S and S-PLUS*, New York, USA: Springer-Verlag.

Pitman, E. J. G. (1937), "Significance tests which may be applied to samples from any populations," *Biometrika*, 29, 322–335.

Postman, M., Huchra, J. P., and Geller, M. J. (1986), "Probes of large-scale structures in the corona borealis region," *Astrophysical Journal*, 92, 1238–1247.

Prim, R. C. (1957), "Shortest connection networks and some generalizations," *Bell System Technical Journal*, 36, 1389–1401.

Proudfoot, J., Goldberg, D., Mann, A., Everitt, B. S., Marks, I., and Gray, J. A. (2003), "Computerized, interactive, multimedia cognitive-behavioural program for anxiety and depression in general practice," *Psychological Medicine*, 33, 217–227.

Quine, S. (1975), *Achievement Orientation of Aboriginal and White Adolescents*, Doctoral Dissertation, Australian National University, Canberra, Australia.

R Development Core Team (2009a), *An Introduction to R*, R Foundation for Statistical Computing, Vienna, Austria, URL `http://www.R-project.org`, ISBN 3-900051-12-7.

R Development Core Team (2009b), *R: A Language and Environment for Statistical Computing*, R Foundation for Statistical Computing, Vienna, Austria, URL `http://www.R-project.org`, ISBN 3-900051-07-0.

R Development Core Team (2009c), *R Data Import/Export*, R Foundation for Statistical Computing, Vienna, Austria, URL `http://www.R-project.org`, ISBN 3-900051-10-0.

R Development Core Team (2009d), *R Installation and Administration*, R Foundation for Statistical Computing, Vienna, Austria, URL `http://www.R-project.org`, ISBN 3-900051-09-7.

R Development Core Team (2009e), *Writing R Extensions*, R Foundation for Statistical Computing, Vienna, Austria, URL `http://www.R-project.org`, ISBN 3-900051-11-9.

Rabe-Hesketh, S. and Skrondal, A. (2008), *Multilevel and Longitudinal Modeling Using Stata*, College Station, Texas, USA: Stata Press, 2nd edition.

Ripley, B. D. (1996), *Pattern Recognition and Neural Networks*, Cambridge, UK: Cambridge University Press, URL `http://www.stats.ox.ac.uk/pub/PRNN/`.

Roeder, K. (1990), "Density estimation with confidence sets exemplified by superclusters and voids in galaxies," *Journal of the American Statistical Association*, 85, 617–624.

Rohlf, F. J. (1970), "Adaptive hierarchical clustering schemes," *Systematic Zoology*, 19, 58–82.

Romesburg, H. C. (1984), *Cluster Analysis for Researchers*, Belmont, CA: Lifetime Learning Publications.

Rubin, D. (1976), "Inference and missing data," *Biometrika*, 63, 581–592.

Sarkar, D. (2008), *Lattice: Multivariate Data Visualization with R*, New York, USA: Springer-Verlag.

Sarkar, D. (2009), ***lattice**: Lattice Graphics*, URL `http://CRAN.R-project.org/package=lattice`, R package version 0.17-22.

Sauerbrei, W. and Royston, P. (1999), "Building multivariable prognostic and diagnostic models: Transformation of the predictors by using fractional polynomials," *Journal of the Royal Statistical Society, Series A*, 162, 71–94.

Schmid, C. F. (1954), *Handbook of Graphic Presentation*, New York: Ronald Press.

Schumacher, M., Basert, G., Bojar, H., Hübner, K., Olschewski, M., Sauerbrei, W., Schmoor, C., Beyerle, C., Neumann, R. L. A., and Rauschecker, H. F. for the German Breast Cancer Study Group (1994), "Randomized $2 \times 2$ trial evaluating hormonal treatment and the duration of chemotherapy in node-positive breast cancer patients," *Journal of Clinical Oncology*, 12, 2086–2093.

Schwarzer, G. (2009), ***meta**: Meta-Analysis*, URL `http://CRAN.R-project.org/package=meta`, R package version 0.9-19.

Schwarzer, G., Carpenter, J. R., and Rücker, G. (2009), *Meta-analysis with R*, New York, USA: Springer-Verlag, forthcoming.

Scott, A. J. and Symons, M. J. (1971), "Clustering methods based on likelihood ratio criteria," *Biometrics*, 27, 387–398.

Scott, D. W. (1992), *Multivariate Density Estimation*, New York, USA: John Wiley & Sons.

Searle, S. R. (1971), *Linear Models*, New York, USA: John Wiley & Sons.

Seeber, G. U. H. (1998), "Poisson regression," in *Encyclopedia of Biostatistics*, eds. P. Armitage and T. Colton, Chichester, UK: John Wiley & Sons.

Shepard, R. N. (1962a), "The analysis of proximities: Multidimensional scaling with unknown distance function Part I," *Psychometrika*, 27, 125–140.

Shepard, R. N. (1962b), "The analysis of proximities: Multidimensional scaling with unknown distance function Part II," *Psychometrika*, 27, 219–246.

Sibson, R. (1979), "Studies in the robustness of multidimensional scaling. Perturbational analysis of classical scaling," *Journal of the Royal Statistical Society, Series B*, 41, 217–229.

Silagy, C. (2003), "Nicotine replacement therapy for smoking cessation (Cochrane Review)," in *The Cochrane Library*, John Wiley & Sons, Issue 4.

Silverman, B. (1986), *Density Estimation*, London, UK: Chapman & Hall/CRC.

Simonoff, J. S. (1996), *Smoothing Methods in Statistics*, New York, USA: Springer-Verlag.

Skrondal, A. and Rabe-Hesketh, S. (2004), *Generalized Latent Variable Modeling: Multilevel, Longitudinal and Structural Equation Models*, Boca Raton, Florida, USA: Chapman & Hall/CRC.

Smith, M. L. (1980), "Publication bias and meta-analysis," *Evaluating Education*, 4, 22–93.

Sokal, R. R. and Rohlf, F. J. (1981), *Biometry*, San Francisco, California, USA: W. H. Freeman, 2nd edition.

Sterlin, T. D. (1959), "Publication decisions and their possible effects on inferences drawn from tests of significance-or vice versa," *Journal of the American Statistical Association*, 54, 30–34.

Stevens, J. (2001), *Applied Multivariate Statistics for the Social Sciences*, Mahwah, New Jersey, USA: Lawrence Erlbaum, 4th edition.

Sutton, A. J. and Abrams, K. R. (2001), "Bayesian methods in meta-analysis and evidence synthesis," *Statistical Methods in Medical Research*, 10, 277–303.

Sutton, A. J., Abrams, K. R., Jones, D. R., and Sheldon, T. A. (2000), *Methods for Meta-Analysis in Medical Research*, Chichester, UK: John Wiley & Sons.

Thall, P. F. and Vail, S. C. (1990), "Some covariance models for longitudinal count data with overdispersion," *Biometrics*, 46, 657–671.

Therneau, T. M., Atkinson, B., and Ripley, B. D. (2009), ***rpart**: Recursive Partitioning*, URL `http://mayoresearch.mayo.edu/mayo/research/biostat/splusfunctions.cfm`, R package version 3.1-43.

Therneau, T. M. and Atkinson, E. J. (1997), "An introduction to recursive partitioning using the rpart routine," Technical Report 61, Section of Biostatistics, Mayo Clinic, Rochester, USA, URL `http://www.mayo.edu/hsr/techrpt/61.pdf`.

Therneau, T. M. and Grambsch, P. M. (2000), *Modeling Survival Data: Extending the Cox Model*, New York, USA: Springer-Verlag.

Therneau, T. M. and Lumley, T. (2009), ***survival**: Survival Analysis, Including Penalised Likelihood*, URL `http://CRAN.R-project.org/package=survival`, R package version 2.35-4.

Timm, N. H. (2002), *Applied Multivariate Analysis*, New York, USA: Springer-Verlag.

Tubb, A., Parker, N. J., and Nickless, G. (1980), "The analysis of Romano-British pottery by atomic absorption spectrophotometry," *Archaeometry*, 22, 153–171.

Tufte, E. R. (1983), *The Visual Display of Quantitative Information*, Cheshire, Connecticut: Graphics Press.

Tukey, J. W. (1953), "The problem of multiple comparisons (unpublished manuscript)," in *The Collected Works of John W. Tukey VIII. Multiple Comparisons: 1948-1983*, New York, USA: Chapman & Hall.

Vanisma, F. and De Greve, J. P. (1972), "Close binary systems before and after mass transfer," *Astrophysics and Space Science*, 87, 377–401.

Venables, W. N. and Ripley, B. D. (2002), *Modern Applied Statistics with S*, New York, USA: Springer-Verlag, 4th edition, URL `http://www.stats.ox.ac.uk/pub/MASS4/`, ISBN 0-387-95457-0.

Wand, M. P. and Jones, M. C. (1995), *Kernel Smoothing*, London, UK: Chapman & Hall/CRC.

Wand, M. P. and Ripley, B. D. (2009), ***KernSmooth**: Functions for Kernel Smoothing for Wand & Jones (1995)*, URL `http://CRAN.R-project.org/package=KernSmooth`, R package version 2.22-22.

Weisberg, S. (2008), ***alr3**: Methods and Data to Accompany Applied Linear Regression 3rd edition*, URL `http://www.stat.umn.edu/alr`, R package version 1.1.7.

Whitehead, A. and Jones, N. M. B. (1994), "A meta-analysis of clinical trials involving different classifications of response into ordered categories," *Statistics in Medicine*, 13, 2503–2515.

Wilkinson, L. (1992), "Graphical displays," *Statistical Methods in Medical Research*, 1, 3–25.

Wood, S. N. (2006), *Generalized Additive Models: An Introduction with R*, Boca Raton, Florida, USA: Chapman & Hall/CRC.

Woodley, W. L., Simpson, J., Biondini, R., and Berkeley, J. (1977), "Rainfall results 1970-75: Florida area cumulus experiment," *Science*, 195, 735–742.

Young, G. and Householder, A. S. (1938), "Discussion of a set of points in terms of their mutual distances," *Psychometrika*, 3, 19–22.

Zeger, S. L. and Liang, K. Y. (1986), "Longitudinal data analysis for discrete and continuous outcomes," *Biometrics*, 42, 121–130.

Zeileis, A. (2004), "Econometric computing with HC and HAC covariance matrix estimators," *Journal of Statistical Software*, 11, 1–17, URL `http://www.jstatsoft.org/v11/i10/`.

Zeileis, A. (2006), "Object-oriented computation of sandwich estimators," *Journal of Statistical Software*, 16, 1–16, URL `http://www.jstatsoft.org/v16/i09/`.

# Index